PRACTICAL ALGEBRA

PETER H. SELBY
Director, Educational Technology
MAN FACTORS, INC.
San Diego, California

John Wiley & Sons, Inc.
New York • London • Sydney • Toronto

Editors: Judy Wilson and Irene Brownstone
Production Manager: Ken Burke
Editorial Supervisor: Lorna Cunkle
Artist: Lorna Cunkle
Composition and Make-up: Meredythe Miller

Library of Congress Cataloging in Publication Data

Selby, Peter H
Practical algebra.

1. Algebra. I. Title.
QA154. 2. S44 512.9'007'7 73-18336
ISBN 0-471-77557-6

Printed in the United States of America

73 74 10 9 8 7 6

For my dear wife, Priscilla . . .

who still can't believe she typed the *whole* thing

To the Reader

This book is intended to acquaint you with the fundamentals of algebra and to show you how these relate to what you already know about arithmetic. Although designed primarily as an introduction for students at the college level, it should also suit the needs of those who studied algebra some time ago and now need a review before continuing in mathematics.

What can you expect to learn from this book? At the conclusion of this program you should find yourself generally familiar with the following topics:

- the basic approach and application of algebra to problem solving;
- the number system (in a much broader way than you have known it from arithmetic);
- monomials and polynomials;
- factoring algebraic expressions;
- how to handle algebraic fractions;
- exponents, roots, and radicals;
- linear and fractional equations;
- functions and graphs
- quadratic equations;
- inequalities;
- ratio, proportion, and variation;
- how to solve word problems.

To use the guide to its greatest benefit, though, you must work enough problems to assure mastery of the concepts. You can measure your mastery by taking the Self-Test at the end of each chapter.

The term "practical" in the title refers to the book's emphasis on techniques and problem solving. Although numerous applications and suggestions for application are included, no attempt has been made to single out any specific applied field to illustrate these algebraic principles. Rather it is the author's hope (and expectation) that each reader will find applications in his own chosen field. Those who continue their study of mathematics will have no difficulty finding use for their knowledge of algebra. Every advanced course you take from this point on will require the use of algebra. Students of engineering, biology, chemistry, psychology, any of the physical sciences, and even sociology and business administration all require a working knowledge of algebra.

As you progress through *Practical Algebra* you will find the chapters broken down into short steps called frames. Each frame presents some new information and asks a question or gives a problem to solve. If your answer is not the same as the answer given below the dashed line, be sure you understand why before you go on to the next frame. You will also find occasional pretests. If you do well on a pretest you may bypass portions of the material. However, this is the place to be very firm with yourself. Don't bypass any topics about which your knowledge is marginal. You will also find occasional tests for self-evaluation. These Self-Tests are your guideposts. If you fail any of these (according to the rating scale established), don't go ahead until you have reviewed appropriate portions of the material.

Finally, your pace through *Practical Algebra* will depend on you. You can go as rapidly as your own aptitude and inclination permit. Keep in mind, however, that the real objective in this kind of self-teaching is learning, not speed.

La Jolla, California
January 1974

Peter H. Selby

SELECTED REFERENCES

Drooyan, I., Wooton, W., *Elementary Algebra for College Students,* 3d ed. (New York: John Wiley & Sons, 1972).

Eulenberg, M. D., Sunko, T. S., *Introductory Algebra: A College Approach* (New York: John Wiley & Sons, 1967).

Rees, P. K., Sparks, F. W., *College Algebra,* 5th ed. (New York: McGraw-Hill, 1967).

Russell, D. S., Collins, M., *Elementary Algebra,* 4th ed. (Boston: Allyn & Bacon, 1971).

Russell, D. S., Lanning, G. E., *Intermediate Algebra,* 2nd ed. (Boston: Allyn & Bacon, 1971).

Rutledge, W. A., Green, S., *Introduction to Algebra for College,* 2nd ed. (Englewood Cliffs, New Jersey: Prentice-Hall, 1968).

Wooton, W., Drooyan, I., *Intermediate Algebra,* 2nd alternate ed. (Belmont, California: Wadsworth Publishing Company, 1968).

REFERENCE CHART FOR SELECTED TEXTBOOKS ON ALGEBRA

	Chapter in This Book	Drooyan and Wooton	Eulenberg and Sunko	Rees and Sparks	Russell and Collins	Rutledge and Green	Wooton and Drooyan	Russell and Lanning
1	Some Basic Concepts	1	1	–	–	3	–	–
2	The Number System	2, 10	2	1	1,2,3	1,2,3	1	1
3	Monomials and Polynomials	4	3	2	3	8	2	2
4	Special Products and Factoring	4	5	3	5	8	–	2
5	Fractions	5	6	4	6	6,7	3	3
6	Exponents, Roots, and Radicals	8	5,8	5	9	10	4	5
7	Linear and Fractional Equations and Formulas	3,6	4	6,9	4,8	5,9,12	5,9	4,8
8	Functions and Graphs	6	7	8	7	11	7,8	6
9	Quadratic Equations	7,9	5	7	10	14	6	7
10	Inequalities	–	4	13	10	16	–	4
11	Ratio, Proportion, and Variation	5	10	14	11	–	–	9
12	Solving Everyday Problems	5	–	–	–	–	5	–

Contents

CHAPTER ONE

Some Basic Concepts

OBJECTIVES

To achieve maximum benefit from this program you need to proceed logically from where you are now in your knowledge of mathematics to where you should be at the end of this program. And since we have assumed that you are familiar with the subject of arithmetic, we will start there.

Algebra is a logical outgrowth of arithmetic and many of the methods of arithmetic are used in algebra, although in modified, expanded, or original form. This chapter will provide a bridge from arithmetic to algebra for the reader who has not studied algebra before. It will also furnish a review for those who, although they perhaps had a first-year high school course in algebra, have largely forgotten what they once knew.

When you complete this chapter you should be able to:

- express the product of factors without the use of multiplication signs;
- identify the literal factors in an algebraic term;
- use letters and symbols to change simple word statements into algebraic expressions
- determine what value of the letter(s) in the denominator of an algebraic fraction would result in an undefined division;
- use parentheses correctly to express multiplication or grouping of terms;
- evaluate algebraic expressions;
- correctly identify terms in an algebraic expression;
- use exponents to indicate repeated multiplication;
- simplify elementary algebraic expressions.

A pretest on this material is provided should you wish to take it. If you find you remember enough from your previous study of algebra to pass the pretest, you may consider skipping Chapter 1 and going directly to Chapter 2. If not, go through the material carefully, studying the parts that seem unfamiliar to you. Keep in mind that your progress through the rest of the book will depend in large measure on how well you remember the fundamentals.

If you decide *not* to take the test, go to frame 1 on page 5.

PRETEST

1. Express the product of the following without using multiplication signs.

   (a) $a \times b \times c$ ____________

   (b) $3 \times k \times m$ ____________

   (c) $\frac{3}{4} \times 8 \times y$ ____________

   (d) $0.5 \times 30 \times q \times t$ ____________

   (e) $2 \times 3 \times st$ ____________ (frame 7)

2. Identify the literal factors in the following expressions.

   (a) $5atk$ ____________

   (b) $7k\left(\frac{y}{t}\right)$ ____________

   (c) $3ab \cdot 2y$ ____________

   (d) $(k)(m)(t)$ ____________

   (e) $0 \cdot 3by$ ____________ (frame 8)

3. Use letters and symbols to change these word statements into algebraic expressions.

   (a) The sum of one-half t and twice t equals twenty. ____________

   (b) Eight times a number (n) minus three times the number equals five more than four times the number. ____________

   (c) The area (A) of a triangle is equal to one-half the base (b) times the height (h).

   ____

   (d) Half of c plus twice d added to five equals eight. ____________

   (frames 9 and 20)

4. What values of the indicated letter in the denominator of each of the following expressions would result in an undefined division?

   (a) $\frac{2}{y-4}$, $y =$ ____________

   (b) $\frac{3b}{4a}$, $a =$ ____________

   (c) $\frac{k}{y-x}$, $x =$ ____________

   (d) $\frac{0.9ky}{7.6cd}$, c or $d =$ ____________ (frame 10)

5. Use parentheses correctly (where applicable) while turning these word statements into algebraic expressions.

(a) Twice the sum of c plus d equals eleven. ________________

(b) k times the sum of x plus y equals p times the quantity z minus t.

____

(c) Three divided by one-half the quantity a plus b equals fourteen.

____

(d) y plus the quantity b minus three equals seven times the quantity four plus c.

____

(e) Three times a number (n), divided by y times the sum of five and the number, is equal to seven. ________________ (frame 19)

6. Evaluate the following expressions.

(a) $3(2+3)-7+\frac{8}{2}=$ ________________

(b) $\frac{(4+2)}{3}-\frac{8}{4}+3\cdot 2=$ ________________

(c) $5-\frac{9}{3}+3(2+1)=$ ________________

(d) $\frac{(3+9)}{(4-1)}-4+(8\div 2)=$ ________________ (frame 23)

7. Evaluate these expressions for $x = 2, y = 3$.

(a) $2(x+y)-\frac{3}{(y-x)}+7=$ ________________

(b) $\frac{9}{y}-\frac{x}{2}+\frac{5xy}{(x+y)}=$ ________________

(c) $\frac{2xy}{4}+\frac{1}{2}(4y-x)=$ ________________

(d) $\frac{x^2y}{4}+\frac{xy^2}{6}+y=$ ________________ (frame 25)

8. How many *terms* are in each of the following expressions?

(a) $4b+\frac{2}{3}cx-3(a-b)$ ________________

(b) $2(c+d)-\frac{k}{m}+3y$ ________________

(c) $c(x)+b^2c$ ________________

(d) $ac(y+x)$ ________________ (frame 26)

9. Write the following expressions using exponents.

(a) $dd + cdd + ccca$ ______________

(b) $mmmy - xx + px$ ______________

(c) $\frac{y}{xx} + yyx - y(xy)$ ______________

(d) $(c + d)(c + d)$ ______________

(e) $cd + de + ef$ ______________ (frame 28)

10. Put into words the meaning of these expressions.

(a) b^2c^3 ______________

(b) $3a^2f$ ______________

(c) 5^2x^3 ______________

(d) $(2y)^2$ ______________

(frame 28)

11. Simplify the following.

(a) $2a + 3b + 3a - b$ ______________

(b) $3ab + 2k - ab + 3$ ______________

(c) $2(a + b) - a + 3b$ ______________

(d) $ax^2 + by + b^2 + 3ax^2$ ______________

(e) $3xy + 3y^2 - 2xy + y^2$ ______________ (frame 30)

Answers to Pretest

1. (a) abc; (b) $3km$; (c) $6y$; (d) $15qt$; (e) $6st$
2. (a) a, t, k; (b) $k, y, \frac{1}{t}$; (c) a, b, y; (d) k, m, t; (e) b, y
3. (a) $\frac{t}{2} + 2t = 20$; (b) $8n - 3n = 4n + 5$; (c) $A = \frac{1}{2}bh$; (d) $\frac{c}{2} + 2d + 5 = 8$
4. (a) $y = 4$; (b) $a = 0$; (c) $x = y$; (d) c or $d = 0$
5. (a) $2(c + d) = 11$; (b) $k(x + y) = p(z - t)$;
   (c) $\frac{3}{\frac{1}{2}(a + b)} = 14$; (d) $y + (b - 3) = 7(4 + c)$; (e) $\frac{3n}{y(5 + n)} = 7$
6. (a) 12; (b) 6; (c) 11; (d) 4
7. (a) 14; (b) 8; (c) 8; (d) 9
8. (a) 3; (b) 3; (c) 2; (d) 1
9. (a) $d^2 + cd^2 + c^3a$; (b) $m^3y - x^2 + px$; (c) $\frac{y}{x^2} + y^2x - xy^2$;
   (d) $(c + d)^2$ or $c^2 + 2cd + d^2$; (e) $cd + de + ef$

10. (a) two factors of b times three factors of c
    (b) three times two factors of a times f
    (c) two factors of five times three factors of x
    (d) two factors of $2y$
11. (a) $5a + 2b$; (b) $2ab + 2k + 3$; (c) $a + 5b$; (d) $4ax^2 + by + b^2$
    (e) $xy + 4y^2$ or $y(x + 4y)$

Instructions: If you made a perfect score or missed no more than two problems, you may consider skipping Chapter 1. However, for any problem where you made an error, review the frame(s) indicated after the problem before continuing. If you then feel that you understand this material thoroughly, go directly to Chapter 2. If you made more than two errors, you should begin with frame 1 of Chapter 1 below.

FROM ARITHMETIC TO ALGEBRA

1. There is nothing mysterious about algebra. It is simply another handy mathematical tool – like arithmetic. The beauty of algebra is that it will provide you with some quick and easy ways to solve problems that may have stumped you before. It will also provide you with a new insight into the world of numbers. Which of these last two advantages of algebra interests you most?

   – – – – – – – – – – – – – – – – – –

   Either answer is fine; it depends on you. Hopefully, you recognize the value of both.

2. Fifty or a hundred years ago studying algebra was not as important as it is now. Today a knowledge of algebra is essential in such diverse fields as engineering, economics, insurance, architecture, statistics, space exploration, and all of the physical sciences.

   Although algebra is really not a difficult subject, it does require that you be willing to learn the meaning of some new words. One of the most important aspects of algebra is the language it uses. To learn its language you must be willing to "listen" carefully when we are discussing any points that may be new to you.

   You will quickly discover that algebra is really just a logical extension of arithmetic. Most of the problems you can solve with algebra are based on operations you learned to perform in arithmetic. The most important of these are the four fundamental operations: addition (+), subtraction (−), multiplication (×), and division (÷). The symbol shown after each word is the one commonly used in arithmetic to indicate (in mathematical language) the operation to be performed.

   Test your memory. Write the name used to indicate the *result* of each of these arithmetical operations.

   (a) The result of addition is called ________.

   (b) The result of subtraction is called ____________________.

   (c) The result of multiplication is called ____________.

   (d) The result of division is called ____________.

- - - - - - - - - - - - - - - - - - -

(a) sum; (b) difference (or remainder); (c) product; (d) quotient
(If you missed these, review them again before you go on.)

3. These same four operations are performed in algebra – with one major difference. In algebra *letters are frequently used to represent numbers.* Why? Simply because we don't always know the numerical value of certain quantities or terms at the outset of a problem. Since we wish to identify the quantity in some way, we use a letter of the alphabet to represent it until its value can be determined. In fact, finding the values of unknown quantities is one of the things we do most frequently in algebra.

   Check the statement below which you think represents the main point we have been discussing above.

   ___(a) Arithmetic uses letters in place of numbers.
   ___(b) In algebra, numbers whose values we don't know are identified by letters.
   ___(c) Algebra differs from arithmetic in its frequent use of letters to represent numbers.

- - - - - - - - - - - - - - - - - - -

(c)

Choice (a) is wrong; it's just the reverse. Choice (b) is not a bad choice, but it is not the main point. Choice (c) is really the best choice.

4. It is this use of letters to represent unknown numbers that makes it possible in algebra to translate long word statements into brief mathematical expressions. We are going to look into the matter of how to do this very shortly. But before we do so it is important that we review certain *axioms of equality,* because these are fundamental to most of our work from this point on.

   Let's start with the word *axiom* itself, a word you may be familiar with from arithmetic (or geometry, if you have studied it). Complete the following sentence.

   An axiom is a basic assumption accepted as true without ________________.

- - - - - - - - - - - - - - - - - - -

proof

5. In both arithmetic and geometry we make use of the following axioms.

   (a) If equals are added to equals, the sums are equal.
   Example: $4 = 6 - 2$; therefore, adding 2 to each side,
   $4 + 2 = (6 - 2) + 2$.

   (b) If equals are subtracted from equals, the differences are equal.
   Example: $6 = 4 + 2$; therefore, subtracting 2 from each side,
   $6 - 2 = (4 + 2) - 2$.

(c) If equals are multiplied by equals, the products are equal.
Example: $\frac{14}{2} = 7$; therefore, multiplying both sides by 2,

$$2 \times \frac{14}{2} = 2 \times 7.$$

(d) If equals are divided by equals, the quotients are equal (provided the divisor is not zero).
Example: $7 \times 2 = 14$; therefore, dividing both sides by 2,
$(7 \times 2) \div 2 = 14 \div 2$.

(e) If $a = 4$ and $b = 4$, then $a = b$.

Test your understanding of these five axioms by completing the following.

(a) If $5 = 7 - 2$, then $5 + 3 =$ ________________.

(b) If $7 = 5 + 2$, then $7 - 4 =$ ________________.

(c) If $\frac{12}{3} = 4$, then $2 \times \frac{12}{3} =$ __________.

(d) If $4 \times 3 = 12$, then $(4 \times 3) \div 2 =$ ______________.

(e) If $x = 21$ and $y = 21$, then $x =$ __________.

- - - - - - - - - - - - - - - - - -

(a) $(7 - 2) + 3$; (b) $(5 + 2) - 4$; (c) 2×4; (d) $12 \div 2$; (e) y

You should be familiar with these axioms already. You must apply them strictly in manipulating algebraic expressions. Many of the difficulties and mistakes of math students are due to their failure to recognize and apply one or more of these axioms. So make sure you know them! We will be referring to them from time to time and checking up to make sure you haven't forgotten them.

ALGEBRAIC REPRESENTATION

6. How does the use of letters, numbers, and mathematical symbols make possible the translation of long word statements into short mathematical sentences or expressions? Let's see.

Example: The sum of five times a number and two times the same number is equal to seven times the number. How can we represent this situation most simply?

Solution: If we let n represent the number we are talking about, we can say the same thing with this short algebraic sentence: $5n + 2n = 7n$

Try this one. Three times a number subtracted from eight times the same number equals five times the number.

Solution: ____________________

- - - - - - - - - - - - - - - - - -

$8n - 3n = 5n$

7. An algebraic statement like $8n - 3n = 5n$ is called an *equation* because it represents two things that are equal to one another. In this example the two quantities that are equal are $(8n - 3n)$ and $5n$.

In this equation we have used an algebraic convention that we should discuss for a moment. In arithmetic we indicate multiplication by use of the "times" sign, $\times$. In algebra there are other ways of expressing the idea of multiplication. Suppose, as in the previous answer, we wish to express the idea of eight times a number. We can do this in any of the following ways:

$$8 \times n \qquad 8 \cdot n \qquad 8(n) \qquad 8n$$

Both the dot and the parentheses are acceptable, but omission of the multiplication sign, as in $8n$ above, is preferred. The times sign ($\times$) is seldom used in algebra since it could be mistaken for the letter x. For two numbers, parentheses are often preferred over a raised dot, which might be confused with a decimal.

Rewrite the following expressions *without using the times sign* to get the corresponding algebraic expressions.

(a) $9 \times a$ ________

(b) $a \times b \times c$ ________

(c) $6 \times y \times s$ ________

(d) $5 \times 7 \times y$ ________________

- - - - - - - - - - - - - - - - - - - -

(a) $9a$; (b) abc; (c) $6ys$; (d) $5 \cdot 7y$, $(5)(7)y$, or $35y$

8. We can use any symbol to represent a number. Traditionally, we use only the letters of the alphabet in addition to the number symbols, called *numerals*. A letter that represents a number is termed a *literal number*. "Literal," in algebra, means letter. As you may recall from arithmetic, when two numbers are multiplied together, the two numbers themselves are called *factors*. Thus, when multiplying in algebra we refer either to *numerical factors* (digits) or *literal factors* (letters).

Examples: In the expression $7xyz$, x, y, and z are the literal factors; 7 is the numerical factor.

In the expression $jk \cdot 3tm$, j, k, t, and m are the literal factors; 3 is the numerical factor.

In the expression $12z(9ak)$, z, a, and k are the literal factors; 12 and 9 are the numerical factors.

We will be discussing the meaning and use of the words *term, factor,* and *expression* in more detail later on; we define them for the present as follows:

> A *term* is either a single number or the product of one or more numerical and/or literal factors. For example, 8, $3z$, and yk are terms. A *factor* is any one of the individual letters or numbers in a term. Thus, 5, c^2, and d are factors of the term $5c^2d$. An *expression* (meaning algebraic expression) consists of one or more terms connected by plus or minus signs.

See if you can identify the literal factors in the following expressions.

(a) $2a$ ____________ (c) $4z(7pk)$ ____________

(b) $3axy$ ____________ (d) $ab \cdot 9ry$ ____________

- - - - - - - - - - - - - - - - - -

(a) a; (b) a, x, y; (c) z, p, k; (d) a, b, r, y

9. Now let us consider the symbols that algebra borrows from arithmetic to indicate the fundamental operations of addition, subtraction, multiplication, and division. We have already discussed the multiplication symbol (×). Just like the multiplication symbol, the division symbol (÷) is rarely used in algebra. More often we use the fraction bar; sometimes we use the colon. Thus, for $x \div y$ we would write

$$\frac{x}{y} \text{ or } x:y$$

Both mean x divided by y. The addition symbol (+) and subtraction symbol (−) are exactly the same in algebra as they are in arithmetic.

Translate the following verbal statements into algebraic expressions using the information given above.

(a) a divided by b ____________

(b) 7 times c times y ____________

(c) the sum of x and ab, divided by the quantity three minus y ____________

(d) two minus $3ky$ plus a, all divided by seven times a times b ____________

- - - - - - - - - - - - - - - - - -

(a) $\frac{a}{b}$ or $a \div b$; (b) $7cy$; (c) $\frac{x + ab}{3 - y}$; (d) $\frac{2 - 3ky + a}{7ab}$

10. One special situation you must watch out for is the case where a letter represents zero. We know from arithmetic that adding or subtracting zero to or from another number does not change the value of an expression. You will recall also that multiplying any number by zero (or zero by another number) gives zero as a result.

What happens when you divide a number by zero? The answer is that *division by zero is meaningless*; it is an undefined operation. These are meaningless expressions;

$$\frac{5}{0}, \frac{x}{0}$$

These meaningless expressions are easy to recognize when you see a zero in the denominator (bottom half) of a fraction. But when the denominator contains a letter (either one or more), you must be very careful to make sure that this letter does not represent zero, or that some value assigned to the letter does not cause the value of the denominator to become zero.

Indicate which values of the letters in the denominators would result in an impossible division. For example, in the expression $\frac{3}{x}$, if $x = 0$ the division would be impossible.

(a) $\frac{8}{c}$ ________ (c) $\frac{6}{x-3}$ ________

(b) $\frac{a}{2b}$ ________ (d) $\frac{8}{xy}$ ________

- -

(a) $c = 0$; (b) $b = 0$; (c) $x = 3$; (d) x or $y = 0$

Note: Division of a number into zero (e.g., $\frac{a}{8}$ where $a = 0$) is *not* meaningless. The fraction equals zero.

11. Before we go on, test yourself to see that you remember these important axioms of equality. Complete the statements below.

(a) If equals are added to ________, the sums are equal.

(b) If equals are ________ by equals, the products are equal.

(c) Expressions that are equal to the same quantity are ________

(d) If equals are subtracted from equals, the differences are ________.

(e) If equals are divided by equals, the ________ are equal (provided the ________ is not zero).

- - - - - - - - - - - - - - - - - -

(a) equals; (b) multiplied; (c) equal to each other; (d) equal;
(e) quotients, denominator

If you missed any of these, return to frame 5 and refresh your memory.

SOME RULES ABOUT INTERCHANGING NUMBERS

In arithmetic we use all of the fundamental operations (adding, subtracting, dividing, multiplying) as they relate to numerical values. Now we need to extend and apply these operations to literal values (letters) used to represent numbers. Such literal values can be used to represent either fixed values (*constants*) or variable numbers (*variables*).

In the paragraphs that follow we discuss some specific rules, or laws, that you will use extensively (and nearly automatically once you are thoroughly familiar with them) throughout your study of algebra and thereafter, no matter what branch of mathematics you study.

You are already familiar with some of these "laws" – although probably not by name – from arithmetic. For example, you doubtless would accept without argument the fact that $6 + 3 = 3 + 6$, or that $3 \cdot 2 = 2 \cdot 3$. (Remember: Get used to seeing and using the dot instead of the times sign for multiplication.) Instead of numerals in the above examples we could have used letters. Thus, $a + b = b + a$ or $a \cdot b = b \cdot a$ (more simply, $ab = ba$).

With this in mind, let's examine each of the specific laws as it relates to any of the four fundamental operations.

12. The first law we are going to talk about is known as the *commutative law for addition.* Stated in words it says that *the sum of two quantities is the same in whatever order they are added.* The example of this that we showed in frame 11 was $6 + 3 = 3 + 6$, or $a + b = b + a$. Here are some other examples:

$$4 + 9 = 9 + 4 \qquad x + y = y + x \qquad 5 + b = b + 5$$

Indicate by the words true or false which of the following is a proper example of the commutative law for addition.

(a) $p + k = k + p$ __________ (c) $a + b = b + c$ __________

(b) $t + 5 = 5 + t$ __________ (d) $7 - z = z - 7$ __________

- - - - - - - - - - - - - - - - - - -

(a) true; (b) true; (c) false; (d) false

13. The last example in frame 12 illustrates that the commutative law for addition does not hold for subtraction. To see this more clearly, suppose in problem (d) we allowed the letter z to represent the numerical value 3. We would then have

$$7 - z = z - 7$$
$$\text{or } 7 - 3 = 3 - 7$$

which is obviously untrue.

We can summarize our conclusions then as follows:

- When *adding,* the order of the numbers *may* be changed. Thus, $b + 9 = 9 + b$.
- When *subtracting,* the order of the numbers *may not* be changed. Thus, $b - 9 \neq 9 - b$. (Note: The symbol $\neq$ means *does not equal.*)

Notice that in the discussion of the commutative law for addition we only used pairs of numbers. The law is defined only for pairs of numbers, not for triples. Adding three numbers is slightly more difficult. For example, if we wish to add $2 + 5 + 8$, we would first add $2 + 5 = 7$, and then add $7 + 8 = 15$. But we could just as well have added $5 + 8 = 13$ and then $2 + 13 = 15$. That is, $(2 + 5) + 8 = 2 + (5 + 8)$. To describe this property we say that addition is associative. The *associative law for addition* states that *the sum of three quantities is the same regardless of the manner in which the partial sums are grouped.* Thus, $3 + 2 + 4 = (3 + 2) + 4 = 3 + (2 + 4)$. Here are some further examples:

$$a + 2 + 3 = (a + 2) + 3 = a + (2 + 3)$$
$$c + d + a = (c + d) + a = c + (d + a)$$
$$x + y + z = (x + y) + z = x + (y + z)$$
$$x + y + z = z + x + y = z + y + x$$

This last example combines the commutative and associative laws and illustrates the somewhat more general fact that *the sum of three numbers is the same regardless of the order in which the addition is performed.*

From what we have learned so far about the commutative and associative laws for addition would you say that the following is true or false?

$$k + t - m = t - k + m$$

- - - - - - - - - - - - - - - - - -

false (The commutative and associative laws for addition do not hold for subtraction.)

14. The commutative law for addition has its counterpart in the *commutative law for multiplication* which states that *the product of two quantities is the same whatever the order of multiplication.* Thus, $a \cdot b = b \cdot a$. If we have three factors, then $a \cdot b \cdot c = a(b \cdot c) = (a \cdot b)c$. This is known as the *associative law for multiplication.* Here are some more examples:

$$2 \cdot 3 \cdot 4 = 3(2 \cdot 4) = (2 \cdot 3)4 = 24$$
$$c \cdot d \cdot f = c(d \cdot f) = (c \cdot d)f = cdf$$
$$8mp = 8(mp) = (8m)p$$

The dots have been omitted from the second and third examples which we can do when using letters as factors, either by themselves or in combination with a number.

Do you think the commutative and associative laws for multiplication hold for division? Let's see. Indicate after each of the following statements of equality whether you think it is true or false.

(a) $8 \div 4 = 4 \div 8$ ____________

(b) $a \div b = b \div a$ ____________

(c) $\frac{1}{9} = \frac{9}{1}$ ____________

(d) $8 \div \frac{4}{2} = \frac{4}{2} \div 8$ ____________

- - - - - - - - - - - - - - - - - -

(a) false; (b) false; (c) false; (d) false

As you can see, the commutative and associative laws hold only for multiplication, not for division.

15. Let us summarize what we have learned about the commutative law as it relates to addition, subtraction, multiplication, and division.

- When *adding,* the order of the numbers *may* be changed. Thus, $b + 9 = 9 + b$.
- When *subtracting,* the order of the numbers *may not* be changed. Thus, $b - 9 \neq 9 - b$.
- When *multiplying,* the order of the numbers *may* be changed. Thus, $9 \cdot b = b \cdot 9$.
- When *dividing,* the order of the numbers *may not* be changed. Thus, $\frac{b}{9} \neq \frac{9}{b}$.

Apply these rules to determine whether the following statements are true or false.

(a) $9 \div 0 = 0 \div 9$ ________

(b) $7(xy) = (7x)y$ ________

(c) $k + r + z = z + k + r$ ________

(d) $c - d = d - c$ ________

(e) $a + b - c = b + a - c$ ________

- - - - - - - - - - - - - - - - - -

(a) false; (b) true; (c) true; (d) false; (e) true

If you doubt the answer to (e) or feel that it violates our rule, notice that we did not change the position of the subtracted number, c. Try substituting numbers for the letters a, b, and c (for example, 4, 3, and 2), and you will find that we get the same answer on either side. As long as we do not change the position of the subtracted number, c, interchanging the positions of the two added numbers, a and b, does not affect the result.

16. Once more, then:

- When *adding* you *may* change the order of the numbers.
- When *subtracting* you *may not* change the order of the numbers.
- When *multiplying* you *may* change the order of the numbers.
- When *dividing* you *may not* change the order of the numbers.

Check the statement below that correctly summarizes what we have just said.

___(a) Changing the order of numbers is permissible when adding or multiplying, but not when subtracting or dividing.
___(b) Changing the order of numbers is permissible with any of the four fundamental operations.
___(c) Changing the order of numbers is never permissible.

- - - - - - - - - - - - - - - - - -

(a)

17. In addition to the commutative and associative laws, there is a third law known as the *distributive law for multiplication.* It states that *the product of an expression of two terms by a single factor is equal to the sum of the products of each term of the expression by the single factor.* As is often the case in mathematics, what takes many words to describe can be stated symbolically in a very concise way.

In the language of mathematics, what the distributive law is saying is that

$$a(b + c) = ab + ac$$

Or, using numbers instead of letters,

$$2(3 + 4) = 2 \cdot 3 + 2 \cdot 4$$

Before considering some further applications of the distributive law you need to be aware that if a number, such as a, is multiplied by itself we write $a \cdot a = a^2$. Also, $a \cdot a \cdot a = a^3$. The exponent (2 or 3, etc.) indicates the number of times the quantity a is used as a factor.

Now here are some more examples of the distributive law:

$$a(a + b) = a^2 + ab$$
$$2b(ab + bc) = 2ab^2 + 2b^2c$$

For more than two terms we use an extended distributive property:

$$2a(2a + 3b - 4ad) = 4a^2 + 6ab - 8a^2d$$
$$3ab(a^2 - 2ad + b) = 3a^3b - 6a^2bd + 3ab^2$$

Apply the extended distributive law to the following multiplication problems. Refer to the examples for help if needed.

(a) $b(c + d + e) =$ ________________

(b) $bc(c + d - 2) =$ ________________

(c) $3x(2 - xy + z) =$ ________________

(d) $4ab(2ab + 3ac - cd) =$ ________________

- - - - - - - - - - - - - - - - - - - -

(a) $bc + bd + be$; (b) $bc^2 + bcd - 2bc$; (c) $6x - 3x^2y + 3xz$,
(d) $8a^2b^2 + 12a^2bc - 4abcd$

18. As a final summary of the commutative, associative, and distributive properties of addition and multiplication we have the following:

The Commutative Law: The result of addition or multiplication is the same in whatever order the terms are added or multiplied.

$$a + b = b + a \qquad ab = ba$$

The Associative Law: The sum of three or more terms or the product of three or more factors is the same in whatever manner they are grouped.

$$a + (b + c) = (a + b) + c = a + b + c$$
$$a(bc) = (ab)c = abc$$

The Distributive Law: The product of an expression of two or more terms multiplied by a single factor is equal to the sum of the products of each term of the expression multiplied by the single factor.

$$a(b + c + d) = ab + ac + ad$$

Apply these laws to determine whether the following statements are true or false. For (i) and (j) complete the operation.

(a) $2 + k = k + 2$ ________________

(b) $c + d = d + e$ ________________

(c) $3 - a = a - 3$ ________________

(d) $2 + c + d = (2 + d) + c$ ______

(e) $3cd = c3d$ ______

(f) $5 \div 3 = 3 \div 5$ ______

(g) $c + k - y = y - k + c$ ______
(If you're in doubt about this one, try substituting numbers for each of the letters.)

(h) $4 + \frac{b}{c} = \frac{b}{c} + 4$ ______

(i) $2(c + d - e) =$ ______

(j) $3mk(2a - 2m + k) =$ ______

(a) true (frame 12); (b) false (frame 12); (c) false (frame 13);
(d) true (frame 13); (e) true (frame 14); (f) false (frame 14);
(g) false (frame 15); (h) true (frame 12); (i) $2c + 2d - 2e$ (frame 17);
(j) $6mka - 6m^2k + 3mk^2$ (frame 17)

For each item you missed, refer to the frame indicated for a review.

WORKING WITH ALGEBRAIC EXPRESSIONS

So far we have examined these elements of algebraic expressions: literal numbers (or values); numerical and literal factors; and the methods of indicating addition, subtraction, multiplication, and division. We have also learned something about three basic laws governing the four fundamental operations and the interchanging of numbers. Now it is time for us to consider some of the methods of "working" with algebraic expressions, that is, rearranging terms, simplifying, and evaluating.

19. In working with the associative law for addition (and elsewhere) we used parentheses to group numbers: $a + (b - c)$. However, we also use parentheses as a convenient method for expressing two (or more) operations.

Parentheses may be used to treat an expression as a single number. Thus, if we wish to double the sum of 3 and x, we write $2(3 + x)$. As you observed in our work with the distributive law for multiplication, this tells us that we must multiply both 3 and x by 2 to get the correct answer. Or suppose we wish to multiply the difference, 9 minus y, by 3. We would write this $(9 - y)3$, or $3(9 - y)$; either is correct.

Using parentheses where needed, express the following algebraically.

(a) the sum of k and twice p ______

(b) twice the sum of s and r ______

(c) ten increased by three times y ______

(d) x divided by the sum of a and b ______

(e) ab, divided by 3 times the sum of s plus 8 ____________________

(f) p plus the quantity (put in parentheses) x minus 7 equals 10 ____________________

- - - - - - - - - - - - - - - - - - -

(a) $k + 2p$; (b) $2(s + r)$; (c) $10 + 3y$; (d) $\frac{x}{a + b}$; (e) $\frac{ab}{3(s + 8}$;
(f) $p + (x - 7) = 10$

20. We will go into the matter of translating word statements into mathematical shorthand in Chapter 12. But since you were doing a little of it in the previous problems, you might like to practice this technique a bit more before going on. If you wish to do so, continue here. If you would rather skip it for now, go to frame 21.

The chief clue to turning words into properly related numbers and letters is to become familiar with the meaning of such phrases as "diminished by," "decreased by," and "difference of." Below is a chart that should help you associate the most common phrases and their corresponding mathematical symbols. It should also help convince you that algebraic representation is generally much more precise than words, which by themselves can often be ambiguous and confusing.

+	−	×	÷	=	x or another letter
sum and plus added to increased by	minus subtract less less than difference remainder decreased by	times product multiplied by	quotient divided by ratio	equals is is equal was equal is as much as	an unknown number

Using the above chart to aid you, express the following algebraically.

(a) twice the sum of b and 18, decreased by 9 ____________________

(b) the produce of a and k decreased by twice the difference of x and y

____

(c) the quotient of 3 and y, minus four times their sum ____________________

(d) 30 decreased by half the product of 12 and c ____________________

(e) the average of 7 and k and z ____________________

---

*Remember from arithmetic: To find an average, divide the *sum* of the terms by the *number* of terms.

(a) $2(b + 18) - 9$; (b) $ak - 2(x - y)$; (c) $\frac{3}{y} - 4(3 + y)$;

(d) $30 - \frac{12c}{2}$ or $30 - \frac{1}{2}(12c)$; (e) $\frac{7 + k + z}{3}$

21. Now that we have discussed addition, subtraction, multiplication, and division, we need to consider the *order* in which these operations are performed. Since we cannot do everything at once, it is important to know what to do first, second, and so on. The rule is as follows:

- Do all multiplications and then all divisions first, as they come up in order, from left to right.
- Do all additions and subtractions second, as they come up in order, from left to right.

For example, note the following expression:

$$6 + 3(2) - \frac{4}{2}$$

Performing the multiplication we get

$$6 + 6 - \frac{4}{2}$$

Next, performing the division gives us

$$6 + 6 - 2$$

Finally, adding and subtracting produces the answer 10. For the moment, the answer is not as important as the procedure.

Examine the expression below and indicate the correct sequence of operations.

$$\frac{3 \cdot 4}{6} + 2(7) - \frac{9}{3} + 2$$

	Operation	Term(s)
(a)	______________	______________
(b)	______________	______________
(c)	______________	______________

(a) multiply, $3 \cdot 4$ and $2(7)$; (b) divide, 12 by 6 and 9 by 3; (c) add and subtract, all terms

22. If you performed the arithmetic correctly in the problem of frame 21 you should have arrived at an answer of 15. But, again, the procedure is what should interest you most at this point.

As you gain practice collecting terms, you may start taking shortcuts by performing several steps at the same time. In the process you no doubt will make some errors. When you do, remind yourself of this one-step-at-a-time procedure. Practice this procedure as you compute the value of the following expressions. Although you are working for an answer, think procedure.

(a) $3 + 4(2) - \frac{12}{4} + 7 =$ __________

(b) $\frac{9}{3} + 5 \cdot 2 - 5 + \frac{2}{3}(9) =$ __________

(c) $12 - \frac{3(4)}{2} - \frac{1}{2} =$ __________

- - - - - - - - - - - - - - - - - -

(a) 15; (b) 14; (c) $5\frac{1}{2}$

23. We have discussed the way parentheses are used in algebraic multiplication. The following examples illustrate ways parentheses help to clarify the *meaning* of an algebraic expression.

- Parentheses can be used to write the product of two numbers without using a multiplication sign. Thus 2 times 3 may be written 2(3), (2)3, or (2)(3).
- Parentheses can be used to group an expression of several terms as a single number. Thus, $2(x + y)$ represents twice the sum of x plus y.
- Parentheses can be used to establish the order of operations in evaluation. Thus, $4(3 + 2)$ means to add the 3 and 2 in parentheses before multiplying by 4. That is, $4(3 + 2) = 4 \cdot 5 = 20$. (Note that if we had written $4 \cdot 3 + 2$ we would have been stating a different problem, since $4 \cdot 3 = 12$, and $12 + 2 = 14$.)

Evaluate the following expressions containing parentheses. Combine terms enclosed in parentheses whenever possible.

(a) $2(3 + 4) - 2 \cdot 3 =$ __________

(b) $6 - \frac{1}{2}(4 - 2) =$ __________

(c) $3 - \frac{1}{3}(4 + 2) =$ __________

(d) $12 - 5(4 - 2) =$ __________

- - - - - - - - - - - - - - - - - -

(a) 8; (b) 5; (c) 1; (d) 2

24. Now we are going to extend this idea by starting out with letters in parentheses and then *substituting* given values in order to *evaluate* (find the value of) the expression. Let's start by evaluating this expression:

$$3(a + b) + 2a - \frac{b}{2} \text{ if } a = 5, b = 4$$

Substituting the numerical values in place of a and b gives

$$3(5 + 4) + 2 \cdot 5 - \frac{4}{2}$$

Adding the numbers inside the parentheses we get

$$3 \cdot 9 + 2 \cdot 5 - \frac{4}{2}$$

Multiplying and dividing gives us $27 + 10 - 2$. And finally, adding and subtracting gives us our answer, 35. There are logical, orderly steps which, when carefully followed, produce correct answers with a minimum of effort.

Evaluate the expressions below. Remember to follow the correct procedure! (Note: Two or more symbols that share the same fraction bar are treated as one term, just as two or more symbols within parentheses are treated as one term.)

(a) $2(a + b) - 12 + 2b$ if $a = 2, b = 4$ ________

(b) $7 + 2k - 3\left(m - \frac{k}{3}\right)$ if $k = 6, m = 4$ ________

(c) $2(x + y) + 3x - \frac{x + y}{2}$ if $x = 4, y = 2$ ________

(d) $2xy + (2x - y) - 10$ if $x = 3, y = 2$ ________

- - - - - - - - - - - - - - - - - - -

(a) 8; (b) 13; (c) 21; (d) 6

If you got these answers without difficulty then you can feel some assurance about your ability to handle problems of this kind. However, if you had to look at the answers and then "juggle" some more to achieve the right answers, be sure to work more problems of this kind.

25. It is time to define a word we have purposely been using rather freely with the thought that you could best acquire its meaning by example. The word is *term.* To clinch its meaning we will define it as follows:

> A *term* is either a single number or letter or the product (or quotient) of several numbers or letters.

Thus, 7, $5a$, xyz, and $\frac{3ay}{2}$ are terms. Each group of numbers or letters we add or subtract is called a term. For example, in the expression $b + c - (x + y) + 2(a + b)$, b, c, $(x + y)$, and $2(a + b)$ are terms.

> A *factor* of a term is one of the numbers or letters multiplied together to comprise a term.

Thus, x, y, and z are factors (literal factors) of the term xyz. Similarly, 3 and $(c + d)$ are factors of the term $3(c + d)$. If we multiply the numbers or letters, each number or letter is called a factor.

Any factor (or group of factors) is called the *coefficient* of the product of the remaining factors. Thus, in the product $3 \cdot 5$, the number 3 is the coefficient of 5, and 5 is the coefficient of 3. In the product $4ab$, 4 is the *numerical coefficient* of ab, $4a$ is the coefficient of b, and ab is the *literal coefficient* of 4. If a letter does not have a coefficient written before it, the coefficient is understood to be 1. Hence a means $1a$, x means $1x$, and k means $1k$.

Study these definitions until you feel clear about them. Then test yourself by completing the following sentences.

(a) Numbers are represented by numerals (such as 1, 2, 3) or by ______________ when their numerical values are not given.

(b) A term is a ______________ number or letter or the ______________ of several numbers or letters.

(c) A factor is one of two or more ______________ being multiplied together to comprise a term.

(d) A literal factor is represented by a ______________.

(e) A numerical factor is represented by a ______________.

(f) Any factor (or factors) is the coefficient of the ______________ factors.

(g) Coefficients are of two kinds: ______________ and ______________ coefficients.

- - - - - - - - - - - - - - - - - - -

(a) letters (Refer to frame 3 if you have forgotten about this.); (b) single, product (or quotient); (c) numbers or letters; (d) letter; (e) numeral; (f) remaining; (g) literal, numerical

26. Don't be afraid of the words *term, factor, literal factor, numerical factor, numerical coefficient,* and *literal coefficient.* Probably the most difficult part of algebra (as pointed out in frame 2) is learning the language. You can't communicate with anyone in French unless you know the vocabulary. It is the same with algebra. Just make up your mind you are going to learn the new words as they come along and you will have won more than half the battle.

Another word we have used frequently before and which we are now ready to define, is the word *expression.*

> An *algebraic expression* is a statement containing one or more terms, some mathematical operations, and symbols of grouping.

Here are some examples of algebraic expressions:

$$3y \qquad 2ab - 5 \qquad x - 2y + 9$$

These are expressions of one, two, and three terms respectively. The first term is called a *monomial,* the second a *binomial,* and the third a *trinomial.* They involve the operations of multiplication, addition, and subtraction, but contain no symbols of grouping.

Indicate in the spaces provided which of the following are algebraic expressions and, if they are, how many terms each has.

	Expression?	No. of Terms
(a) $3xyz$	______	______
(b) $3a + b - c$	______	______
(c) $3ak + \frac{y}{2} + 8$	______	______

	Expression?	No. of Terms
(d) $7ab - y(2 + z)$	________	________
(e) $6(a - b) + cy$	________	________

- - - - - - - - - - - - - - - - - -

(a) yes, 1; (b) yes, 3; (c) yes, 3; (d) yes, 2; (e) yes, 2

Did you have any trouble with the figures in parentheses, such as $(2 + z)$ in problem (d) or $(a - b)$ in problem (e)? Remember that the figures in parentheses are treated as one number!

27. Before we leave our present discussion of factors, there is an interesting and important concept with which you should be familiar since it appears regularly in algebraic equations: the repeated multiplying of a factor by itself. We touched on this in frame 17 but only in connection with our discussion of the distributive law for multiplication.

If we wish to multiply $2 \cdot 2 \cdot 2$, for example, we may express this in shorter form by writing 2^3. The numeral 3, written to the right and slightly above the factor 2, is called an *exponent* and indicates the number of times 2 is to be taken as a factor. Similarly, the short way of writing $a \cdot a$ or aa, is a^2. The expression a^2 is read "a squared" or "a to the second power." Similarly, 2^3 is read "2 cubed" or "2 to the third power." The term x^4 would mean that four x's are to be multiplied together. If a figure or letter has no exponent written at its upper right, the exponent is understood to be 1. Thus, y means y^1 and 4 means 4^1.

Write the following expressions using exponents.

(a) aaa ________

(b) bbb ________

(c) k ________

(d) 7777 ________

(e) mm ________

(f) $yyyyy$ ________

(g) $(a + b)(a + b)$ ________

- - - - - - - - - - - - - - - - - -

(a) a^3; (b) b^3; (c) k^1; (d) 7^4; (e) m^2; (f) y^5; (g) $(a + b)^2$

28. Frequently we have more than one factor in an algebraic term. For example, suppose we have the product bx^3. The question naturally arises, "Does the exponent 3 apply to the b as well as to the x?" The answer is no; it applies only the factor it follows. If we wanted the exponent to apply to both b and x, we would write $(bx)^3$. In bx^3, then, what we have is b to the first power multiplied by x to the third power, or $bxxx$.

When you are dealing with exponents it is a good idea to ask yourself what the exponent means. For example, what is the meaning of a^2? The exponent 2 means a is to be taken as a factor twice. To expand your understanding of the meaning of exponents, answer the following questions.

(a) What is the meaning of n^3? ________________

(b) In the expression $5y^3$, which is to be cubed, y or $5y$?

____

(c) What is the coefficient of y in the expression $6by$? ________ What is the exponent of y in this expression? ________

(d) How can the expression $4yyy$ be written using exponents? ________

(e) How can the expression $2xxxyy$ be written using exponents? ________

(f) What does the expression m^3n^2 mean? ________________

____

(g) What does the expression $4 \cdot 7^2$ mean? ________________

____

- - - - - - - - - - - - - - - - - -

(a) The n is to be taken as a factor three times.
(b) Just the y; $(5y)^3$ would mean three factors of $5y$.
(c) $6b$, 1
(d) $4y^3$
(e) $2x^3y^2$
(f) The m is to be taken as a factor three times and n twice; thus, *mmmnn*.
(g) The 4 is to be taken as a factor once and 7 taken as a factor twice; that is, $4 \cdot 7 \cdot 7$.

29. We have used the words exponents and power but there is another word associated with multiple factors, and that is the word base. The *base* is simply the *repeated factor.* Therefore, when you see the term x^3 it means that x, the base, is to be taken as a factor 3 times. In the term 2^3, the base is 2 and is to be taken as a factor 3 times. Thus, $2^3 = 8$, or

$$\text{base}^{\text{exponent}} = \text{power}$$

Therefore, when we speak of raising a number, say 2, to the third power, we may think of this as 2^3. What we mean, however, is the result of the multiplication $2 \cdot 2 \cdot 2$, or 8.

To summarize:

> An *exponent* is a number that indicates how many times another number, the *base*, is to be used as a repeated factor. The *power* is the answer obtained.

Identify the base, exponent, and power in $x^2 = y$. ________________

____

- - - - - - - - - - - - - - - - - -

The base is x, the exponent is 2, and the power is y.

30. Exponents are a very handy mathematical tool and are used frequently in algebraic expressions of all kinds. They can be used to express the number of square yards of carpet you need for your living room or to state Einstein's mass-energy relationship, $E = mc^2$.

Now we are going to turn our attention from exponents to another algebraic activity: how to combine like and unlike terms in an algebraic expression. *Like terms are terms that have the same literal parts.* The process of combining the like terms in an expression is known as *simplifying* the expression. Let's start with a familiar example.

Just as we do not expect to add apples and oranges and obtain any kind of meaningful result, only like terms can be combined. Thus, 2 apples plus 3 apples is 5 apples; 3 feet plus 4 feet is 7 feet. Similarly, we can combine the terms $3a$ and $4a$ to get $7a$ because the literal parts (letters) are the same. This is called simplifying the expression $3a + 4a$.

On the other hand, we cannot simplify the expression $3a + 5b$ because we cannot add the unlike terms $3a$ and $5b$ (apples and oranges). There is no way we can combine unlike terms (that is, terms having different literal parts).

Can the expression $4a + 3b + 6a - b + 7$ be simplified? (Yes/No) If so, what is the new expression you get when you combine like terms? ____________________

Yes; $10a + 2b + 7$

31. *Like (similar) terms can be combined.* Remember, like terms have the same literal coefficient. Also, these literal coefficients must be of the same power (that is, they must have the same exponents). Here are two examples of the combining of like terms.

$$6a + 2a^2 - 3a - a^2 = 3a + a^2$$
$$2ab + cd^2 + cd - ab = ab + cd + cd^2$$

In the second example the terms containing c and d cannot be combined because they do not have the same exponents.

Simplify the following expressions where possible.

(a) $3a + 4a - 7 =$ ____________________

(b) $4a - 3b + 2a + 6b =$ ____________________

(c) $2k + 4k - 8c - 8 =$ ____________________

(d) $2x^2 - 3y^2 + 4x^2 - 13 =$ ____________________

(e) $x^2 - y^2 + 7x^2 + 2y^2 =$ ____________________

(f) $2xy - 3ak + 3xy + 4ak =$ ____________________

(a) $7a - 7$; (b) $6a + 3b$; (c) $6k - 8c - 8$; (d) $6x^2 - 3y^2 - 13$;
(e) $8x^2 + y^2$; (f) $5xy + ak$

SELF-TEST

The following problems review the algebraic methods covered in this chapter. If your answer does not agree with that given, check your work once more, then refer (if necessary) to the frame indicated and review the explanations given there.

1. Express the product of the following without using multiplication signs.

   (a) $a \times b \times c$ ____________ (d) $0.5 \times 40 \times t$ ____________

   (b) $3 \times c \times d$ ____________ (e) $3 \times 4 \times dy$ ____________

   (c) $\frac{2}{5} \times 15 \times q$ ____________ (frame 7)

2. Identify the literal factors in the following expressions.

   (a) $7apb$ ____________ (d) $(x)(k)(t)$ ____________

   (b) $3k\left(\frac{p}{y}\right)$ ____________ (e) $0.3kz$ ____________

   (c) $4ad \cdot 2g$ ____________ (frame 8)

3. Use letters and symbols to change these word statements into algebraic expressions.

   (a) The sum of one-half x and one x equals 12. ____________

   (b) Twice d plus half of b added to 3 equals nine. ____________

   (c) Ten times a number (n) minus three times the number equals 7 more than four times the number. ____________

   (d) The area (A) of a triangle is equal to one-half the base (b) times the height (h).

   ____

   (frames 9 and 20)

4. What values of the indicated letter in the denominator of each of the following expressions would result in an impossible division?

   (a) $\frac{7}{x-3}$, $x =$ ____________

   (b) $\frac{3c}{4d}$, $d =$ ____________

   (c) $\frac{w}{a-b}$, $b =$ ____________

   (d) $\frac{5ky}{27nm}$, n or $m =$ ____________ (frame 10)

5. Use parentheses to express the following relationships.

   (a) Twice the sum of $c + d$ equals 7. ____________

(b) b divided by the sum of b and a, plus twice qp equals 9. ____________

(c) Two added to one-third the quantity of y minus z equals $2z$.

____

(d) a plus half the quantity of y minus 2 equals 13. ____________

(e) Three times a number (n), divided by y times the sum of 1 and the number, is equal to 7. ____________ (frame 19)

6. Find the value of the following expressions. (Remember the order of operations: multiplication and division first, addition and subtraction last. Terms enclosed in parentheses should be combined wherever possible.)

(a) $2(3+2)-7+\frac{9}{3}=$ ________

(b) $\frac{(6+4)}{2}-\frac{6}{3}+2\cdot 4=$ ________

(c) $7-\frac{12}{4}+2(3+1)=$ ________

(d) $\frac{(9+3)}{4-1}-3+(6\div 2)=$ ________ (frame 23)

7. Evaluate these expressions for $a=2$, $c=3$.

(a) $2(a+c)-\frac{3}{(c-a)}+7=$ ________

(b) $\frac{9}{c}-\frac{a}{2}+\frac{5ac}{(a+c)}$ ________

(c) $\frac{2ac}{4}+\frac{1}{2}(4c-a)=$ ________

(d) $\frac{a^2c}{4}+\frac{ac^2}{6}+c=$ ________ (frame 24)

8. How many terms are in each of the following expressions?

(a) $4z+\frac{1}{3}\cdot bx-3(k-z)$ ________

(b) $3(c+d)-\frac{y}{x}+2z$ ________

(c) $c(d)+b^2c-\frac{dx}{3}$ ________

(d) $bc+cd-de(y+x)$ ________ (frame 25)

9. Write the following expressions using exponents where appropriate.

(a) $cc+acc+bbbc$ ____________

(b) $mmmy-xx+mx$ ____________

(c) $\frac{y}{mm} + xyy - m(my)$ ____________________

(d) $(b - c)(b - c)$ ____________________

(e) $ab + bc + cd$ ____________________ (frame 27)

10. What do the following expressions mean?

(a) a^2b^3 ____________________

(b) $7d^2e$ ____________________

(c) 3^2x^3 ____________________

(d) $(4y)^2$ ____________________

(frame 28)

11. Simplify the following.

(a) $2a + 3b + 3a - b$ ____________________

(b) $3ab + 2k - ab + 3$ ____________________

(c) $2(a + b) - a + 3b$ ____________________

(d) $ax^2 + by + b^2 + 3ax^2$ ____________________

(e) $3xy + 3y^2 - 2xy + y^2$ ____________________ (frame 30)

Answers to Self-Test

1. (a) abc; (b) $3cd$; (c) $6q$; (d) $20t$; (e) $12dy$
2. (a) a, p, b; (b) $k, p, \frac{1}{y}$; (c) a, d, g; (d) x, k, t; (e) k, z
3. (a) $\frac{x}{2} + x = 12$; (b) $2d + \frac{b}{2} + 3 = 9$; (c) $10n - 3n = 4n + 7$;
   (d) $A = \frac{1}{2}bh$ or $\frac{bh}{2}$
4. (a) $x = 3$; (b) $d = 0$; (c) $b = a$; (d) n or $m = 0$
5. (a) $2(c + d) = 7$; (b) $\frac{b}{b + a} + 2qp = 9$; (c) $2 + \frac{1}{3}(y - z) = 2z$;
   (d) $a + \frac{1}{2}(y - 2) = 13$; (e) $\frac{3n}{y(1 + n)} = 7$
6. (a) 6; (b) 11; (c) 12; (d) 4
7. (a) 14; (b) 8; (c) 8; (d) 9
8. (a) 3; (b) 3; (c) 3; (d) 3
9. (a) $c^2 + ac^2 + b^3c$; (b) $m^3y - x^2 + mx$; (c) $\frac{y}{m^2} + xy^2 - m^2y$; (d) $(b - c)^2$;
   (e) $ab + bc + cd$
10. (a) two factors of a times three factors of b; (b) 7 times two factors of d times e; (c) two factors of 3 times three factors of x; (d) two factors of $4y$
11. (a) $5a + 2b$; (b) $2ab + 2k + 3$; (c) $a + 5b$; (d) $4ax^2 + by + b^2$;
    (e) $xy + 4y^2$

CHAPTER TWO

The Number System

OBJECTIVES

This chapter deals with the system of numbers we use in algebra and nearly every other branch of mathematics. It extends the concept of numbers beyond that of arithmetic, principally by introducing the subject of negative numbers and discussing their use.

When you have completed the work in this chapter you should be able to:

- state the absolute values of signed numbers and compute the difference of the absolute values;
- recognize numerical values on the number line and compute the distances between number pairs;
- add, subtract, multiply, and divide signed numbers, keeping proper track of the signs;
- raise signed numbers to higher powers;
- perform complex operations with signed numbers, such as combined multiplication and division;
- evaluate expressions containing signed numbers.

THE REAL NUMBER SYSTEM

1. The numbers you first learned to count with are called *natural numbers* or *counting numbers.* They are whole numbers greater than zero. Later you learned how to add, subtract, multiply, and divide these numbers. You soon found, however, that while some divisions (such as $8 \div 4$ or $9 \div 3$) resulted in a whole number as a quotient, others (such as $5 \div 2$ or $7 \div 3$) did not. A new class of numbers, known as *fractions,* had to be introduced to give meaning to the results of such divisions.

   Another name for the natural numbers is ________________ numbers.

   - - - - - - - - - - - - - - - - - -

   counting

2. The number system of arithmetic consists of whole numbers, zero, and fractions. There is, however, a handicap to this sytem of numbers. We cannot, for example, subtract a large number from a smaller number and arrive at an answer that is either

zero or greater. To overcome this handicap mathematicians invented *negative numbers* or numbers that are less than zero. For every positive number there exists a number that is the negative (opposite) of the positive number.

The opposite of a positive number is known as a ______________________.

- - - - - - - - - - - - - - - - - - - -

negative number

3. We began in arithmetic with the set of whole numbers (and zero). When we consider the corresponding positive and negative numbers we have what is known as the set of all *integers,* as shown below.

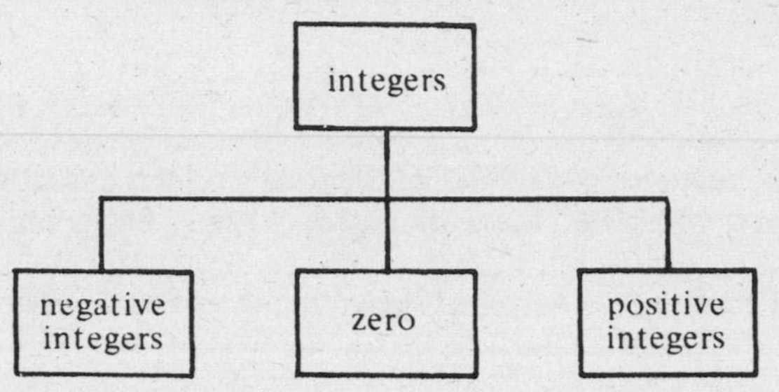

Another name for the set of positive and negative whole numbers and zero is

______________________.

- - - - - - - - - - - - - - - - - - - -

integers

4. If we include the entire collection of integers and positive and negative fractions, we have what is known as the set of *rational numbers.* This is a diagram of the rational number system.

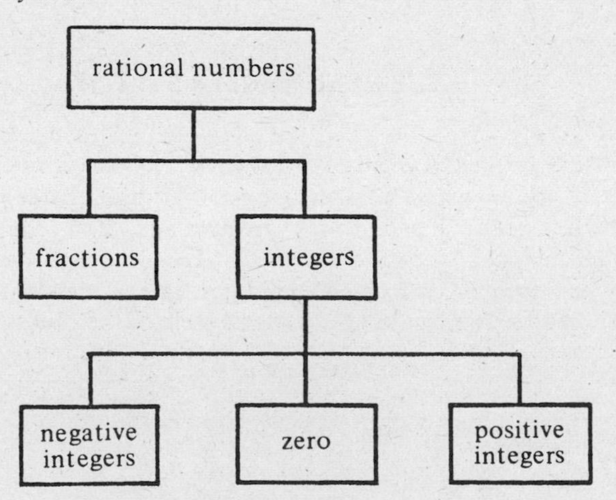

A rational number is a number that can be expressed as the quotient of two integers, the word "rational" being derived from the word "ratio."

The set of rational numbers includes both ______________ and ______________.

- - - - - - - - - - - - - - - - - -

integers, fractions

5. The counterpart to rational numbers is *irrational numbers*—numbers that cannot be expressed as the ratio of two integers. Since there is nothing you need to know about irrational numbers at the moment—other than the fact that they exist—we will mention only two examples of such numbers: π (the ratio of the circumference of a circle to its diameter), and $\sqrt{2}$ (the square root of 2). Neither can be expressed as an ordinary fraction.

The rational numbers together with the set of irrational numbers comprise the entire family of *real numbers,* as shown in our completed diagram.

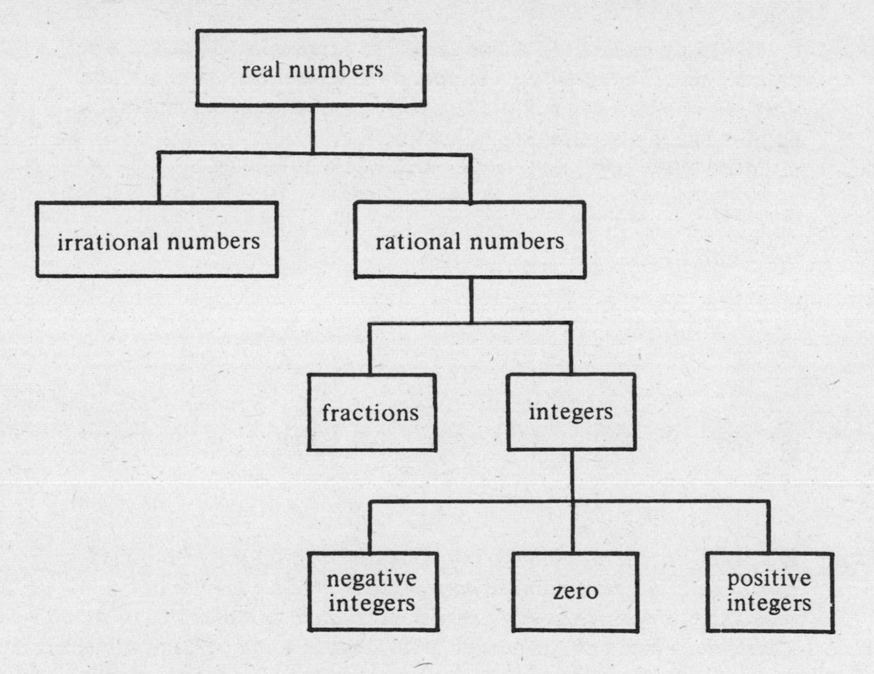

An irrational number is one that cannot be expressed as the ______________ of two integers.

- - - - - - - - - - - - - - - - - -

ratio

6. The diagram in frame 5 represents all the elements of the real number system and therefore all of the numbers with which we will be concerned in our study of algebra. Some of the names may be new to you, but the only really new element in the rational number set is the negative integers. In order to refer to our positive and negative numbers properly we will call them *signed* numbers. Keep in mind, however, that while zero is neither positive nor negative we include it with the signed

numbers. "Signed numbers" is, therefore, another name for ______________________

__________________________________.

- - - - - - - - - - - - - - - - - - - -

positive and negative numbers.

SIGNED NUMBERS

7. You are already familiar with several kinds of signed numbers. A thermometer, for example, has a scale containing both positive and negative numbers. We usually refer to these as numbers above zero or below zero. Above-zero numbers are positive, indicated by a plus (+) sign. Below-zero numbers are negative, indicated by a minus (−) sign.

   If +15 represents 15° above zero, −15 represents 15° below zero. Plus and minus numbers on a temperature scale therefore constitute a set of opposites. We will use this concept of a number scale or number line in discussing signed numbers.

   Write in the opposite term for each of the following.

   (a) profit ____________________

   (b) south ____________________

   (c) faster ____________________

   (d) lower ____________________

- - - - - - - - - - - - - - - - - - - -

(a) loss; (b) north; (c) slower; (d) higher

+100 –
+90 –
+80 –
+70 –
+60 –
+50 –
+40 –
+30 –
+20 –
+10 –
0 –
−10 –
−20 –
−30 –
−40 –
−50 –

8. Here is an important point to remember when working with signed numbers: *A negative number must be indicated by the use of a minus sign; a positive number may be indicated by a plus sign or by no sign at all.* Therefore, a plus sign—or no sign at all—means a number is positive; a minus sign means a number is negative.

   To develop further this idea of opposite values, answer the following questions.

   (a) If we let $+75^\circ$ represent 75° *west* longitude, what does -45° represent?

   ____

   (b) If +1492 represents the year 1492 *A.D.*, what does −1492 stand for?

   ____

   (c) Which is colder, -15° or -25°? ________________

(d) What is the difference in dollars between a *gain* of $10 and a *loss* of $5?

____

(e) If +12 means 12 feet *above* sea level, how would you indicate 6 feet *below* sea level? __________

- - - - - - - - - - - - - - - - - - -

(a) 45° east longitude; (b) 1492 B.C.; (c) -25°; (d) $15; (e) −6

9. The figure below is called a *number line* or *number scale.* (We will use the terms interchangeably.)

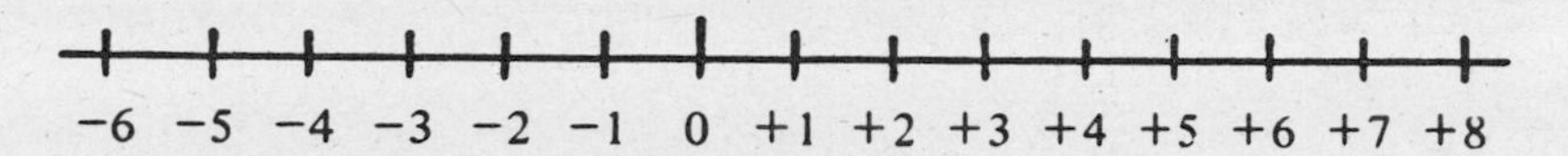

The representation of real numbers as points on a line is an important concept in establishing the connection between arithmetic and geometry. The most important fact about this representation is that every point corresponds to one and only one real number and that every real number corresponds to one and only one point.

In order to be able to write simple and meaningful rules for working with signed numbers we need to give a name to the distance between a number and zero on the number line. We call this distance the *absolute value* of the number.

By looking at the number scale we see that +3 and −3 must have the same absolute value since they are each three units from zero. Hence, +3 and −3 both have an absolute value of 3. Similarly, +5 and −5 are five units from zero and therefore the absolute value of each is 5. Notice also that the absolute value of −5 is greater than the absolute value of +3, even though the number +3 is larger than the number −5.

The concept of absolute value occurs so frequently that a special symbol is used to represent the absolute value of a number. We use $|n|$ to mean the absolute value of the number n. Thus, $|4| = 4$ is read "the absolute value of 4 is 4," and $|-6| = 6$ is read "the absolute value of −6 is 6."

When using the number scale we will consider the absolute value to mean the distance between a number and zero without regard to the direction from zero. Although we will give another definition of absolute value later, this is the only one we need at present.

Practice applying these concepts by filling in the correct answers to the following problems.

(a) The absolute values of 25 and −8 are __________ and __________ .

(b) The absolute values of 10 and −10 are __________ and __________ .

(c) $|12|$ equals __________ and $|-30|$ equals __________ .

- - - - - - - - - - - - - - - - - - -

(a) 25 and 8; (b) 10 and 10; (c) 12 and 30

10. Let us now consider the two uses of signs. In arithmetic the plus (+) sign always indicates addition and the minus (−) sign subtraction. They are called *signs of operation* because they tell us what mathematical operation to perform on the numbers. In algebra, as you have discovered, plus and minus signs may also be used to indicate that numbers are positive or negative. When signs are used in this way they are called *signs of quality* or *signs of condition.* To avoid confusion we will adopt the practice of writing $^{-}3$ when we mean negative three. (Note that the position of the minus sign is higher than you are used to finding it.) Parentheses are also used to help avoid confusion. Thus, the expression $+3 + (^{-}4)$ means that we are to add negative four to positive three.

Write negative 5 added to negative 8. ____________________

- - - - - - - - - - - - - - - - - - -

$^{-}5 + {}^{-}8$ or $^{-}5 + (^{-}8)$

NUMBER SCALES

11. As yet we haven't talked about how to work with signed numbers. We have only been getting acquainted with some new ideas. To explain how to add and subtract positive and negative numbers we will make use of number scales. There are two kinds of number scales—vertical and horizontal. The *vertical number scale* shown to the right is similar to the temperature scale in frame 7. Note again that the absolute values of the positive and negative numbers are the same for numbers that are the same distance from zero.

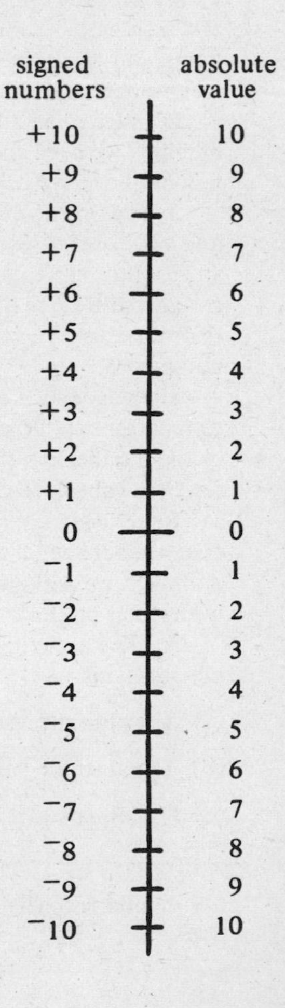

Similarly, in the *horizontal number scale* shown below the positive and negative numbers appear on opposite sides of the origin (zero). The absolute values of the positive numbers increase to the right and the absolute values of the negative numbers increase to the left.

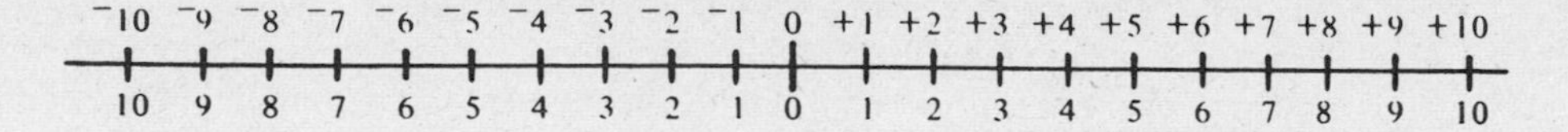

On both scales the positive numbers represent equally spaced distances from the origin in one direction, and the negative numbers similar distances from the origin in the opposite direction.

The two kinds of scales we have been discussing are known as the

____________________ and the ____________________ number scales.

\- - - - - - - - - - - - - - - - - - - -

vertical, horizontal

12. Number scales help us understand the meaning of signed numbers and also provide a convenient way of visualizing which of two signed numbers is the greater. This helps us solve problems involving signed numbers. But what do we mean by "greater"?

By *greater* we mean higher (on a vertical number scale) or further to the right (on a horizontal number scale). We have chosen the positive direction or positive sense of the vertical scale to be upward and the positive direction of the horizontal number scale to be to the right. On the scale shown below, for example, we say that b is greater than a but less than c.

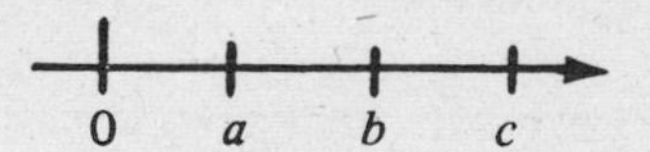

When we speak of one number being greater (or less) than another, therefore, we are referring to its location on the number scale relative to the location of the other number. And when we compare two numbers, we determine the value of one number with respect to the other by taking into account the absolute values and the signs of the numbers.

It is not unusual to find that the absolute value of one number is less than that of another although its relative value is the greater of the two. For example, $+10^\circ$ is greater (higher on the temperature scale) than -30°, yet its absolute value (10) is numerically less than 30. In fact any positive number is greater than any negative number.

In each of the following instances, decide whether or not you feel it would be necessary to distinguish between the absolute value of a number and its relative value. Indicate your answer by writing no or yes in the spaces provided.

(a) To determine if the number is greater than some other number. ________

(b) To determine if the number is greater or less than zero. ________

(c) To determine if the number is on a vertical or horizontal number line. ________

- - - - - - - - - - - - - - - - - -

(a) Yes. This is important to understand. When we speak of "greater than" or "less than" we are referring to the relative positions of two numbers on the number line. The only way we can (at the moment) locate a number is by considering both its absolute value *and* its sign. What we need to know is the value of each number, and this includes the sign. For example, -2 is greater than -5 in terms of their relative values (that is, on the number line). However, since $|-2| = 2$ and $|-5| = 5$, 2 is less than 5 in terms of their absolute values. But "greater than" or "less than" refers to position on the number scale, hence we must consider the signs of both numbers.
(b) Yes. Since absolute values are neither negative nor positive, they do not indicate whether a number is greater or less than zero. We need to know its relative value to determine this.
(c) No. Vertical and horizontal number scales are essentially the same except that on the vertical scale relative values increase upward whereas on the horizontal scale they increase to the right.

13. We have defined the absolute value of a number as its distance from zero on the number line, regardless of direction. It would be clear from this definition that we also are saying "regardless of sign." Thus, we can also consider absolute value as the value of a number when its sign is ignored. This is a somewhat more operational definition that you should be aware of when working problems involving signed numbers and their absolute values, particularly when you don't have a number scale in front of you to help you visualize the situation.

When we speak of the *value of a number,* we mean its relative or positional value on the number scale, and this includes *both its distance and its direction from zero.*

Using the horizontal number line shown below, answer the following questions.

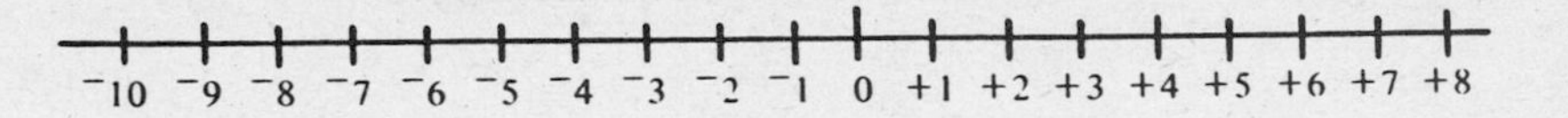

(a) Which is greater, +2 or $^{-}9$? ________

(b) Is zero greater than or less than $^{-}7$? ________

(c) How many units apart are 8 and $^{-}4$? ________

(d) Is $^{-}3$ greater than or less than $^{-}4$? ________

(e) If you start at +3 and count five spaces to the left, what number will you stop at? ________

(f) How would you express in mathematical symbols what you did in problem (e)?

____

(a) $+2$; (b) greater; (c) 12 units; (d) greater; (e) $^{-}2$; (f) $+3-5={}^{-}2$

14. Let's review some of the important concepts regarding signed numbers.

- *Signed numbers:* either positive or negative (indicated by the preceding plus or minus sign), or zero.
- *The idea of opposites, of which positive and negative numbers are but one kind:* up or down, hot or cold, above or beneath, gain or loss, east or west.
- *Absolute value of a number:* its distance from zero on a number scale; the value when its sign is ignored.
- *Greater:* higher or further to the right on a number scale.
- *The value of a number:* its location on a number scale, considering both the distance and the direction from zero.
- *The use of number scales to show how positive and negative numbers are related to each other and how intervals between them are measured:* vertical and horizontal number scales of uniform units and increasing upward (in the case of vertical scales) or to the right (in the case of horizontal scales).

Now we must learn how to add, subtract, multiply, and divide signed numbers by using these concepts.

COMBINING SIGNED NUMBERS

15. We will consider first the subject of addition. When we say "*add*" in algebra we mean *combine.* Thus, when we speak of adding two numbers we really mean we are combining them to obtain a single number that represents the total or combination of the two. It is a method of counting the total number of units in both numbers, taken together.

For example, if you first gain four pounds and then lose six pounds, the combined result (or algebraic sum) is a loss of two pounds. To help visualize this we will use the horizontal number scale.

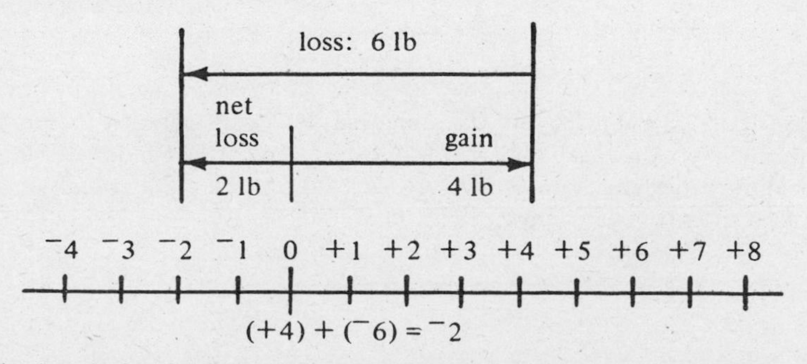

The algebraic solution is shown below the number scale. This is the method you will soon be using almost exclusively.

What would be the result (other than your own annoyance) if after gaining four pounds you gain another six pounds? Use the number scale below to diagram the steps involved in combining +4 and +6. Show the algebraic solution.

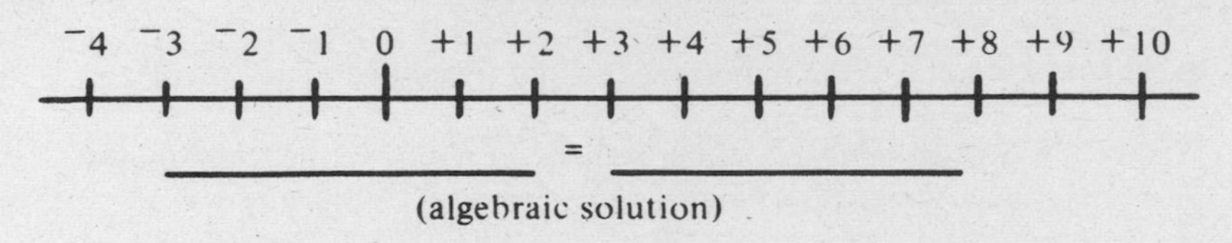

________________ = ________________

(algebraic solution)

- - - - - - - - - - - - - - - - - -

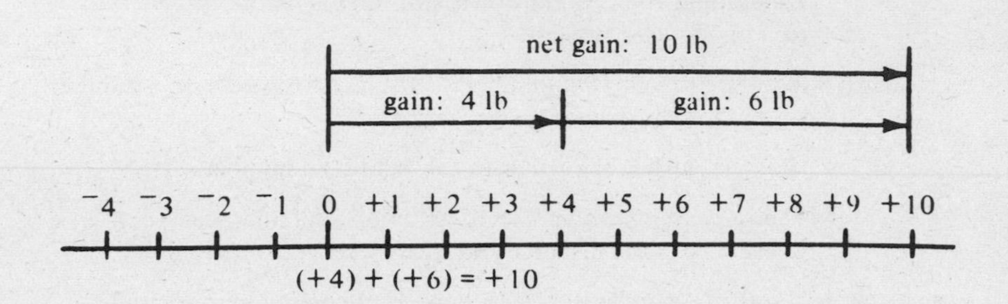

16. Number scales can help us visualize what happens when we combine signed numbers. However, it would be rather awkward if we had to draw a number scale every time we wanted to add two numbers algebraically. To avoid having to rely on number scales, two rules (partially illustrated by the preceding two problems) have been developed. The first is given below.

> *Rule 1:* To add two numbers with like signs, add their absolute values and prefix (place in front of the answer) their common sign.

The problem you just solved in frame 15 is an example of this rule. The absolute values of the two numbers were 4 and 6 and their common sign was +. Adding the numbers gave us 10 and prefixing the common sign gave us +10 as an answer. Using symbols we would have: $^+4 + ^+6 = |4| + |6| = 10$.

This procedure is not quite as essential with positive numbers. What would happen if both numbers were negative? In this case we would get

$$^-4 + (^-6) = {}^-(|^-4| + |^-6|) = {}^-(4 + 6) = {}^-10$$

Apply Rule 1 to the following problem.

$(^-4) + (^-5) =$ ______________________________

17. Having added numbers with the same signs let's consider now the problem of adding two numbers that have different (opposite) signs. If the numbers have the same absolute value, their sum is zero. (Check this on the number scale.) Otherwise, follow the rule given below.

Rule 2: To add two numbers with opposite signs, subtract the smaller absolute value from the larger and prefix the sign of the number whose absolute value is larger.

This sounds more complicated than it really is. Suppose that in your dieting you lose 6 pounds after gaining 4. The algebraic solution looks like this:

$$(+4) + (^-6) = {}^-2$$
$$\text{or } (+4) + (^-6) = {}^-(|^-6| - |4|) = {}^-(6 - 4) = {}^-2$$

Putting this into words, the rule tells us to subtract first the smaller absolute value, 4, from the larger absolute value, 6, which gives us 2. It then tells us to prefix the sign of the number with the larger absolute value. Since $^-6$ has the larger absolute value and its sign is minus, we get $^-2$ as our answer.

Try the same thing but with a different set of numbers.

$(^-7) + (+8) =$ ________________________________

- - - - - - - - - - - - - - - - - -

$(^-7) + (+8) = +(|8| - |^-7|) = +(8 - 7) = +1$

18. Let's see if you learned the two rules for algebraic addition which we discussed in frames 16 and 17. Fill in the proper words in the blank spaces below.

(a) *Rule 1:* To add two numbers with like signs, ________________ (add/subtract) their absolute values and prefix the common sign.

Example: $(^-3) + (^-7) = {}^-10$

(b) *Rule 2:* To add two numbers with opposite signs, ________________ (add/subtract) the ________________ (smaller/larger) absolute value from the ________________ (smaller/larger) and prefix the sign of the number with the ________________ (smaller/larger) absolute value.

Example: $(^-8) + (+5) = {}^-(|^-8| - |5|) = {}^-(8 - 5) = {}^-3$

- - - - - - - - - - - - - - - - - -

(a) add; (b) subtract, smaller, larger, larger

19. Practice using these rules by applying them in the following problems. If you are in doubt, check your results by using a number scale such as the one in frame 13.

(a) $(+8) + (+4) =$ ____________ (d) $(+10) + (^-8) =$ ____________

(b) $(^-7) + (^-2) =$ ____________ (e) $(^-6) + (+1) =$ ____________

(c) $(^-3) + (+10) =$ ____________ (f) $(+4a) + (^-2a) =$ ____________

- - - - - - - - - - - - - - - - - -

(a) $+12$; (b) $^-9$; (c) $+7$; (d) $+2$; (e) $^-5$; (f) $+2a$

The last problem contained literal coefficients in the two terms being combined but this should not have given you any trouble. Just handle them as you learned in Chapter 1 and apply your rules for signs as you would any pair of numbers. You will be doing a lot more adding (combining) of algebraic terms before long.

20. You should be aware of another mathematical custom that will simplify your work.

> When writing additions horizontally (as we usually do in algebra) you may omit the signs of operation and the parentheses and use only the signs of quality. Also, if the first signed number is positive, its + sign may be omitted.

For example, instead of writing $(+3) + (+8) + (^-9)$ write $3 + 8 - 9$. Notice that first we eliminated all parentheses. Then we dropped the signs of operation (the two plus signs between the parentheses), putting in their place the signs of quality appearing before each number. Finally, since the sign in front of the first number was plus, we dropped it.

Rewrite the following, using a minimum of signs and parentheses.

(a) $(+8) + (^-4) + (+2) =$ ____________________

(b) $(+2) + (+3) + (^-8) =$ ____________________

(c) $(+8) + (^-6) + (^-4) =$ ____________________

(d) $(+2) + (^-4) + (+3) + (^-6) =$ ____________________

(e) $(^-2) + (^-4) + (^+3) =$ ____________________

- - - - - - - - - - - - - - - - - - - -

(a) $8 - 4 + 2$; (b) $2 + 3 - 8$; (c) $8 - 6 - 4$; (d) $2 - 4 + 3 - 6$;
(e) $^-2 - 4 + 3$

21. The problems in frame 20—especially (d)—raise another question: How do we combine a series of positive and negative values? The answer is this: Add all the positive numbers (the sum will be positive), add all the negative numbers (the sum will be negative), then use the rule of addition of unlike signed numbers to arrive at the final answer. Let's apply this to problem (d) in frame 20.

Add the positive numbers:
$$\begin{array}{r} 2 \\ \underline{3} \\ 5 \end{array}$$

Add the negative numbers:
$$\begin{array}{r} ^-4 \\ \underline{^-6} \\ ^-10 \end{array}$$

Add their sums: $5 - 10 = {}^-5$

The advantage of adding the positive and negative numbers separately may not be obvious in such a simple problem. However, consider the problem shown below. Notice that it contains decimal fractions.

$$20.2 + 3.7 - 12.9 + 1.1 - 6.4$$

Work out the solution to this problem step by step below.

(a) Adding the positive numbers gives us____________.

(b) Adding the negative numbers gives us____________.

(c) Adding the two sums gives us ____________.

\- - - - - - - - - - - - - - - - -

(a) +25.0
(b) −19.3
(c) + 5.7

SUBTRACTING SIGNED NUMBERS

22. Now let us consider the problem of subtracting signed numbers. Algebraic subtraction can best be visualized by considering it as finding the *directed distance* from one position to another on the number scale. This approach includes two concepts: the *interval* (distance) between two positions, and the *direction* of change (positive or negative). This is different from arithmetic where we were concerned only with positive numbers. Because we worked only with numbers greater than zero in arithmetic we did not subtract a larger number from a smaller one. In algebra we do. In fact, negative numbers were "invented" so we could perform this kind of subtraction.

In problem (e) of frame 13 you subtracted 5 from 3. Now let's subtract 3 from 5 using the vertical number scale shown to the right. Note how it illustrates the concepts of scale distance and direction. To subtract 3 from 5, count the number of spaces between +3 (the number being subtracted, called the *subtrahend*) and +5 (the number from which 3 is being subtracted, called the *minuend*). The scale distance is 2 and the direction (from 3 to 5) is upward (positive), hence the difference is +2.

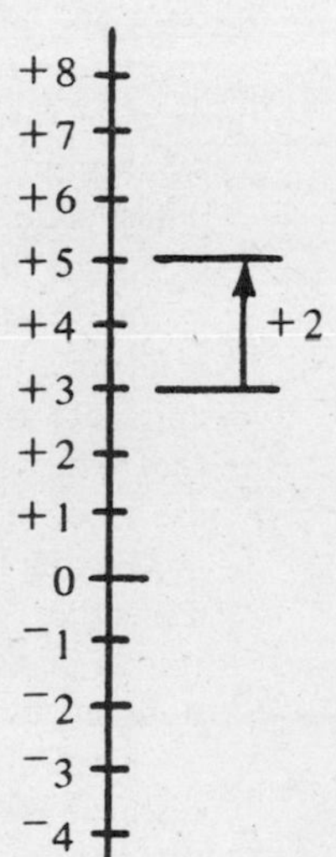

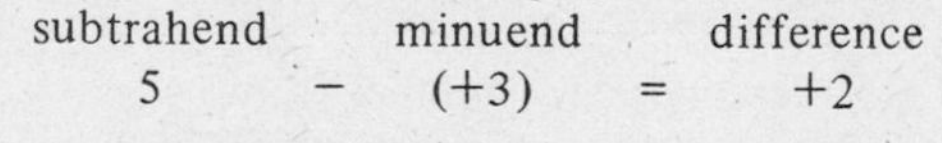

Subtract 5 from 3 and explain algebraically how you arrived at your answer

\- - - - - - - - - - - - - - - - -

To subtract 5 from 3 count the number of spaces between +5 (the subtrahend) and +3 (the minuend). The distance is 2. Since the direction is downward (negative) the difference is $^-2$.

This explanation may seem rather wordy. The words represent a reasoning process that is important. If you follow the reasoning you will not have difficulty later.

23. So far we have been concerned with subtracting one positive number from another positive number. Now let us subtract a negative number from a positive number.

For example, to subtract ¯3 from +5, count from ¯3 to +5. The distance is 8 and the direction is upward (positive). The difference, therefore, is +8, as shown on the vertical scale. Remember: Always count from the subtrahend to the minuend; this determines the direction in which you are counting and therefore the sign of the answer.

Write out below, in a horizontal line, the algebraic solution to the subtraction performed above.

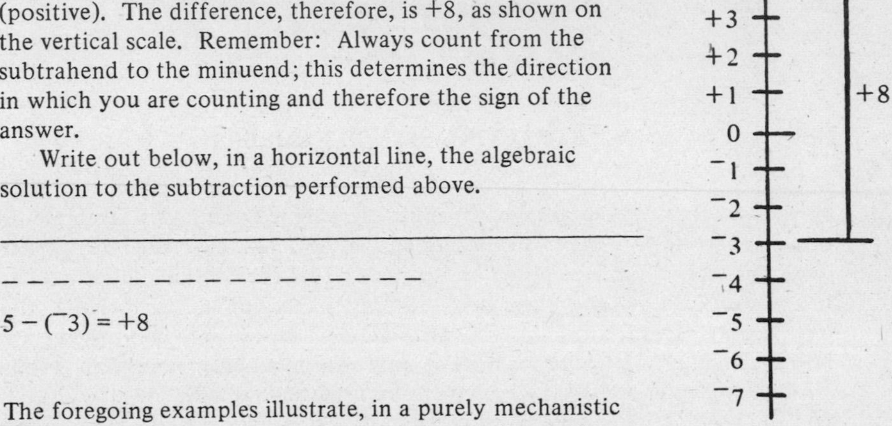

____

- - - - - - - - - - - - - - - - - - - -

5 − (¯3) = +8

24. The foregoing examples illustrate, in a purely mechanistic way, how we can use the number scales to perform subtraction. But the reason why it works out as it does can be seen more readily if we remember that the process of subtraction involves finding a number which, when added to the subtrahend, produces the minuend. For example, what number added to ¯3 will equal 5? (Or, ¯3 + ? = 5.) The answer is, of course, +8. Putting it in more general terms, $a - b = c$ means the same as finding the value of c such that $a = b + c$.

After we have examined each of the four different subtraction combinations we will summarize the procedures and state a definition or rule. It is important, therefore, that you write each subtraction in proper form when you are asked to do so.

Having learned how to subtract a negative number from a positive one, we will now reverse the procedure and subtract a positive number from a negative one, using the same numbers as before.

To subtract +3 from ¯5, count from +3 to ¯5. The distance is 8 and the direction is downward, hence the difference is ¯8, as shown to the right. Write this operation in the form $a - b = c$.

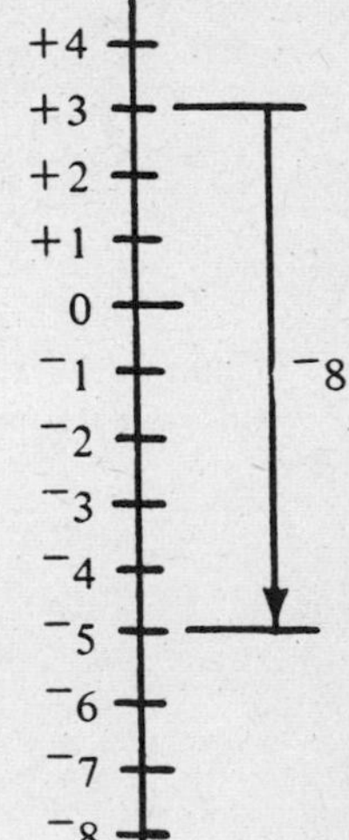

____

- - - - - - - - - - - - - - - - - - - -

¯5 − (+3) = ¯8

25. So far we have subtracted a positive number from a positive, a negative from a positive, and a positive from a negative. Let us now consider the one remaining combination, subtracting a negative number from another negative number.

Using our familiar absolute values 3 and 5, we will subtract $^{-}3$ from $^{-}5$. Using the number scale shown to the right we count from $^{-}3$ to $^{-}5$. The distance is 2 units and the direction is downward (negative), hence the difference is $^{-}2$. Write this problem in the form $a - b = c$.

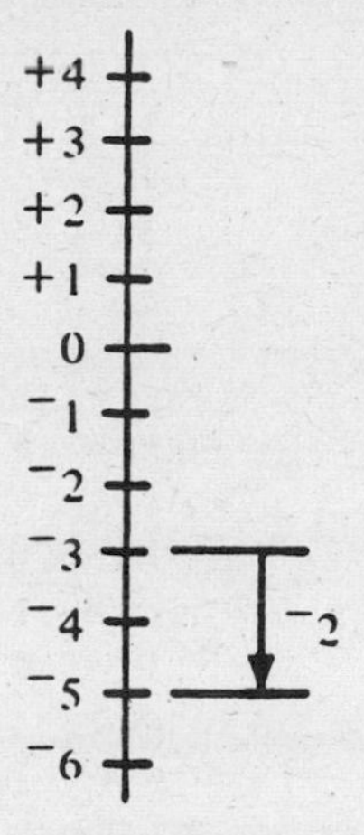

____

- - - - - - - - - - - - - - - - - -

$^{-}5 - (^{-}3) = {}^{-}2$

26. The preceding examples have illustrated the new and therefore unfamiliar process of subtracting various combinations of signed numbers. It is time for us to establish a general rule for subtraction based on the concept that subtraction and addition are inverse (opposite) processes.

Rule 3: To subtract a number, add (using the correct rule for addition) the negative (opposite) of that number. Using symbols, $a - b = a + (^{-}b)$.

Let's review how this rule works. Below is a summary of what we have covered.

	examples from our discussion of the number scale	the same problem worked out by the definition of subtraction
frame 22	$+5 - (+3) = +2$	$+5 - (+3) = +5 + (^{-}3) = 2$
frame 23	$+5 - (^{-}3) = +8$	$+5 - (^{-}3) = +5 + (+3) = 8$
frame 24	$^{-}5 - (+3) = {}^{-}8$	$^{-}5 - (+3) = {}^{-}5 + (^{-}3) = {}^{-}8$
frame 25	$^{-}5 - (^{-}3) = {}^{-}2$	$^{-}5 - (^{-}3) = {}^{-}5 + (+3) = {}^{-}2$

You can see that our rule for subtraction works and thus provides a simple method of subtracting.

Use the rule of subtraction to solve the following problems.

Example: $(^{-}3) - (+5) = {}^{-}3 + (^{-}5) = {}^{-}8$

(a) $(+7) - (^{-}4) =$ ____________________ = ____________________

(b) $(^{-}8) - (^{-}2) =$ ____________________ = ____________________

(c) $(7) - (3) =$ ____________________ = ____________________

(d) $1 - 4 =$ ____________________ = ____________________

(e) $1 - (^{-}4) =$ ____________________ = ____________________

(a) $(+7) - ({}^-4) = +7 + (+4) = 11$; (b) $({}^-8) - ({}^-2) = {}^-8 + (+2) = -6$;
(c) $(7) - (3) = 7 + ({}^-3) = 4$; (d) $1 - 4 = 1 + ({}^-4) = {}^-3$;
(e) $1 - ({}^-4) = 1 + (+4) = 5$

27. Don't let the omission of the plus signs and parentheses in some of these problems confuse you. Remember, if there is no sign in front of a number it is positive. Remember, too, that parentheses are needed only when there is a possibility of confusion between signs of operation (adding or subtracting) and signs of quality (positive or negative). If no confusion is likely, the parentheses may be omitted.

If you are still having difficulty subtracting signed numbers you would do well to return to frame 22 and re-read the material to this point. Before we leave the subject of the addition and subtraction of signed numbers, let us state our rules once more.

Addition

Rule 1: To add two signed numbers having like signs, add their absolute values and prefix their common sign. Thus, ${}^-9 + ({}^-7) = {}^-16$.

Rule 2: To add two numbers having opposite signs, find the difference of their absolute values and prefix the sign of the number having the larger absolute value. Thus, $9 + ({}^-7) = +2$.

Subtraction

Rule 3: To subtract a number, add its opposite (its additive inverse). Thus, $9 - (+7) = 9 + ({}^-7) = +2$.

MULTIPLYING SIGNED NUMBERS

28. Since you now understand the difference between the minus (−) symbol as a sign of quality and as a sign of operation, we shall discontinue the practice of raising it above the usual position to indicate a negative number. We will continue to use parentheses, however, to show that the minus sign is associated with a number.

Having learned something about how to add and subtract signed numbers let's consider the matter of multiplying such numbers.

Rule 4: When multiplying two signed numbers, if the signs are the same the product will be positive. Thus, $3 \cdot 4 = 12$ or $(-3)(-4) = 12$.

Regardless of whether the signs are positive or negative, if they are the same the product will be positive.

Apply this rule in solving the following problems.

(a) $(+9)(+4) =$ ____________ (e) $(+2)(8) =$ ____________

(b) $(-7)(-3) =$ ____________ (f) $(3)(+12) =$ ____________

(c) $(5)(6) =$ ____________ (g) $7(5) =$ ____________

(d) $7 \cdot 8 =$ ____________

- - - - - - - - - - - - - - - - - -

(a) 36; (b) 21; (c) 30; (d) 56; (e) 16; (f) 36; (g) 35

29. The inclusion or omission of plus signs and parentheses in the previous problems was intended to give you practice in recognizing the facts brought out in the preceding frame: when the numbers in a multiplication are positive, a plus sign need not be used, and parentheses are necessary only if the numbers are negative. As you continue with algebra you will find that the plus sign is nearly always omitted except when its omission might introduce doubt as to the quality of a number (i.e., whether it is positive or negative).

The second rule for multiplication (fifth rule for signed numbers) follows.

> *Rule 5:* When multiplying two signed numbers, if the signs of the numbers are different, the product will be negative. Thus, $(-3)(4) = -12$ or $(3)(-4) = -12$.

Regardless of the order in which the positive and negative factors appear, the combination of two unlike signs produces a negative product.

Apply this rule in the following problems.

(a) $(5)(-5) =$ ____________ (d) $(-1)(+1) =$ ____________

(b) $(-7)(2) =$ ____________ (e) $6(-2) =$ ____________

(c) $(0.5)(-8) =$ ____________ (f) $-4 \cdot 4 =$ ____________

- - - - - - - - - - - - - - - - - -

(a) -25; (b) -14; (c) -4; (d) -1; (e) -12; (f) -16

The purpose of problem (c) was to remind you that not all multiplying is done with integers.

30. Now let us consider what happens when we wish to multiply more than two signed numbers. We could, of course, form the product of two numbers at a time until we had used up all the factors, then apply the sign we obtained for the last product. However, there is a rule that will simplify the procedure.

> *Rule 6:* Regardless of the number of factors, the product of more than two numbers is always negative if there are an odd number of negative factors, and positive if there are an even number of negative factors.

Thus, $(2)(-3)(4)(-5) = +120$ (even number of negatives); whereas $(-2)(+3)(-4)(-5) = -120$ (odd number of negatives). The product of two or more numbers is zero if any of the numbers (factors) is zero. (See frame 10, Chapter 1.) Thus, $(-2)(0)(3)(-4)(5) = 0$.

Apply Rule 6 in solving the following problems.

(a) $(1)(-3)(-2)(-5) =$ ____________ (d) $(-2)^3 =$ ____________

(b) $(-7)(4)(6)(-2) =$ ____________ (e) $(8)(-8) =$ ____________

(c) $(5)(-5)(5) =$ ____________ (f) $(-2)(-2)(-2) =$ ____________

- - - - - - - - - - - - - - - - - -

(a) -30; (b) 336; (c) -125; (d) -8; (e) -64; (f) -8

31. Did you notice anything special about problems (d) and (f) in frame 30? They are the same problem written differently. However, writing the problem in exponential form, $(-2)^3$, gives us a clue to another small variation of Rule 6. This variation of the rule relates to exponents.

> Odd powers of negative numbers are negative; even powers of negative numbers are positive.

Let's see what this means. Looking at problem (d) in frame 30 again, it is apparent that the exponent 3 is an odd (not even) power of -2, a negative number. This means that there will be an odd number of negative factors—three to be exact. According to our rule, therefore, the product should be negative. Thus, $(-2)(-2)(-2) = -8$. If the exponent is an even number (such as 4) then there are an even number of negative factors and the answer is positive. Thus, $(-2)^4 = (-2)(-2)(-2)(-2) = +16$.

Apply the rule for exponents to the following problems.

(a) $(-4)^2 =$ ____________ (d) $(-1)^9 =$ ____________

(b) $(-3)^2 =$ ____________ (e) $(-a)^4 =$ ____________

(c) $(-5)^3 =$ ____________ (f) $(-7)^1 =$ ____________

- - - - - - - - - - - - - - - - - -

(a) $+16$; (b) $+9$; (c) -125; (d) -1; (e) $+a^4$; (f) -7

Did $(-a)^4$ give you any difficulty? It shouldn't have. Just treat it as you would any other number. You might find it helpful to write out the problem in the form $(-a)(-a)(-a)(-a)$ and multiply across just to prove to yourself that the rule holds.

DIVIDING SIGNED NUMBERS

32. Although we have discussed the procedures for adding, subtracting, and multiplying signed numbers and given you rules governing these operations, you have had to accept them without any real proof that they are valid. By "proof" we mean logical, mathematical proof rather than simply what would appear reasonable from a common-sense viewpoint. However, there is not sufficient time in a course of this kind for such proofs. Nor have we taken a rigorous approach to the number system. So you will have to accept these rules on faith for the present. It is perfectly possible to prove all of them, but it requires a totally different approach than the one we have used. Most college algebra textbooks do include these proofs.

Now let us consider the division of signed numbers. Perhaps you can discover your own rule in this case. See if you can work out the correct signs for the answers to the following problems.

(a) $6 \div 3 =$ ________ (c) $6 \div -3 =$ ________

(b) $(-6) \div (-3) =$ ________ (d) $(-6) \div 3 =$ ________

- - - - - - - - - - - - - - - - - -

(a) $\frac{+6}{+3} = +2$; (b) $\frac{-6}{-3} = +2$; (c) $\frac{+6}{-3} = -2$; (d) $\frac{-6}{+3} = -2$

It is a little easier to visualize the signs if these divisions are written in fractional form.

33. The method of determining the sign of the quotient in a division is based on the rule of signs in multiplication. The rule follows. Observe how it relates to the problems of frame 32.

> In dividing two signed numbers, if the signs are alike the quotient will be positive; if the signs are different the quotient will be negative. .

This rule reads very much like our rule for signs in multiplication. In both multiplication *and* division, if the two numbers involved have the same sign, the result will be positive; if they have different signs the result will be negative.

Use the division rule in solving these problems.

(a) $\frac{-4}{-2} =$ ________ (d) $\frac{+25}{+5} =$ ________

(b) $\frac{-9}{+3} =$ ________ (e) $\frac{-1}{-1} =$ ________

(c) $\frac{+16}{-4} =$ ________ (f) $\frac{-2}{0} =$ ________

- - - - - - - - - - - - - - - - - -

(a) +2; (b) −3; (c) −4; (d) +5; (e) +1; (f) no value

Remember, division by zero is meaningless. There is no value that satisfies the requirements of problem (f) or, in fact, any case where the divisor is zero.

34. Since we often combine multiplication and division in algebra problems, we should look at one or two situations of this kind.

Example: Multiply and then divide as indicated.

$$\frac{(-4)(+6)}{(-2)(-2)}$$

Multiplying the factors of the numerator and denominator we get

$$\frac{-24}{+4}$$

or −6 for an answer.

Example: Multiply and then divide as indicated.

$$\frac{(-1)(-7)(+6)}{(+2)(-1)(+3)} = \frac{+42}{-6} = -7$$

The trick is always to multiply the factors in the numerator together to get one term, do the same in the denominator, and then perform the final division to obtain the answer—always paying careful attention to the signs.

Perform the multiplications and divisions called for in the following problems.

(a) $\frac{(-9)(-2)}{(+6)(+3)} =$ __________ (c) $\frac{(-4)(+2)(-4)}{(-2)^3} =$ __________

(b) $\frac{(5)(-5)(+2)}{(-1)(-10)} =$ __________ (d) $\frac{(-3)(12)}{(+3)(-3)(2)} =$ __________

(a) 1; (b) −5; (c) −4; (d) +2

EVALUATING EXPRESSIONS CONTAINING SIGNED NUMBERS

35. There is just one more aspect of positive and negative numbers that we need to consider: evaluating expressions containing signed numbers. The rule consists of two steps.

Rule 7: (a) To evaluate an expression containing signed numbers, first substitute the values given for the letters, enclosing them in parentheses.
(b) Perform the indicated operations in the correct order. If possible, simplify inside parentheses first; next, apply any exponents; and finally, follow the procedural rules you learned in frame 21.

For example, let's evaluate the expression $2x - 5y = ?$ for $x = -3$, $y = +2$. The first step is to substitute the given values for x and y, enclosing terms in parentheses.

$$2x - 5y = 2(-3) - 5(+2) = ?$$

Then perform the indicated operations in the correct order.

$$-6 - 10 = -16$$

Evaluate the following expression for $a = 3$, $b = -4$.

$$ab - \frac{3b}{a} = ?$$

(a) Substitute the given values: ____________________

(b) Perform the indicated operations: ____________________

(a) $(3)(-4) - \frac{3(-4)}{3}$; (b) $-12 - \frac{(-12)}{3} = -12 - (-4)$ or $-12 + 4 = -8$

Notice that the second step had to be subdivided further. Don't hesitate to do this when necessary. That is, rewrite your expression each time you perform a separate operation. *Don't try to perform more than one operation at a time!* Most mistakes made in algebra are a result of trying to perform more than one operation at a time.

36. Here is a more difficult problem. Evaluate the following expression for $a = 2$ and $b = -3$.

$$\frac{2ab - 4b^2}{ab}$$

Substituting, we get

$$\frac{2(2)(-3) - 4(-3)^2}{2(-3)}$$

Applying the exponent

$$\frac{2(2)(-3) - 4(9)}{2(-3)}$$

Performing multiplications

$$\frac{-12 - 36}{-6}$$

Finally, combining terms in numerator and dividing

$$\frac{-48}{-6} = 8$$

Evaluating algebraic expressions is good practice in handling positive and negative numbers. Using the rules and the previous examples as a guide, evaluate the following expressions.

(a) $3xy^2$, for $x = 1, y = -2$ ____________________

(b) $2x^2 - y^2$, for $x = -2, y = 3$ ____________________

(c) $ab^2 + 4$, for $a = -3, b = -4$ ____________________

(d) $x^2 + x^3 - x$, for $x = -4$ ____________________

(e) $\dfrac{ab + 8}{a^2 + b}$, for $a = -3, b = -4$ ____________________

- -

(a) $3(1)(-2)^2 = 3(4) = 12$; (b) $2(-2)^2 - (3)^2 = 2(4) - 9 = 8 - 9 = -1$; (c) $(-3)(-4)^2 + 4 = -3(16) + 4 = -48 + 4 = -44$; (d) $(-4)^2 + (-4)^3 - (-4) = 16 + (-64) + 4 = 20 - 64 = -44$; (e) $\dfrac{(-3)(-4) + 8}{(-3)^2 + (-4)} = \dfrac{(-3)(-4) + 8}{9 - 4} = \dfrac{12 + 8}{5} = 4$

SELF-TEST

Although we have now covered the fundamentals of signed numbers and the rules governing their applications, you should complete the self-test that follows. As with any aspect of mathematics, skill and confidence come only with practice. If you find you are still having difficulty with any particular kind of problem, turn to the frames referenced and review the explanations given there.

1. Another name for the natural numbers is ______________ numbers. (frame 1)

2. The opposite of a positive number is a ______________ number. (frame 2)

3. Another name for the set of positive and negative whole numbers (and zero) is ______________. (frame 3)

4. The collective term for integers and fractions is ______________ numbers. (frame 4)

5. Draw a diagram of the real number system. (frame 5)

6. What does a thermometer reading of +45 mean? ______________ (frame 7)

7. What does a thermometer reading of −12 mean? ______________ (frame 7

8. If +25 means a grain of $25, what does −25 mean? ______________ (frame 8

9. If +10 means 10 steps forward, what does −10 mean? ______________ (frame 8

10. If −18 means 18 inches below the water line, what does +18 mean? ______________ (frame 8

11. If +3 means a positive force of 3*g* (gravity units), what does −3 mean? ______________ (frame 8

12. Write the absolute value of each pair of numbers and the difference of their absolute values.

	absolute values	difference
(a) +21 and −16	____________	_____
(b) −17 and −9	____________	_____
(c) −35 and +25	____________	_____
(d) $+4\frac{1}{2}$ and 3	____________	_____

(frame 9)

13. Use the number line shown below to answer these questions.

$^{-}10 \quad ^{-}9 \quad ^{-}8 \quad ^{-}7 \quad ^{-}6 \quad ^{-}5 \quad ^{-}4 \quad ^{-}3 \quad ^{-}2 \quad ^{-}1 \quad 0 \quad +1 \quad +2 \quad +3 \quad +4 \quad +5 \quad +6 \quad +7 \quad +8$

(a) Is −14 greater than −15? __________

(b) In counting from −3 to +5 is your direction of travel positive or negative?

____

(c) Which is greater, $-\frac{1}{2}$ or 0? __________

(d) How many units apart are −3 and +5 on the number line? ______ (frame 13)

14. Perform the following additions.

(a) (+9) + (−5) = ____________

(b) (−4) + (−13) = __________

(c) (−15) + (+11) = __________

(d) (+3) + (−3) = __________ (frame 16)

15. Express the following terms with a minimum number of symbols and then add them.

(a) (+3) + (+7) + (−1) + (−5) = ________________ = _____

(b) (−13) + (−2) + (+30) + (−6) = ________________ = _____

(c) (+3) +(−3) + (+11) +(+4) = ________________ = _____

(d) (−1) + (−2) + (−3) − (−4) = ________________ = _____

(frame 20)

16. Combine terms.

(a) (+4) − (−2) = _____

(b) (−7) − (−9) − (+4) = _____

(c) (+18) − (+3) − (+9) = ______

(d) 8 − 4 + 2 − 12 = ______ (frame 26)

17. Perform the following multiplications.

(a) $(+3)(-7) =$ ________

(b) $(-9)(-8) =$ ________

(c) $(+6)(+\frac{1}{2}) =$ ________

(d) $(-a)(+a) =$ ________ (frame 28)

18. Multiply as indicated.

(a) $(-3)(+3)(-4) =$ ______

(b) $(-5)(-7)(-1) =$ ______

(c) $(+2)(+2)(-2) =$ ______

(d) $(+y)(-y)(+y)(-y) =$ ______

(frame 30)

19. Perform the indicated operations.

(a) $(-3)^3 =$ ______

(b) $(-7)^2 =$ ______

(c) $(-y)^3 =$ ______

(d) $(-5)^4 =$ ______ (frame 31)

20. Divide as indicated.

(a) $\frac{-24}{+8} =$ ______

(b) $\frac{+120}{-12} =$ ______

(c) $\frac{+2.5}{-0.5} =$ ______

(d) $\frac{-15}{-6} =$ ______ (frame 33)

21. Multiply and divide as indicated. (Remember: Multiply in factors in numerator and denominator before attempting to divide.)

(a) $\frac{(+2)(-3)}{(-1)(-4)} =$ ________

(b) $\frac{(-9)(-8)}{(-3)(+4)} =$ ________

(c) $\frac{(4)(5)}{(-2)^2} =$ ________

(d) $\frac{(-6)(+5)(-2)}{(-1)(-12)} =$ ________

(frame 34)

22. Evaluate the following.

(a) abc^2, for $a = +1$, $b = +2$, $c = -3$ ____________

(b) $x^2 - 2xy + y^2$, for $x = -2$, $y = +3$ ____________

(c) $b(b^3 - 2)$, for $b = -2$ ____________

(d) $\frac{2x - 3y}{4z}$, for $x = -2$, $y = -1$, $z = +3$ ____________ (frame 35)

Answers to Self-Test

1. counting
2. negative
3. integers
4. rational
5. see frame 5
6. 45° above zero
7. 12° below zero
8. a loss of $25
9. 10 steps backward
10. 18 inches above the water line
11. a negative force of $3g$
12.

	absolute values	difference
(a)	21 and 16	5
(b)	17 and 9	8
(c)	35 and 25	10
(d)	$4\frac{1}{2}$ and 3	$1\frac{1}{2}$

13. (a) Yes; (b) Positive; (c) 0; (d) 8
14. (a) +4; (b) −17; (c) −4; (d) 0
15. (a) $3 + 7 - 1 - 5 = +4$; (b) $-13 - 2 + 30 - 6 = +9$ (c) $3 - 3 + 11 + 4 = +15$; (d) $-1 - 2 - 3 + 4 = -2$
16. (a) +6; (b) −2; (c) +6; (d) −6
17. (a) −21; (b) +72; (c) +3; (d) $-a^2$
18. (a) +36; (b) −35; (c) −8; (d) $+y^4$
19. (a) −27; (b) +49; (c) $-y^3$; (d) +625
20. (a) −3; (b) −10; (c) −5; (d) $+2\frac{1}{2}$
21. (a) $-1\frac{1}{2}$; (b) −6; (c) +5; (d) +5
22. (a) +18; (b) +25; (c) +20; (d) $-\frac{1}{12}$

If your answers do not agree with any of those given above, be sure to check your work before referring to the appropriate frames for review. Re-read any portions of this chapter you feel are necessary to reinforce your understanding before going on to Chapter 3.

CHAPTER THREE

Monomials and Polynomials

OBJECTIVES

Because you will encounter a great many algebraic expressions containing one or more terms, it is important that you be able to recognize such expressions as being (or not being) proper polynomials, and also that you know how to work with them. By now you have probably come to realize that by "work with them" we mean add, subtract, multiply divide, evaluate, and otherwise manipulate them in the solution of algebraic problems. It is the purpose of this chapter to teach you to do just that. Specifically, when you have finished this chapter you will be able to:

- distinguish between an algebraic expression and a true polynomial;
- distinguish between a monomial, a binomial, a trinomial, and a polynomial of more than three terms;
- recognize like terms and be able to combine them;
- simplify and combine monomial terms;
- add polynomials;
- remove grouping symbols and simplify polynomials;
- multiply and divide polynomials;
- multiply powers of the same base;
- find the power of the power of a base;
- divide powers.

KINDS OF POLYNOMIALS

1. We are going to begin this chapter by reviewing some words we discussed in Chapter 1. To follow the discussion of monomials and polynomials you must remember such words as term, factor, and expression.

   A *term* is either a single number or the product of a numerical coefficient and one or more literal coefficients. Thus, 3, $5c^2d$, $7nk$, and $4xyz$ are terms.

A *factor* is any one (or group) of the individual letters or numbers being multiplied in a term. Thus, 5, c, and d are factors of the term $5c^2d$.

Numbers in a term are called *numerical factors,* whereas letters are called *literal factors.*

An algebraic *expression* consists of one or more terms together with some symbols of operation (that is, symbols that tell us to add, subtract, multiply or divide). Thus $3a - 7b + (\frac{1}{2})cd^2 - 7(x + y)$ is an algebraic expression.

See if you can answer the following questions about the expression $3a - 7b + (\frac{1}{2})cd^2 - 7(x + y)$.

(a) How many terms does it have? ____________

(b) How many factors are listed in the third term? ____________

(c) How many factors are listed in the fourth term? ____________

(d) How many numerical factors are listed in the entire expression? ____________

- - - - - - - - - - - - - - - - - -

(a) four; (b) three; (c) two; (d) four

If you had any difficulty getting the correct answers, re-read frames 26 and 27 of Chapter 1.

2. We reviewed the terminology because we will need it in discussing polynomials. In simplest words, a *polynomial* is the sum of one or more terms, each of which is the product of a collection of numbers and letters. There is one further restriction: the exponents of letters must be positive integers. This means that fractional and negative exponents are excluded. It also means—although the reason for this may not be clear (or seem important) to you until later—that polynomials do not have variables (letters) in any denominator nor any variables under a radical sign.

The expressions $6x$, $3y^2 - 6$, and $5a^2 - 3a - 7$ are all polynomials. However,

$$\frac{3}{x}, \frac{x-3}{5x}, \frac{7}{x} - \frac{2}{3x^2}, \text{ and } 2\sqrt{x}$$

are algebraic expressions but they *are not* polynomials. Three of these expressions contain a variable (x) in the denominator and one of them contains a variable under the radical sign (the square root of x).

Indicate which of the following is a polynomial and which simply an expression.

(a) $3n^2k + \frac{1}{2}mv^2$ ____________

(b) $7xy^2(3z)$ ____________

(c) $\frac{2}{b} - 7$ ____________

(d) $y^2 - 3\sqrt{x} + 4$ ____________

(e) $k + \frac{3}{\sqrt{x}}$ ____________

- - - - - - - - - - - - - - - - - -

(a) polynomial; (b) polynomial; (c) expression (because the variable b occurs in the denominator of the first term); (d) expression (because the second term contains the variable x under a radical sign); (e) expression (because the second term contains $\sqrt{x}$)

3. Remember, a polynomial cannot contain a negative or a fractional exponent, and this means that it cannot have a variable (such as a letter x or y) in the denominator nor can it contain a variable under a radical sign. You will understand better the relationship between these two concepts when you get to Chapter 6.

Because many of the polynomials we work with in algebra contain either one, two, or three terms, for convenience they are given separate names. Thus, polynomials that contain only one term (such as $5x^2y^3$) are called *monomials.* Those with two terms (such as $2a^2 - 3bc$) are called *binomials. Trinomials,* as you would expect, are those containing three terms. There is no special name given to polynomials having more than three terms.

To identify an expression as a monomial, binomial, trinomial, or just plain polynomial (if it contains more than three terms), follow the procedure given in frame 2.

(1) Look for plus or minus signs separating the terms. (Although we defined a polynomial as the sum of one or more terms, any one of those terms could be a negative term.)

(2) Make sure there are no variables in the denominators of any of the terms, nor any variables under the radical sign ($\sqrt{\ }$).

(3) Count the number of terms, and name the expression accordingly.

Identify each of the following expressions as monomial, binomial, trinomial, or polynomial (more than three terms).

(a) $a^2 + b^2$ ____________________

(b) $ak^3 - \left(\frac{a-b}{2}\right) + 14k$ ____________________

(c) $ak(x^2 - y^2)$ ____________________

(d) $ax^3 + bx^2 + cx + d$ ____________________

- - - - - - - - - - - - - - - - - -

(a) binomial; (b) trinomial; (c) monomial; (d) polynomial

Note that the second term in problem (b) is enclosed in parentheses to indicate that it is to be treated as a single factor. Even if we had not used parentheses, the fraction bar would have had the same effect as parentheses. This is accepted practice in algebra: a fraction bar has the effect of grouping the letters and numbers above and below it into a single term. We could, therefore, have written the expressions as

$$ak^3 - \frac{a-b}{2} + 14k$$

Use of parentheses avoids any possible confusion.

4. Because it is customary to write a term in its simplest form, we usually multiply all the numerals together and write the product in front as a single number. It is then known as the *numerical coefficient,* a phrase we introduced briefly in Chapter 1, frame 25. Similarly, all the factors involving the same letter are brought together and written as some power of the letter. For example, $7 \cdot c \cdot d \cdot 2 \cdot c \cdot c \cdot d = 14c^3d^2$.
   Simplify the following expression.

   $4 \cdot m \cdot m \cdot 2 \cdot k \cdot 3 \cdot m \cdot p =$ ____________

$24m^3kp$

5. In Chapter 1, frame 30, we defined *like terms* as terms having the same literal parts (the same letters, that is). Like terms differ, therefore, only in their numerical coefficients. But for the literal parts to be the same, they must also have the same exponents. Thus, $3ac^2$ and $5ac^2$ are like terms, but $2a^3b$ and $4ab^3$ are not like terms because their literal parts are not identical. Here are some other examples of unlike terms:

   $3ab$ and $4ky$ $\qquad$ $5x^2y$ and $4xy^2$ $\qquad$ $9abc$ and $7bcd$

   Write down the like terms (if any) in the following polynomials.

(a) $a + 3a - x + z$ ____________

(b) $7 - 3y + 3$ ____________

(c) $2x^2y + xy^2 - xy$ ____________

(d) $3a^2b^2 - 6a^2b^2 + a^2b^2c$ ____________

(e) $2ac + 3ab - 4bc + ac$ ____________

(f) $2(a + b) + 3(a - b) - (a + b)$ ____________

(a) a and $3a$; (b) 7 and 3; (c) none; (d) $3a^2b^2$ and $-6a^2b^2$; (e) $2ac$ and $+ac$; (f) $2(a + b)$ and $-(a + b)$

ADDING LIKE TERMS

6. Now that you have learned to recognize like terms you can learn how to add them. The process is simple.

   To add like terms, add their numerical coefficients and keep the common literal coefficient. Thus, $2x^2 + 5x^2 - 3x_2 =$ $(2 + 5 - 3)x^2$ or $4x^2$.

If the terms you are adding contain parentheses and signs of operation, remember (Chapter 2, frame 20), you may omit the signs of operation and the parentheses and use only the signs of quality. A plus sign before the first term also may be dropped. For example, $(+2y) + (-3y) + (+5y)$ may be written $2y - 3y + 5y$. Adding these like terms gives us $(2 - 3 + 5)y$ or $4y$.

To simplify an expression composed of like and unlike terms, add the like terms first and then write down the unlike terms. For example, $2a + 3a + 4ay - 4a + 7z - 4 = (2 + 3 - 4)a + 4ay + 7z - 4 = a + 4ay + 7z - 4$.

Simplify the following polynomials.

Example: $3xy + 2x^2y - xy + 4x^2y = (3 - 1)xy + (2 + 4)x^2y = 2xy + 6x^2y$

(a) $4ab - 6bc + 3bc - 2ab =$ ________________ = ____________

(b) $(+3a) - (+2a) + (-4a) =$ ________________ = ____________

(c) $x^2 - y^2 + 4 - 3y^2 =$ ________________ = ____________

(d) $rw - 2rw + 7 - 3 + y =$ ________________ = ____________

(e) $abc + bac + acb =$ ________________ = ____________

- - - - - - - - - - - - - - - - - -

(a) $(4 - 2)ab + (-6 + 3)bc = 2ab - 3bc$; (b) $3a - 2a - 4a = -3a$;
(c) $x^2 + (-1 - 3)y^2 + 4 = x^2 - 4y^2 + 4$; (d) $(1 - 2)rw + (7 - 3) + y = -rw + 4 + y$;
(e) $(1 + 1 + 1)abc = 3abc$ or $3bac$ or $3acb$

Problem (e) may have given you some concern. The purpose of this problem was to remind you that the *order* of the factors is not important; as long as the literal coefficients consist of the same letters, the terms are like terms. It is customary to arrange the letters of a term in alphabetical order, but not necessary from a mathematical point of view.

7. Since most of the polynomials we work with will contain various powers of the same letters, you will save time and avoid confusion if you arrange the terms in descending powers of one of the letters. For example, rearranging the terms $x + 4 - 3x^2 + 4x^3$ in the order of descending powers of x gives $4x^3 - 3x^2 + x + 4$.

Try arranging the following polynomial in convenient order.

$3ax^2y - 4bxy^2 + 7x^3 - 2ab$ ____________________

- - - - - - - - - - - - - - - - - -

The best order would be $7x^3 + 3ax^2y - 4bxy^2 - 2ab$ (that is, descending power of x). It would appear that you could select descending powers of either x or y. However, since the letter x appears in more terms than the letter y, x is the better choice.

8. It is best to arrange algebraic expressions in descending powers of the letter that appears most frequently. Practice this rule by rearranging the following polynomials in what you consider the best sequence.

(a) $4b^2 + 3a^2 - 7ab + 14$ (in terms of a) ____________________

(b) $3xy - y^2 + 2x^3 - 5x^2$ (in terms of x) ____________________

(c) $3z^2 - 4 - z^2 + 2z$ ________________

(d) $abc^2 + ab^2c + a^2bc$ (in terms of a) ________________

(e) $7c^2 - 3ac + dc^3 + 11$ (in terms of c) ________________

- - - - - - - - - - - - - - - - - - -

(a) $3a^2 - 7ab + 4b^2 + 14$; (b) $2x^3 - 5x^2 + 3xy - y^2$; (c) $3z^2 - z^2 + 2z - 4$;
(d) $a^2bc + ab^2c + abc^2$; (e) $dc^3 + 7c^2 - 3ac + 11$

ADDING POLYNOMIALS

9. Once you know how to arrange polynomials in convenient order you can use this skill to assist you when adding them.

Example: $(3x - y) + (2y + x)$

Removing parentheses and grouping like terms we get

$$(3x + x) + (2y - y)$$

Combining like terms gives $4x + y$ as the sum of these two binomials.

Example: $(3x^2y + x - 2xy^2 + 4) + (-3x + 2y + 5xy^2 - 7)$

Removing parentheses and grouping like terms we get

$$(3x^2y) + (x - 3x) + (-2xy^2 + 5xy^2) + (2y) + (4 - 7)$$

Combining like terms gives $3x^2y - 2x + 3xy^2 + 2y - 3$.

Addition of polynomials can be simplified—particularly when three or more polynomials are involved—by arranging the work in columns containing like terms. Thus, the second example above could be written:

$$\begin{array}{r} 3x^2y + x - 2xy^2 + 4 \qquad\quad \\ -3x + 5xy^2 - 7 + 2y \\ \hline 3x^2y - 2x + 3xy^2 - 3 + 2y \end{array}$$

Use either procedure to add the following polynomials.

(a) $(3a^2 + ab + c) + (2c - 4a^2 - ab) =$ ________________

(b) $(2x - y) + (3y - x) + (3x + 7) =$ ________________

(c) $(x^3 + 2x) + (x^2 - 4) + (x - 2x^2) =$ ________________

(d) $(10x^2y + 3xy^2 - 3xy) + (5xy + x^2 - y^2) =$ ________________

- - - - - - - - - - - - - - - - - - -

(a) $-a^2 + 3c$; (b) $4x + 2y + 7$; (c) $x^3 - x^2 + 3x - 4$;
(d) $10x^2y + 3xy^2 + 2xy + x^2 - y^2$

SUBTRACTING POLYNOMIALS

10. Before attempting to subtract all types of polynomials, you first need to know how to subtract monomials. In Chapter 2, frame 26, you learned this rule: *To subtract a number, add its opposite.* Thus, to subtract $-3x$, add $+3x$; to subtract $+2y^2$, add $-2y^2$.

Example: Subtract $-5a$ from $+2a$.

We write this first as $(+2a) - (-5a)$. Then, since $5a$ is the opposite of $-5a$, write $(+2a) + (+5a)$. Finally, we get $2a + 5a = 7a$.

Example: $(-5x^2) - (+2x^2)$

Rewrite the problem as $(-5x^2) + (-2x^2)$. Then simplify to get $-5x^2 - 2x^2 = -7x^2$.

Perform the subtractions indicated below.

(a) $(+3k) - (-7k) =$ ________________ = ________________

(b) $(-2z^2) - (-5z^2) =$ ________________ = ________________

(c) $(-7x^2y) - (+2x^2y) =$ ________________ = ________________

(d) $(+11ab) - (+4ab) =$ ________________ = ________________

- - - - - - - - - - - - - - - - - -

(a) $(+3k) + (+7k) = 3k + 7k = 10k$
(b) $(-2z^2) + (+5z^2) = -2z^2 + 5z^2 = 3z^2$
(c) $(-7x^2y) + (-2x^2y) = -7x^2y - 2x^2y = -9x^2y$
(d) $(+11ab) + (-4ab) = 11ab - 4ab = 7ab$

11. Now let us consider an expression with more terms.

$$(+7ab) - (3ab) - (-ab) - (-2ab)$$

Each minus sign before a parenthesis indicates a subtraction. Therefore we *add* the opposite of each number in parentheses. For the problem above this gives us $(+7ab) + (-3ab) + (+ab) + (+2ab)$ or simply $7ab - 3ab + ab + 2ab = 7ab$.

We are now ready to combine the operations of addition and subtraction with monomials.

Example: $(+2bc) - (-3bc) + (-10bc) - (+5bc)$

Changing subtraction to addition we get:

$$(+2bc) + (+3bc) + (-10bc) + (-5bc)$$

Finally we simplify and combine terms.

$$2bc + 3bc - 10bc - 5bc = -10bc$$

Use this procedure to combine the following like terms:

(a) $(-3xy) - (+xy) + (+4xy) - (-5xy) =$ ________________________

____

(b) $(+7ak^2) + (-ak^2) - (+3ak^2) - (-2ak^2) =$ ____________________

____

(c) $(-2abc) - (-3abc) - (-4abc) - (-abc) =$ ____________________

____

- - - - - - - - - - - - - - - - - -

(a) $(-3xy) + (-xy) + (+4xy) + (+5xy) = -3xy - xy + 4xy + 5xy = 5xy$
(b) $(+7ak^2) + (-ak^2) + (-3ak^2) + (+2ak^2) = 7ak^2 - ak^2 - 3ak^2 + 2ak^2 = 5ak^2$
(c) $(-2abc) + (+3abc) + (+4abc) + (+abc) = -2abc + 3abc + 4abc + abc = 6abc$

12. So far we have subtracted only monomials. Now we are ready to proceed to polynomials of more than one term.

> To subtract two polynomials, change each sign in the subtrahend to its opposite and proceed as in the addition of polynomials.

Subtraction, like addition, can be performed either horizontally or vertically.

Example of horizontal subtraction: $(5x^2 - 2xy + 3) - (4x^2 + x - 2)$

First we rewrite the problem:

$$(5x^2 - 2xy + 3) + (-4x^2 - x + 2)$$

Then remove parentheses and combine like terms:

$$5x^2 - 2xy + 3 - 4x^2 - x + 2 = x^2 - 2xy - x + 5$$

Example of vertical subtraction:

$$\begin{array}{rl} & 5x^2 - 2xy + 3 \\ \text{(subtract)} & \underline{4x^2 \qquad\quad - 2 + x} \end{array} \qquad \begin{array}{rl} & 5x^2 - 2xy + 3 \\ \text{(add)} & \underline{-4x^2 \qquad\quad + 2 - x} \\ & x^2 - 2xy + 5 - x \end{array}$$

Use whichever procedure seems easier to you in arranging and subtracting the following as indicated.

(a) $3x - 2y + 6$ from $5x + 4y + 8 =$ ____________________

(b) $4ab - b^2 + 2a^2$ from $3a^2 + ab - b^2 =$ ____________________

(c) $x^3 - 2x + y - 3x^2$ from $x^2 - 3x^3 + 4x =$ ____________________

(d) $mv^2 - 4 + 3av$ from $2av - 3mv^2 - 4 =$ ____________________

- - - - - - - - - - - - - - - - - -

(a) $2x + 6y + 2$; (b) $a^2 - 3ab$; (c) $-4x^3 + 4x^2 + 6x - y$; (d) $-4mv^2 - av$

If you used the horizontal method of subtraction you should have remembered to enclose each of the polynomials in parentheses. This is most important because the subtraction symbol (−) before the subtrahend has the effect of changing the sign of every term in the subtrahend when the minus sign and parentheses are removed. Only by enclosing the subtrahend in parentheses can we indicate that the subtraction symbol applies to every term.

13. The example we used in frame 12 to illustrate the subtraction of one trinomial from another was $(5x^2 - 2xy + 3) - (4x^2 + x - 2)$. Following our rule, we changed the subtraction sign to a sign of addition, then wrote the additive inverse (opposite) of each term in the subtrahend. This gave us the sum $(5x^2 - 2xy + 3) + (-4x^2 - x + 2)$.

The result would have been exactly the same had we simply removed the parentheses around the first polynomial, dropped the minus sign in front of the second polynomial, and changed the sign of each of its terms. This would have given us $5x^2 - 2xy + 3 - 4x^2 - x + 2$ directly, after which we would combine like terms to get the final answer. The signs of the terms in the first polynomial (the minuend) were unchanged because there was a plus sign (understood) in front of it. However, all the signs in the second polynomial (the subtrahend) were reversed because there was a minus sign before it. We can summarize this as follows:

Rule 1: When removing parentheses preceded by a plus sign, do not change the signs of the enclosed terms.

Rule 2: When removing parentheses preceded by a minus sign, drop the minus sign and parentheses and change the sign of each enclosed term.

Below are some examples of the foregoing rules.

Rule 1

$$4b + (3b - 7) = 4b + 3b - 7 = 7b - 7$$

$$(10a^2 - 3a) + (-5 - 5a^2) = 10a^2 - 3a - 5 - 5a^2 = 5a^2 - 3a - 5$$

Rule 2

$$(2xy - y^2 + 3) - (2y^2 + xy) = 2xy - y^2 + 3 - 2y^2 - xy = xy - 3y^2 + 3$$

$$(2ab - 3cd) - (-3ab + 2cd) = 2ab - 3cd + 3ab - 2cd = 5ab - 5cd$$

Rules 1 and 2 (combined)

$$(3m - 2k) - (7m + 4k) + (m - k) = 3m - 2k - 7m - 4k + m - k = -3m - 7k$$

Remove parentheses and combine like terms in the following.

(a) $(3a - 4b) - (9a + 2b) =$ ________________ = ________________

(b) $(5xy + 3z) - (z + 2xy) =$ ________________ = ________________

(c) $(x^2 + y^2) - (2y^2 - x^2) + (w) =$ ________________ = ________________

(d) $(a - b + c) + (a + b - c) =$ ________________ = ________________

(e) $(-k^2 + m - 2n) - (n - 2m + 3k^2) =$ ________________ = ________________

(f) $(3 - a + b) + (7a) - (5 + 3b) =$ ________________ = ________________

(a) $3a - 4b - 9a - 2b = -6a - 6b$; (b) $5xy + 3z - z - 2xy = 3xy + 2z$;
(c) $x^2 + y^2 - 2y^2 + x^2 + w = 2x^2 - y^2 + w$; (d) $a - b + c + a + b - c = 2a$;
(e) $-k^2 + m - 2n - n + 2m - 3k^2 = -4k^2 + 3m - 3n$;
(f) $3 - a + b + 7a - 5 - 3b = 6a - 2b - 2$

USE OF GROUPING SYMBOLS

14. You have learned how to add and subtract monomials and polynomials. You have also learned the correct procedure for removing parentheses that are preceded either by a plus or a minus sign. However, there are grouping symbols other than parentheses that are used in mathematics and you should be aware of these. The grouping symbols most commonly used are these:

parentheses: ( ) as in $2(2x - y)$

brackets: [ ] as in $3 - [2 + (x - y)]$

braces: { } as in $z - \{4 + [y - (2 + x)]\}$

bar: ——— as in the fraction $\frac{3 - 4x}{y}$.

We have already encountered parentheses and fraction bars. The rule for removing all of these grouping symbols is this:

> When more than one set of grouping symbols is used, remove one set at a time, beginning with the innermost set, following the same rule of signs you learned for parentheses (frame 13).

Example: $2x - \{3 - [7 + (4x - 5)]\}$

Remove parentheses: $2x - \{3 - [7 + 4x - 5]\}$
Remove brackets: $2x - \{3 - 7 - 4x + 5\}$
Remove braces: $2x - 3 + 7 + 4x - 5$
Combine like terms: $6x - 1$

Use this procedure to simplify the following expression.

$$3z - \{7w - [8 - (2z - 3 - w)] - 4\}$$

Remove parentheses: $3z - \{7w - [8 - 2z + 3 + w] - 4\}$
Remove brackets: $3z - \{7w - 8 + 2z - 3 - w - 4\}$
Remove braces: $3z - 7w + 8 - 2z + 3 + w + 4$
Combine like terms: $z - 6w + 15$

Remember: When you remove a grouping symbol that is preceded by a minus sign, you must drop that minus sign and change the sign of every term within the grouping symbol.

15. The rule for removing grouping symbols was applied three times in the problem you just completed. Repeated applications such as this are not at all unusual in working with algebraic expressions, so practice is important. With this thought in mind, practice simplifying the following expressions, using the same procedures you applied in frame 14.

(a) $4 + \{2x - [3y + (4 - 5x - 5y)]\} =$

(b) $2a - \{4 - [3b - (5a + 7 - b)] + 2\} =$

(c) $xy - \{3 - [(2xy + 5) - 5xy] + 4\} =$

(d) $\{[-(2a - 3b) - 6 + 4a] - 2\} + 5b =$

(e) $x^2y + \{3 - [4x + (2x^2y - 7)]\} + x =$

(a) $4 + \{2x - [3y + 4 - 5x - 5y]\} = 4 + \{2x - 3y - 4 + 5x + 5y\} =$
$4 + 2x - 3y - 4 + 5x + 5y = 7x + 2y$

(b) $2a - \{4 - [3b - 5a - 7 + b] + 2\} = 2a - \{4 - 3b + 5a + 7 - b + 2\} =$
$2a - 4 + 3b - 5a - 7 + b - 2 = -3a + 4b - 13$

(c) $xy - \{3 - [2xy + 5 - 5xy] + 4\} = xy - \{3 - 2xy - 5 + 5xy + 4\} =$
$xy - 3 + 2xy + 5 - 5xy - 4 = -2xy - 2$

(d) $\{[-2a + 3b - 6 + 4a] - 2\} + 5b = \{-2a + 3b - 6 + 4a - 2\} + 5b =$
$-2a + 3b - 6 + 4a - 2 + 5b = 2a + 8b - 8$

(e) $x^2y + \{3 - [4x + 2x^2y - 7]\} + x = x^2y + \{3 - 4x - 2x^2y + 7\} + x =$
$x^2y + 3 - 4x - 2x^2y + 7 + x = -x^2y - 3x + 10$

Don't consider your answers correct unless you followed the correct procedure. It is the procedure that is most important at this point. Form the habit of taking one step at a time, changing signs as required. You will have occasion to use and remove grouping symbols frequently when adding and subtracting polynomials or solving long equations.

MULTIPLYING MONOMIALS AND POLYNOMIALS

16. In Chapter 1, frame 27, you learned that an exponent is a small numeral placed to the right and slightly above another numeral (or letter) to indicate how many times that number (or letter) is to be taken as a factor. Thus, 2^3 is a convenient way of writing $2 \cdot 2 \cdot 2$, a^3 is an abbreviated way of writing aaa, and $(3x)^3$ is a short way of writing $(3x)(3x)(3x)$. We will now discuss the multiplication of monomials.

Example: $(3a)(-4a^2)$

Multiply numerical coefficients: $(3)(-4) = -12$
Multiply literal coefficients: $(a)(a^2) = (a)(a)(a) = a^3$
Multiply the results: $(-12)(a^3) = -12a^3$

Observe that in our answer the literal factor a has an exponent of 3, indicating that the base, a, appeared as a factor three times. Note also that the exponent 3 is the *sum* of the exponents of a in the two monomials. This leads to the following rule:

> To multiply powers of the same base, keep the base and add the exponents.

Thus, $3^2 \cdot 3^3 = 3^5$; $x^4 \cdot x^5 = x^9$; $a^2 \cdot a \cdot a^3 = a^6$; $5^x \cdot 5^y = 5^{x+y}$. Or, returning to monomials (which usually are a combination of numbers *and* letters) we have these examples:

$$2x^2 \cdot 3x^3 = 6x^5 \qquad -a^4 \cdot 5a^3 = -5a^7 \qquad 3ab \cdot a^2 b \cdot 4b^3 = 12a^3 b^5$$

Multiply the following monomials using the procedure shown above.

(a) $c \cdot c^2 =$ __________ (d) $2a^2(-3ab) =$ __________

(b) $b^2 \cdot b \cdot 2b^6 =$ __________ (e) $x^2 x^3 x^4 =$ __________

(c) $2^a \cdot 2^b =$ __________ (f) $-2a \cdot 7ab \cdot b^2 =$ __________

- - - - - - - - - - - - - - - - - - -

(a) c^3; (b) $2b^9$; (c) 2^{a+b}; (d) $-6a^3b$; (e) x^9; (f) $-14a^2b^3$

17. Closely related to the procedure for multiplying powers of the same base by adding their exponents is that of raising a power to a higher power. For example, we call x^2 a *power* because it consists of a base with an exponent. Suppose we wish to raise x^2 to the third power, which we write as $(x^2)^3$. We first must remember that the exponent 3 means use the term x^2 as a factor three times: $(x^2)(x^2)(x^2)$. We have just learned that when multiplying powers of the same base we add their exponents. Therefore,

$$(x^2)(x^2)(x^2) = x^{2+2+2} = x^6$$

Notice that we get the same result if we simply multiply the two exponents (3 and 2) together. Since this is always true we can state the following rule:

> To find the power of a power of a base, keep the base and multiply the exponents.

Further examples: $(4^3)^4 = 4^{12}$ $\quad (k)^7 = k^7$

$(a^5)^3 = a^{15}$ $\quad (m^2)^3(n^3)^2 = m^6 n^6$

$(y^a)^b = y^{ab}$ $\quad (x^3)^4 x^2 x^3 = x^{17}$

Solve the following.

(a) $(a^4)^2 =$ __________ (d) $(k^4)^4 k^3 =$ __________

(b) $(x^3)^5 =$ __________ (e) $(b^5)^2(b^2)^4 =$ __________

(c) $(w^b)^c =$ __________ (f) $t^3(t^3)^3 =$ __________

- - - - - - - - - - - - - - - - - - -

(a) a^8; (b) x^{15}; (c) w^{bc}; (d) $k^{16}k^3 = k^{19}$; (e) $b^{10}b^8 = b^{18}$;
(f) $t^3 t^9 = t^{12}$

18. In Chapter 1, frame 17, we learned from the distributive law for multiplication that $a(b + c) = ab + ac$. Now let us suppose we wish to multiply 6 times 17. We can perform this multiplication in the conventional way

$$\begin{array}{r} 17 \\ \underline{6} \\ 102 \end{array}$$

Or, using the distributive law, we could do it like this:

$$\begin{aligned} 6(10 + 7) &= 6 \cdot 10 + 6 \cdot 7 \\ &= 60 + 42 \\ &= 102 \end{aligned}$$

In a practical sense, then, the distributive law tells us that if we wish to multiply the sum of two numbers by a third number, we can multiply each number of the sum by the third number and then add the products. The result will be exactly the same as if we had found the sum of the two numbers first and *then* multiplied by the third number.

Why are we concerned with this now? In algebra many numbers (algebraic expressions) are already divided into several parts. In fact, this essentially is what polynomials are. The distributive law, therefore, provides the following rule as it relates to multiplying algebraic expressions:

> To multiply a polynomial by a monomial, multiply each term of the polynomial by the monomial.

See if you can apply this rule to the following problem

$3(a + b - c) =$ ________________

$3a + 3b - 3c$

19. Here are some further examples of multiplication of polynomials by monomials using the distributive law:

$$\begin{aligned} 7(3 + x) &= 21 + 7x \\ x(4 - y) &= 4x - xy \\ -2a(3a + ab - b) &= -6a^2 - 2a^2b + 2ab \\ 3(x^2 + y - 8) &= 3x^2 + 3y - 24 \\ -ak(a + k) &= -a^2k - ak^2 \end{aligned}$$

Although we usually perform these multiplications horizontally they could, of course, be performed vertically. Thus, $-2a(3a + ab - b)$ could be written

$$\begin{array}{rr} & 3a + ab - b \\ \text{multiplied by} & \underline{\quad -2a \quad} \\ \text{equals} & -6a^2 - 2a^2b + 2ab \end{array}$$

Vertical multiplication is not really necessary with monomials. It is nearly essential, however, when multiplying polynomials of two or more terms.

Perform the following multiplications using the horizontal method.

(a) $-2(-a + 9) =$ ______

(b) $ry(r - y) =$ ______

(c) $-k^2(ak - 3 + a) =$ ______

(d) $2x^2y(x^2 - xy + 3y) =$ ______

(e) $-d(a - b - c + d) =$ ______

(f) $\frac{a}{2}(2a - 4 + 3b) =$ ______

- - - - - - - - - - - - - - - - - -

(a) $2a - 18$; (b) $r^2y - ry^2$; (c) $-ak^3 + 3k^2 - ak^2$;

(d) $2x^4y - 2x^3y^2 + 6x^2y^2$; (e) $-ad + bd + cd - d^2$; (f) $a^2 - 2a + \frac{3ab}{2}$

Note: Watch your signs! If you are in doubt about how to handle them, review the rule (Chapter 2, frame 29) for multiplying signed numbers.

20. Now let us consider the problem of multiplying a binomial by a binomial.

Example: Multiply $(2x + 3)$ by $(2 + x)$.

Arrange terms in same order:	$2x + 3$
	$x + 2$
Multiply upper binomial by	$2x^2 + 3x$
lower left and right terms:	$+ 4x + 6$
Add the like terms:	$2x^2 + 7x + 6$

In this case we multiplied each term of one binomial by each term of the other binomial, placing like terms under one another and adding.

Multiply $(4 - 2y)$ by $(y + 3)$. Show all your work so you can check your procedure. Your procedure must be correct as well as your answer.

- - - - - - - - - - - - - - - - - -

Arrange terms in same order:	$-2y + 4$
	$y + 3$
Multiply:	$-2y^2 + 4y$
	$- 6y + 12$
Add like terms:	$-2y^2 - 2y + 12$

21. Here is a slightly harder problem.

Example: Multiply $(r^2 - 3r + 5)$ by $(5r - 2)$.

$$\begin{array}{r} r^2 - 3r + 5 \\ 5r - 2 \\ \hline 5r^3 - 15r^2 + 25r \qquad \\ - 2r^2 + 6r - 10 \\ \hline 5r^3 - 17r^2 + 31r - 10 \end{array}$$

In this case we multiplied a trinomial by a binomial. The procedure was, however, exactly the same as that used in frame 20. In fact, regardless of the number of terms in either factor, the procedure is identical. It may look odd at times when few of the terms seem to match, but don't let this bother you. Just stick to the established procedure and you will arrive at the correct answer, no matter how many terms the polynomials may contain.

Perform the following multiplications, remembering to arrange the terms in order of descending powers where necessary.

(a) $(2a - 3b)(a + b) =$ ____________

(b) $(3m + 1)(4 - m) =$ ____________

(c) $(2k - 3 + k^2)(k - 4) =$ ____________

(d) $(x^2 + xy + y^2)(x + y) =$ ____________

(e) $(y + k + 2)(y + 1) =$ ____________

(f) $(x^2 + 1)(x + 2) =$ ____________

- - - - - - - - - - - - - - - - - - - -

(a) $2a^2 - ab - 3b^2$; (b) $-3m^2 + 11m + 4$; (c) $k^3 - 2k^2 - 11k + 12$;
(d) $x^3 + 2x^2y + 2xy^2 + y^3$

(e) solution:

$$\begin{array}{r} y + k + 2 \\ y + 1 \qquad \\ \hline y^2 + ky + 2y \qquad\qquad \\ + y + k + 2 \\ \hline y^2 + ky + 3y + k + 2 \end{array}$$

answer: $y^2 + ky + 3y + k + 2$

(f) solution:

$$\begin{array}{llll} x^2 & + 1 & & \\ x & + 2 & & \\ \hline x^3 & & + x & \\ & + 2x^2 & & + 2 \\ \hline x^3 & + 2x^2 & + x & + 2 \end{array}$$

answer: $x^3 + 2x^2 + x + 2$

(The trick here is to leave spaces for the missing powers of x.)

DIVIDING MONOMIALS AND POLYNOMIALS

22. Now that we have considered the matter of multiplying monomials and polynomials of more than one term, it is time to give some attention to how to divide them. We will start with the smallest of the monomials, namely, powers of a single letter.

Example: $\dfrac{x^5}{x^3}$

Assuming that $x \neq 0$, we can write this as

$$\frac{xxxxx}{xxx}$$

Written this way you will recognize at once that the three x's in the denominator will cancel out with (or, more correctly, divide into) three of the x's in the numerator.

$$\frac{\overset{1}{\cancel{x}}\overset{1}{\cancel{x}}\overset{1}{\cancel{x}}xx}{\underset{1}{\cancel{x}}\underset{1}{\cancel{x}}\underset{1}{\cancel{x}}}$$

This leaves only two factors of x, or x^2, in the numerator. Notice that the answer, x^2, would have been the same had we subtracted the exponent of the denominator (3) from the exponent of the numerator (5). Thus,

$$\frac{x^5}{x^3} = x^2, x \neq 0$$

What is the answer to the following division?

$$\frac{x^4}{x^7} = \_\_\_\_\_\_ \quad (x \neq 0)$$

$$\frac{1}{x^3}$$

If you got the correct answer you probably reasoned that the four x's in the numerator cancelled with four of the x's in the denominator.

$$\frac{xxxx}{xxxxxxx} = \frac{1}{xxx} = \frac{1}{x^3}$$

23. There is one other possible combination of powers in the numerator and denominator (other than the numerator being a higher power than the denominator or vice versa): the exponents could be equal. Since any number, letter, or combination of numbers and letters (that is not zero) divided by itself equals one (unity), then if $x \neq 0$

$$\frac{x}{x} \text{ or } \frac{x^2}{x^2} \text{ or } \frac{x^5}{x^5} \text{ or } \frac{x^{15}}{x^{15}} = 1.$$

The foregoing conclusions allow us to state three rules for dividing powers.

Rule 1: If the exponent of the numerator is larger than that of the denominator, find the quotient as follows:

$$\frac{x^a}{x^b} = x^{a-b} \qquad \frac{x^5}{x^3} = x^2 \qquad (x \neq 0)$$

Rule 2: If the exponent of the numerator is smaller than that of the denominator, find the quotient as follows:

$$\frac{x^a}{x^b} = \frac{1}{x^{b-a}} \qquad \frac{x^3}{x^5} = \frac{1}{x^2} \qquad (x \neq 0)$$

Rule 3: If the exponents of the numerator and denominator are equal, the quotient is 1.

$$\frac{x^a}{x^a} = 1 \qquad (x \neq 0)$$

With the above three rules in mind, solve the following:

$\frac{k^5}{k^9} =$ ________

- - - - - - - - - - - - - - - - - - -

$\frac{1}{k^4}$

24. If you have studied algebra before, you may find that you have a strong tendency to write $\frac{1}{k^4}$ as k^{-4}. This is equally correct since, by definition, $k^{-4} = \frac{1}{k^4}$. Putting this in somewhat more general terms we can say that if $a \neq 0$

$$a^{-n} = \frac{1}{a^n}$$

Perform the following divisions of the same base.

(a) $\frac{x^8}{x^6} =$ ________ if $x \neq 0$

(b) $\frac{4^3}{4^5} =$ ________

(c) $\frac{a^3}{a^3} =$ ________ if $a \neq 0$

(d) $\frac{x^{2a}}{x^a} =$ ________ if $x \neq 0$

(e) $\frac{b^5}{b^7} =$ ________ if $b \neq 0$

(f) $\frac{-7^3}{7^3} =$ ________

(g) $\frac{8^3}{8^7} =$ ________

(h) $\frac{m^a}{m^{2a}} =$ ________ if $m \neq 0$

- - - - - - - - - - - - - - - - - - -

(a) x^2; (b) $\frac{1}{4^2}$; (c) 1; (d) x^a; (e) $\frac{1}{b^2}$; (f) -1; (g) $\frac{1}{8^4}$; (h) $\frac{1}{m^a}$

25. Now we need to extend the technique we have been using to the division of monomials in general. Instead of just one exponential factor in the dividend or divisor, there may be several. And these may consist of a combination of letters and numbers.

Example: $\frac{-48x^3y^4}{8xy^2} = -6x^2y^2$

Dividing 8 into -48 gave -6, dividing x into x^3 gave x^2, and y^2 into y^4 left y^2 as the final factor in the quotient.

Study the following examples until you see clearly how the answer was obtained. (Assume that none of the variables is zero.)

$$\frac{24a^2}{6a} = 4a \qquad \frac{-18b^3d^7}{3bd} = -6b^2d^6$$

$$\frac{-12k^3}{-12k^3} = 1 \qquad \frac{15x^2}{-5x^3} = \frac{-3}{x}$$

$$\frac{-5a^2bc}{25ab^2c^2} = -\frac{a}{5bc} \qquad \frac{9abc}{-9a^2bc} = -\frac{1}{a}$$

Solve the following. Refer to frame 33, Chapter 2, if you are hazy about handling signs when dividing.

(a) $\frac{17a^2y^3}{-17ay} =$ ________ (c) $\frac{36xy^3}{6x^2y} =$ ________

(b) $\frac{7abc^2}{14ab} =$ ________ (d) $\frac{-28m^3n}{-7mn^3} =$ ________

- - - - - - - - - - - - - - - - - - - -

(a) $-ay^2$; (b) $\frac{c^2}{2}$; (c) $\frac{6y^2}{x}$; (d) $\frac{4m^2}{n^2}$

26. The last division we will consider here is dividing a polynomial by a monomial. The rule is simple:

To divide a polynomial by a monomial, divide each term of the polynomial by the monomial.

Examples: $\frac{9x + 6}{3} = 3x + 2$

$$\frac{8x^3 - 4x^2}{2x} = \frac{8x^3}{2x} - \frac{4x^2}{2x} = 4x^2 - 2x$$

$$\frac{7k - ak^2}{k} = 7 - ak$$

Remember, divide each term of the polynomial, one at a time, by the monomial. Here again it will help if you arrange the terms of the polynomial in descending order of powers of one letter, if they are not already in that order. The fact that the polynomial may be a lengthy one does not make the problem any more difficult, just longer.

Example: $\frac{9x^5 - 27x^4 + 18x^3 - 3x^2 + 6x}{3x} = 3x^4 - 9x^3 + 6x^2 - x + 2$

Perform the following divisions. Assume no denominators are zero. Don't forget to rearrange terms where necessary.

(a) $\frac{4k^3 - 2k^2 + 12k - 6}{2} =$ ________________________________

(b) $\frac{25a + 15a^3 - 10a^2}{5a} =$ ________________________________

(c) $\frac{7m^4n^3 - 14m^5n^4 + 21m^3n^2}{7mn} =$ ________________________________

(d) $\dfrac{9x^2y - 36xy^2}{-9xy} =$ ______________

(e) $\dfrac{cd - cdk}{cd} =$ ______________

- - - - - - - - - - - - - - - - - -

(a) $2k^3 - k^2 + 6k - 3$; (b) $3a^2 - 2a + 5$; (c) $-2m^4n^3 + m^3n^2 + 3m^2n$;
(d) $-x + 4y$; (e) $1 - k$

It cannot be repeated too often that becoming familiar with the *language* of algebra is most important. If a word or phrase (such as *polynomial* or *literal coefficient*) has no meaning to you, look it up at once. Remember: You can't understand French until you have learned the vocabulary, and algebra is in many ways a new language to you.

SELF-TEST

1. How many terms are in each of the following algebraic expressions?

(a) x^2y ______________

(b) $abc^2 - abc + c^2$ ______________

(c) $\dfrac{k^2mn}{2} + 4k^2$ ______________

(d) $\dfrac{x^3y^2}{4} + \dfrac{3}{xy}$ ______________

(frame 1)

2. Indicate which of the algebraic expressions in problem 1 above are polynomials.

(a) ______________

(b) ______________

(c) ______________

(d) ______________

(frame 2)

3. Identify each of the following expressions as either a binomial, a trinomial, or a polynomial.

(a) $ac^2 + 2c - 3$ ______________

(b) $x^3y^2 - 7xy + 7$ ______________

(c) $x^3 + 2x^2 - 4x - 12$ ______________

(d) $4 + \dfrac{a^2}{2}$ ______________

(frame 3)

4. Identify the like terms (if any) in the following.

(a) $2xy + yz - 4xyz + 3yz$ ______________

(b) $a^2b^2 + a^2 - b^2 + 2a^2$ ______________

(c) $xy + xz - yz$ ______________

(d) $2(x + y) - 3(x - y) - 4(x + y)$ ______________

(frame 5)

5. Simplify and add the following monomial terms.

(a) $2a + 4c - 6a + 2c + b =$ ______

(b) $3xy - y^2 - 2x^2 + 4y - 3x^2 =$ ______

(c) $2x^2 + 3 - 4x^2 - 5x - 2x =$ ______

(d) $2 + 7 - 4 + k =$ ______ (frame 6)

6. Rearrange these polynomials in order of descending powers, combining terms if possible.

(a) $7 - x^3 + 2x - 4x^2$ ______

(b) $xy - y^2 + 3x^2 - 2$ ______

(c) $4k^2 - 5 - 2k^2 + 7k$ ______

(d) $xy^3 + 2y - 4 - 3y^2$ ______ (frames 7-8)

7. Add the following polynomials.

(a) $(4x^2 - 3 + 2x) + (5x - 2x^2) =$ ______

(b) $(x + y) + (2x - y) + (9 + y) =$ ______

(c) $(3a^2b - ab + b^2) + (3b^2 - a^2b + 2ab) =$ ______

(d) $(k^3 + 8) + (2k - 1) + (3k - 4k^3) =$ ______ (frame 9)

8. Combine the following like terms.

(a) $(7ak) + (-ak) - (+3ak) =$ ______

(b) $(-3xy) - (-7xy) + (-4xy) =$ ______

(c) $(ab^2) - (3ab^2) + (8ab^2) - (-4ab^2) =$ ______

(d) $(7z) - (-7z) + (-7z) - (+7z) =$ ______ (frame 11)

9. Arrange in order and subtract.

(a) $2a + 3b - 4$ from $6a - b + 2 =$ ______

(b) $3 - z^2 - 4z$ from $9z + 6z^2 - 1 =$ ______

(c) $x - 2y$ from $5y - x^2 + 3x - 9 =$ ______

(d) $2a - 3a^2 + ab$ from $6a^2 - 5a =$ ______ (frame 12)

10. Simplify the following in two steps—first removing parentheses, then combining like terms.

(a) $(4xy - 3y^2) - (xy + y^2) =$ ______ $=$ ______

(b) $(ab^2 + 2a^2b) - (3ab^2 - 7) =$ ______ $=$

______

(c) $(2q - 3k) - (4k - 3q) + (-k) =$ ______________________ $=$

____

(d) $(a + b) - (b - c) - (c - d) =$ ______________________ $=$

______________ (frame 13)

11. Remove the grouping symbols and simplify.

(a) $3xy - \{4 + [2x - (xy + 7)] - 9x\} =$ ______________

(b) $-12 + \{2ak + [(3 - ak) + 4] - k^2\} =$ ______________

(c) $\{[-(x^2 - y^2) - 4] + xy\} - 3 =$ ______________

(d) $\{3w - [(w + 4k) - 8] - 7k\} =$ ______________ (frames 14–15)

12. Perform the following multiplications.

(a) $a^3 \cdot a^6 =$ ________

(b) $x^4 \cdot x^6 \cdot x =$ ________

(c) $3^a \cdot 3^b \cdot 3^c =$ ________

(d) $-2b^2c \cdot 4b^2c(b^3) =$ ________ (frame 16)

13. Perform the indicated operations.

(a) $(x^3)^3 =$ ________

(b) $(t^2)^2 \cdot t^4 =$ ________

(c) $(a^3)^2 \cdot a^2 =$ ________

(d) $(a^b)^c =$ ________ (frame 17)

14. Perform the following multiplications.

(a) $ak(a - k) =$ ______________

(b) $-2y(y - x + 3) =$ ______________

(c) $xy^2(-x + y - z) =$ ______________

(d) $\frac{b}{3}(9b - 6 + 12a) =$ ______________ (frame 19)

15. Perform the following multiplications, rearranging terms where necessary.

(a) $(3k - 2m)$ by $(k + 3m) =$ ______________

(b) $(3 + a^2 - 2a)$ by $(a + 2) =$ ______________

(c) $(xy - 2x^2 + y^2)$ by $(x - y) =$ ______________

(d) $(3ab + 2a^2 - 4)$ by $(2a + b) =$ ______________ (frame 21)

16. Perform these divisions. Assume there are no zeros in the denominators.

(a) $\frac{a^8}{a^5} =$ ________

(b) $\frac{x^5}{x^6} =$ ________

(c) $\frac{y^5z^7}{y^2z^9} =$ ________

(d) $\frac{18xy}{-3x^2y^2} =$ ________ (frame 22)

17. Perform the following divisions. Assume there are no zeros in the denominators.

(a) $\frac{25m^3}{5m} =$ ________ (c) $\frac{81k^2m^2}{-9k^4m^6} =$ ________

(b) $\frac{-49a^5}{7a^2} =$ ________ (d) $\frac{-16ab}{4a^2b^2} =$ ________ (frame 25)

18. Perform the following divisions. Assume there are no zeros in the denominators.

(a) $\frac{3c^3 - c^2 + 4c}{c} =$ ____________________

(b) $\frac{16x - 4x^3 + 12x^2}{2x} =$ ____________________

(c) $\frac{11m^3k^2 + 33mk^3 - 22mk}{11mk} =$ ____________________

(d) $\frac{3by^2 - 4by + b^2y}{by} =$ ____________________ (frame 26)

Answers to Self-Test

1. (a) 1; (b) 3; (c) 2; (d) 2
2. (a) not a polynomial; (b) polynomial; (c) polynomial; (d) not a polynomial
3. (a) trinomial; (b) trinomial; (c) polynomial of more than three terms; (d) binomial
4. (a) yz and $3yz$; (b) a^2 and $2a^2$; (c) none; (d) $2(x + y)$ and $-4(x + y)$
5. (a) $-4a + 6c + b$; (b) $-5x^2 + 3xy - y^2 + 4y$; (c) $-2x^2 - 7x + 3$; (d) $k + 5$
6. (a) $-x^3 - 4x^2 + 2x + 7$; (b) $3x^2 + xy - y^2 - 2$; (c) $2k^2 + 7k - 5$; (d) $xy^3 - 3y^2 + 2y - 4$
7. (a) $2x^2 + 7x - 3$; (b) $3x + y + 9$; (c) $2a^2b + ab + 4b^2$; (d) $-3k^3 + 5k + 7$
8. (a) $3ak$; (b) 0; (c) $10ab^2$; (d) 0
9. (a) $4a - 4b + 6$; (b) $7z^2 + 13z - 4$; (c) $-x^2 + 2x + 7y - 9$; (d) $9a^2 - 7a - ab$
10. (a) $4xy - 3y^2 - xy - y^2 = (3xy - 4y^2)$
    (b) $ab^2 + 2a^2b - 3ab^2 + 7 = 2a^2b - 2ab^2 + 7$
    (c) $2q - 3k - 4k + 3q - k = 5q - 8k$
    (d) $a + b - b + c - c + d = a + d$
11. (a) $7x + 4xy + 3$; (b) $ak - k^2 - 5$; (c) $-x^2 + xy + y^2 - 7$; (d) $2w - 11k + 8$
12. (a) a^9; (b) x^{11}; (c) 3^{a+b+c}; (d) $-8b^7c^2$
13. (a) x^9; (b) t^8; (c) a^8; (d) a^{bc}
14. (a) $a^2k - ak^2$; (b) $-2y^2 + 2xy - 6y$; (c) $-x^2y^2 + xy^3 - xy^2z$; (d) $3b^2 - 2b + 4ab$
15. (a) $3k^2 + 7km - 6m^2$; (b) $a^3 - a + 6$; (c) $-2x^3 + 3x^2y - y^3$; (d) $4a^3 + 8a^2b + 3ab^2 - 8a - 4b$
16. (a) a^3; (b) $\frac{1}{x}$; (c) $\frac{y^3}{z^2}$; (d) $\frac{-6}{xy}$
17. (a) $5m^2$; (b) $-7a^3$; (c) $\frac{-9}{k^2m^4}$; (d) $\frac{-4}{ab}$
18. (a) $3c^2 - c + 4$; (b) $8 - 2x^2 + 6x$; (c) $m^2k + 3k^2 - 2$; (d) $3y - 4 + b$

REVIEW TEST 1

Chapter 1

1. If equals are added to equals the sums are ____________.

2. Expressions that are equal to the same quantity are equal to ____________.

3. The law that says that the sum of two quantities is the same in whatever order they are added is known as the ____________ (commutative/associative/distributive) law for addition.

4. The fundamental operations of algebra (adding, subtracting, multiplying, and dividing) are basically the same as they are for arithmetic. (True / False)

5. A letter used to represent a number is called a ____________ term (or coefficient).

6. Show the three ways commonly used in algebra to indicate the multiplication of a times b. ____________

7. *Factor* is the name given to numbers or letters being ____________ (added/multiplied) together.

8. Plus (+) and minus (−) signs are used as signs of operation in algebra (just as they are in arithmetic) as well as being used as signs of quality. (True / False)

9. Subtraction is a commutative process because $a - b$ is the same as b − a. (True / False)

10. In the following expression, insert parentheses in such a way as to indicate that $3y - 6$ is to be treated as a single quantity.

    $$2x - 3y - 6 + z$$

11. Identify the literal factors in the expression $6xy(z)$. ____________

12. Use letters and symbols to change the following word statement into an algebraic expression.

    Twice a number (n) plus half the number is equal to two less than five times the number. ____________

13. What value of the letter in the denominator of the following fraction would result in an impossible division?

    $$\frac{3ky}{a-2}, a = \_\_\_\_\_\_$$

14. Evaluate the expression $3(3 - 1) - \frac{6}{3} + 4$. ____________

15. Which number is the exponent in the following expression?

    $4^3 = 64$ ________

16. Evaluate the following expression for $m = 3$, $k = 4$.

    $3(m + 2k) - \frac{6k}{3} =$ ________

17. How many separate terms (not factors) are in the following algebraic expression?

    $\frac{a}{b} + 2(4 - c) - xy^3 + 7$ ____________

18. Write the following expression using exponents.

    $xyy + kkk - \frac{z}{xx} =$ ________________________

19. What does the expression $2d^2y^3$ mean in *words*? ____________________________

    ____

20. Simplify $2a + 3b + a - 2b + 4$. ________________________

Chapter 2

21. The opposite of a positive number is a ________________________ number.

22. The term *integers* refers to the set of p_______________ and n_______________ numbers and z________ .

23. The *rational* numbers consist of the i_______________ and f_______________ .

24. Another name for the natural numbers is c_______________ numbers.

25. If +10 means 10 miles per hour over the speed limit, what does −5 mean?

    ____

26. Give the absolute values of the numbers +17 and −3 and the difference between their absolute values.

    Absolute values _______________ Difference ________

27. What does a thermometer reading of $-18°$ mean? ________________________

28. In counting from −7 to +1 is your direction of travel positive or negative?

    ____

29. Perform the following addition.

    $(-3) + (+5) =$ ____________

30. Express the following terms using a minimum number of symbols, then add them.

$$(-7) + (+3) + (-4) + (-2) = \_\_\_\_\_\_\_\_\_\_ = \_\_\_\_\_$$

31. Combine terms. $(+2) - (-3) + (-7) =$ __________

32. Multiply. $(+2a)(-3b) =$ __________

33. Multiply. $(-4)(+2)(-3) =$ __________

34. When writing additions horizontally, it is correct to omit the signs of operation and the parentheses and use only the signs of quality. (True / False)

35. Perform the indicated operation. $(-4)^3 =$ __________

36. Odd powers of negative numbers are positive. (True / False)

37. Divide as indicated.

$$\frac{-36}{-9} = \_\_\_\_\_\_\_\_\_\_$$

38. Which is greater, $+0.01$ or -19? __________

39. Multiply and divide as indicated.

$$\frac{(-3)(+4)}{(+1)(-2)} = \_\_\_\_\_\_\_\_\_\_$$

40. Evaluate the following expression for $x = -2$, $y = 3$.

$$x^2y + 2x - 3y + 6 = \_\_\_\_\_\_\_\_\_\_$$

Chapter 3

41. Is the following expression a binomial, a trinomial, or a polynomial?

$$ax^3 - \frac{x^2(1-k)}{4} + 7y^2k - 7z \quad \_\_\_\_\_\_\_\_\_\_$$

42. Which of the following are *like* terms?

$$2x^2y + 3xy - 7x^2y + 4x^2 \quad \_\_\_\_\_\_\_\_\_\_$$

43. Add the following like terms.

$$4xy - 7xy + xy = \_\_\_\_\_\_\_\_\_\_$$

44. Combine the like terms.

$$2k - 3ab + 4 - y + 7ab - ab = \_\_\_\_\_\_\_\_\_\_$$

45. Add the following terms.

$$7zyx - 2xyz + zxy + 7 = \_\_\_\_\_\_\_\_\_\_$$

46. Arrange in order of descending powers.

$12 - 2x^2 - 5x + 3x^3 =$ ______________

47. Add $(3c - 4a + 2b)$ and $(7b + 2a - c)$ horizontally, in three steps.

____

48. Add these polynomials vertically.

$3b + 2c - 4a \qquad 2a - b + 3c \qquad c - 5a + 4b$

49. Subtract $4n + 2k - 3m$ from $8n - 5k - m$, horizontally. ______________

____

50. Rearrange terms and subtract (vertically or horizontally).

$x - 8 + x^3 - 2x^2$ from $4x^2 - 4 + 3x^3 - 2x$

51. Perform the following subtraction vertically.

$(9xy + 4xz - 7zy) - (5xz + 3xy - zy)$

52. Simplify by removing grouping symbols and combining like terms.

$7x - \{9 - [4 - (x + y)] + 3y\} =$ ______________

____

53. Perform the following multiplication.

$3x^2y^3(-4xy^2) =$ ______________

54. Multiply. $(x^3)^3 \cdot x^2 \cdot x =$ ________

55. Multiply. $-3mk(m + 2k - 4) =$ ______________

56. Multiply. $3y^2 - 2 + 4y$ by $2 + y =$ ______________

57. Divide. k^7 by $k^3 =$ ________

58. Perform the following division; use positive exponents.

$\frac{y^3}{y^8} =$ ________

59. Simplify.

$$\frac{-36a^6b^2}{-9a^3b^5} = \underline{\hspace{4cm}}$$

60. Divide. $7b^3 + 28b^2 - 14b$ by $7b$ = ____________________

Answers to Review Test 1

Chapter 1

1. equal
2. each other
3. commutative
4. True
5. literal
6. $a \cdot b$, $(a)(b)$, ab
7. multiplied
8. True
9. False
10. $2x - (3y - 6) + z$
11. x, y, and z
12. $2n + \frac{n}{2} = 5n - 2$
13. $a = 2$
14. $3(2) - 2 + 4 = 8$
15. 3
16. 25
17. four
18. $xy^2 + k^3 - \frac{z}{x^2}$
19. two times two factors of d, times three factors of y, or $2ddyyy$
20. $3a + b + 4$

Chapter 2

21. negative
22. positive, negative, zero
23. integers, fractions
24. counting
25. 5 miles per hour under the speed limit
26. 17, 3, 14
27. eighteen degrees below zero
28. positive
29. +2
30. $-7 + 3 - 4 - 2 = -10$
31. $2 + 3 - 7 = -2$
32. $-6ab$
33. +24
34. True
35. −64

36. False
37. +4
38. +0.01
39. +6
40. 5

Chapter 3

41. a polynomial
42. $2x^2y$ and $-7x^2y$
43. $-2xy$
44. $3ab + 2k - y + 4$
45. $6xyz + 7$
46. $3x^3 - 2x^2 - 5x + 12$
47. $3c - 4a + 2b + 7b + 2a - c = (2a - 4a) + (7b + 2b) + (3c - c) = -2a + 9b + 2c$
48.
$$\begin{array}{r} -4a + 3b + 2c \\ 2a - b + 3c \\ \underline{-5a + 4b + c} \\ -7a + 6b + 6c \end{array}$$
49. $8n - 5k - m - 4n - 2k + 3m = (8n - 4n) + (-5k - 2k) + (3m - m) = 4n - 7k + 2m$
50. $2x^3 + 6x^2 - 3x + 4$
51.
$$\begin{array}{rr} & 9xy + 4xz - 7zy \\ \text{(subtract)} & \underline{3xy + 5xz - zy} \\ & 6xy - xz - 6zy \end{array}$$
52. $7x - \{9 - [4 - x - y] + 3y\} = 7x - \{9 - 4 + x + y + 3y\} =$
$7x - 9 + 4 - x - y - 3y = 6x - 4y - 5$
53. $-12x^3y^5$
54. x^{12}
55. $-3m^2k - 6mk^2 + 12mk$
56
$$\begin{array}{r} 3y^2 + 4y - 2 \\ \underline{y + 2} \\ 3y^3 + 4y^2 - 2y \phantom{{}-4} \\ \underline{+ 6y^2 + 8y - 4} \\ 3y^3 + 10y^2 + 6y - 4 \end{array}$$
57. k^4
58. $\frac{y^3}{y^8} = \frac{1}{y^5}$
59. $\frac{4a^3}{b^3}$
60. $\frac{7b^3 + 28b^2 - 14b}{7b} = b^2 + 4b - 2$

If you missed *any* of the above problems, be sure to review the applicable portions of the first three chapters. If you missed more than three problems from any one chapter, you would do well to review the entire chapter before continuing. In mathematics–more than in nearly any other subject–later portions of the work are highly dependent on the information presented earlier. You will avoid needless confusion later on if you will take time now to make sure you understand and remember what we have covered thus far.

CHAPTER FOUR

Special Products and Factoring

OBJECTIVES

In this chapter you will learn more about the operations of multiplying, dividing, and separating products into their factors. When you complete the chapter you will be able to

- express a number or algebraic expression as a product of prime factors;
- recognize a prime number;
- square a monomial;
- multiply and divide factors;
- factor linear algebraic expressions;
- find the principal square roots of monomials;
- find the products of binomials by inspection;
- factor trinomials;
- square binomials;
- factor polynomials completely;
- recognize perfect square expressions;
- factor the difference of two squares.

WHAT FACTORING INVOLVES

1. In Chapter 1 we reviewed the basic terminology and procedures of arithmetic and saw how these carry over into algebra. In Chapter 3 we reviewed further the meanings of such words as product, factor, and quotient. You have practiced the techniques of multiplication and division as they relate to powers, monomials, and polynomials in general. In this chapter we will devote most of our attention to the process known as factoring and to methods for finding certain kinds of special products.

   Factoring is expressing a number (or algebraic expression) as a product of certain factors. This activity generally involves finding one factor of a product and then using it as a divisor to find the other factors. The process of factoring is, therefore, the reverse of multiplication. For example, since $3 \cdot 8 = 24$, we can say that

3 and 8 are *factors* of the product 24. Or, since $a(a + 4) = a^2 + 4a$, we can say that a and $(a + 4)$ are factors of $a^2 + 4^a$. Since we know what the factors are to start with, this kind of "factoring" is obvious. It is not as easy when we have to factor a rather lengthy polynomial whose factors we do not know.

To make sure you know what we mean by factoring, as we have discussed it so far, check the answer below which you consider a correct definition of factoring.

___(a) Finding the divisor to form the quotient.
___(b) Finding the product.
___(c) Finding the numbers that form a product.

– – – – – – – – – – – – – – – – – – –

(a) You are partially correct, but take another look. Suppose we are seeking the factors of the known product 15. The procedure would be to try various numbers as divisors until we finally found one that would divide into 15 an even number of times. Thus, we would find that 3 divided into 15 five times. That is, $15 \div 3 = 5$. But of the two numbers (15 and 3) only one of them, 3, is a *factor*; the other is a *product*. You would find the other factor by dividing 3 into 15, giving you 5. Using this procedure, then, you have factored 15. But the answer you selected is a roundabout way of describing the process of factoring. Answer (c) is better.
(b) Incorrect. Finding a product consists of multiplying two or more factors together; this is just the reverse of factoring. Factoring is analyzing a given product to determine what numbers (the factors) could be multiplied together to produce the product.
(c) Correct. The *numbers* are the factors.

2. To make sure there is no confusion in your mind about the relationship between the three processes of multiplication, division, and factoring, let's analyze them.

Multiplication	*Division*	*Factoring*
factor X factor = product	product ÷ factor = factor	product = factor X factor
$3 \times 7 = 21$	$21 \div 3 = 7$	$21 = 3 \times 7$
$a(a + 3) = a^2 + 3a$	$(a^2 + 3a) \div a = a + 3$	$a^2 + 3a = a(a + 3)$

In the first example of factoring it is assumed that we know the product 21 and that we wish to find the two (or more) factors which, when multiplied together, will produce the given number. Our procedure is to try to divide 21 by various numbers until we finally find one (either 3 or 7, in this case) that will divide into it an even number of times. The number we divided by would then represent one factor, and the quotient (when it was divided into 21) would be the other factor. If we divided by 3, then 7 would be the quotient; if we divided by 7, 3 would be the quotient. In either case 3 and 7 would emerge as the factors of 21.

In the second example the factoring task consisted of finding the component multipliers of a binomial $(a^2 + 3a)$ to discover what literal or numerical factors they have in common. It is apparent that the only factor they share is a, hence dividing the binomial by a gives us the second factor.

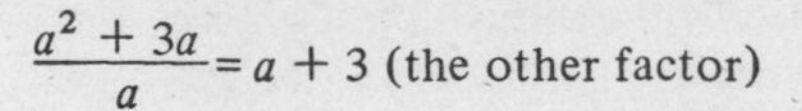

What are the two factors of $3x^2 - 6xy$? ____________________

- - - - - - - - - - - - - - - - -

$3x$ and $x - 2y$

If you followed the reasoning process discussed in frame 2, you first recognized that 3 and x were common to both terms. You therefore divided the binomial by $3x$ to find the other factor, namely $x - 2y$. Notice that we are seeking the largest common factor.

3. By now you should see that factoring really is a special kind of division in which both the divisor and quotient are to be found. *It is the divisor and the quotient that are the factors of a product.* It is important, therefore, that you be able to multiply and divide with speed and accuracy if you are going to do much factoring. The following exercises will help give you the facility you need. (See frame 26, Chapter 3 if you need help on the division problems.)

(a) $3(a + b) =$ ____________________

(b) $2x(x - 3y) =$ ____________________

(c) $x^2(1 - 3x) =$ ____________________

(d) $a^2b(b + 2a) =$ ____________________

(e) $a(x + y - z) =$ ____________________

(f) $-5y^3(3x - 2y + 4) =$ ____________________

(g) $(6x - 3y) \div 3 =$ ____________________

(h) $(8a^2 + 6a) \div 2a =$ ____________________

(i) $10k^3 - 4k^2 + 2k) \div 2k =$ ____________________

(j) $(ab + ac + ad) \div a =$ ____________________

- - - - - - - - - - - - - - - - - -

(a) $3a + 3b$; (b) $2x^2 - 6xy$; (c) $x^2 - 3x^3$; (d) $a^2b^2 + 2a^3b$;
(e) $ax + ay - az$; (f) $-15xy^3 + 10y^4 - 20y^3$; (g) $2x - y$; (h) $4a + 3$;
(i) $5k^2 - 2k + 1$; (j) $b + c + d$

4. Before considering further the subject of factoring, several related concepts should be noticed.

Two factors of any number are 1 and the number itself. Thus, 1 and 9 are factors of 9; 1 and b are factors of b. However, when we speak of "factoring a number," we usually mean to find its factors other than 1 and the number itself.

A prime number is a whole number greater than 1 which has no whole number factors except 1 and itself. Thus, 2, 3, 7, 11, and 17 are examples of prime numbers. Such numbers as 4, 6, 9, 12, and 15 are not prime numbers because they have whole number factors other than 1 and themselves; 9 is divisible by 3, 12 is divisible by 2, 3, 4, and 6, etc.

Monomials need not be factored further since they are already prime expressions. Other polynomials should be factored until they can be factored no further (that is, until all factors are prime expressions).

Answer the following.

(a) Which of these are prime numbers: 7, 13, 18, 27, 31? ______

(b) Name all the factors of 24 (other than 1 and 24). ______

(c) Can you factor this expression further?

$2x^2(3xy - 4)$ ______

- - - - - - - - - - - - - - - - - -

(a) 7, 13, and 31; (b) 2, 3, 4, 6, 8, 12; (c) No (because there are no other factors common to *both* terms of the binomial)

FACTORING POLYNOMIALS

5. You will recall that we referred to factoring as a special kind of division in which both the divisor and quotient are found. Hence, when you factor a number (or polynomial) you not only find the quotient, you also find the divisor. And it is the divisor you must find first since you cannot divide without it.

In frame 2 you learned that to factor a polynomial requires that you first identify the *common monomial factors* (that is, the literal and numerical factors common to all terms of the polynomial). In the expression $3x^2 - 6xy$, 3 and x are the common monomial factors, so we use $3x$ as a divisor and find $x - 2y$ to be the quotient. In order to factor a polynomial it is important to find both the common factors *and* the highest common factors.

Example: Find the factors of the polynomial $6x^5 + 12x^4 - 18x^3$.

Inspection of the three terms reveals that all three numerical coefficients could be divided by 3 and all the literal coefficients by x. However, 6 is the highest (that is, largest) numerical factor and x^3 is the highest literal factor. Taken together, $6x^3$ represents the *highest common monomial factor* of the polynomial. Dividing by this factor gives us the other factor, $x^2 + 2x - 3$. (Notice that we are merely reversing the distributive law for multiplication.)

What is the highest common monomial factor in this polynomial?

$12x^4y^2 + 24xy^3 - 36x^3y + 6x^2y^4$ ______

- - - - - - - - - - - - - - - - - -

$6xy$

6. Keep in mind that in selecting the highest common monomial factor of an expression, we must choose the highest common numerical factor and the smallest exponent found for any common literal factor appearing in each term of the polynomial.

Factor the following where possible.

(a) $2x + 6$ ____________

(b) $a + 3$ ____________

(c) $3x - 9$ ____________

(d) $7 + 3a$ ____________

(e) $r + ry - r^2$ ____________

(f) $5k^3 - 15k^2 + 35k$ ____________

(g) $a^2bc - ab^2c + abc^2$ ____________

(h) $6m^3 + 12m^2 - 9m + 1$ ____________

(i) $y^3 - 2y + 4$ ____________

(j) $10x^3y^2 - 5x^2y^3$ ____________

- - - - - - - - - - - - - - - - - -

(a) $2(x + 3)$; (b) not possible (prime); (c) $3(x - 3)$; (d) not possible (prime); (e) $r(1 + y - r)$; (f) $5k(k^2 - 3k + 7)$; (g) $abc(a - b + c)$ (h) not possible (prime); (i) not possible (prime); (j) $5x^2y^2(2x - y)$

SQUARING MONOMIALS

7. In order to be able to find some of the special products that we will consider later, you should be able to square monomials quickly and easily. You can learn to do so with relatively little practice. First, however, let's review some of the procedures you will apply to squaring monomials.

In Chapter 2, frame 31, you learned that even powers of negative numbers are positive (all powers of positive numbers are positive). Thus, $(-4)^2 = +16$. You also learned (Chapter 3, frame 17) that to find the power of a power of a base you keep the base and multiply the exponents. Thus, $(x^3)^2 = x^6$. Combining these two ideas enables us to square monomials.

Example: Square the term $(-3x^2y^3)$. We write this as $(-3x^2y^3)^2$.

Squaring the numerical factor we get $(-3)^2 = 9$.
Squaring the literal factors gives us $(x^2y^3)^2 = x^4y^6$. The final result is $9x^4y^6$.

From this example we can derive a general rule:

> To square a monomial, square its numerical coefficient, keep each literal factor, and double the exponent of each literal factor. Thus, $(4ab^3)^2 = (4ab^3)(4ab^3) = 16a^2b^6$.

Work the following problems. (Remember to square both numerator and denominator of a fraction.)

(a) $(2mk^2)^2$ = ____________ (d) $(-5a^3b^4)^2$ = ____________

(b) $(-1)^2$ = ____________ (e) $(-ak)^2$ = ____________

(c) $\left(\frac{1}{2}gt\right)^2$ = ____________ (f) $\left(\frac{2}{5}a^3b\right)^2$ = ____________

- - - - - - - - - - - - - - - - - -

(a) $4m^2k^4$; (b) 1; (c) $\frac{1}{4}g^2t^2$; (d) $25a^6b^8$; (e) a^2k^2; (f) $\frac{4}{25}a^6b^2$

FINDING MONOMIAL SQUARE ROOTS

8. The opposite of squaring a monomial is finding its square root. When a number can be written as the product of two equal factors (such as $25 = 5 \cdot 5$) either of the factors is called the *square root.* Thus, 5 is a square root of 25. However, since squaring either a positive or negative number produces a positive result, every positive number has two square roots which are equal in absolute value but opposite in sign, or quality. Thus, the square roots of 25 are +5 and −5; the square roots of a^2 are $+a$ and $-a$. (The square root of a negative number is not defined in the real number system.)

The positive square root of a number is called the *principal square root.* The symbol $\sqrt{\ }$ is used to indicate the principal (positive) square root of a number and is known as the *radical sign.* Thus, $\sqrt{64} = 8$. Similarly, the principal square root of a fraction is the square root of its numerator divided by the square root of its denominator. Thus, $\sqrt{\frac{25}{49}} = \frac{5}{7}$. The combination of the radical sign together with the number under it is called a *radical.* Thus, $\sqrt{64}$ and $\sqrt{\frac{25}{49}}$ are called radicals. The number under the radical sign is known as the *radicand.* If there is a plus sign or no sign at all before the radicand, the positive root is desired. When there is a minus sign before the radical, the negative root is indicated. Thus, $\sqrt{16} = 4$, whereas $-\sqrt{36} = -6$.

Supply the missing terms in each of the following.

(a) A square root of a number is one of its ______ __________ _______.

(b) The positive root is called the ____________ square root.

(c) The symbol $\sqrt{\ }$ is called the ____________ sign.

(d) Every positive number has __________ (how many) square roots.

(e) The number under the radical sign is called the ____________.

(f) The combination of the radical sign together with the number under it is known as the ____________.

- - - - - - - - - - - - - - - - - -

(a) two equal factors; (b) principal; (c) radical; (d) two; (e) radicand; (f) radical

9. Now that you are familiar with the basic concepts, symbols, and terminology of square roots, it's time to see how these apply to monomials. We will start with the following rule:

> To find the principal square root of a monomial, find the principal square root of its numerical coefficient, keep each literal factor, and use half the exponent of each literal factor.

Let's apply this rule to finding the principal square root of $36a^4b^6$. We write this $\sqrt{36a^4b^6}$. Taking the square root of the numerical coefficient gives us $\sqrt{36} = 6$. Keeping each base with half its exponent we get $\sqrt{a^4b^6} = a^2b^3$. Combining these gives $\sqrt{36a^4b^6} = 6a^2b^3$. The problem checks: $6a^2b^3 \cdot 6a^2b^3 = 36a^4b^6$.

Further Examples:

$\sqrt{64k^6m^{12}} = 8k^3m^6$ $\sqrt{81x^2y^2z^2} = 9xyz$ $\sqrt{\dfrac{16b^8}{9a^4}} = \dfrac{4b^4}{3a^2}$

Find the principal square roots of the following monomials. Check your results by showing that they fulfill the requirements of the rule above. Watch your decimal point in problem (d).

(a) $\sqrt{100} =$ ______________

(b) $\sqrt{x^2y^6z^4} =$ ______________

(c) $\sqrt{\frac{1}{4}k^4m^2} =$ ______________

(d) $\sqrt{0.04} =$ ______________

(e) $\sqrt{\dfrac{49}{81}} =$ ______________

(f) $\sqrt{\dfrac{a^4b^6}{c^6d^8}} =$ ______________

(g) $\sqrt{\dfrac{25x^{12}y^{16}}{400z^8}} =$ ______________

(h) $\sqrt{81a^{2x}} =$ ______________

- - - - - - - - - - - - - - - - - -

(a) 10; (b) xy^3z^2; (c) $\frac{1}{2}k^2m$; (d) 0.2; (e) $\frac{7}{9}$; (f) $\dfrac{a^2b^3}{c^3d^4}$;
(g) $\dfrac{5x^6y^8}{20z^4}$; (h) $9a^x$

Did you check by multiplying your answer by itself to make certain the product was the original problem? Checking is useful and satisfying; it reaffirms the concepts or procedures involved in a problem. Checking should become a habit with you.

MULTIPLYING BINOMIALS

10. Now it is time for us to consider further the matter of multiplying binomials. In Chapter 3, frame 20, you learned how to vertically multiply two binomials to get their product. This was a first and important step in learning how to multiply larger polynomials together. However, in many cases binomials can be multiplied quickly and more simply by means of horizontal multiplication. Here is an example to show you why.

Example: Multiply $(2x + 3)$ by $(x + 2)$.

Vertical Method		*Procedure*
$2x + 3$	(1)	Multiply each term of the upper binomial by each term of the lower binomial. (The arrows show how the inner and outer products are formed.)
$x + 2$		
$2x^2 + 3x$		
$+ 4x + 6$	(2)	Add like terms.
$2x^2 + 7x + 6$ (answer)		

Horizontal Method		*Procedure*
$4x$	(1)	Multiply first terms: $2x \cdot x = 2x^2$
$(2x + 3)(x + 2)$	(2)	Add product of inner and outer terms: $3x + 4x = 7x$
$3x$	(3)	Multiply last terms: $3 \cdot 2 = 6$
answer: $2x^2 + 7x + 6$	(4)	Combine terms: $2x^2 + 7x + 6$

What advantages do you see in the new (horizontal) shorter method? ______

______

- - - - - - - - - - - - - - - - - -

There are several advantages to this shorter method that may have occurred to you:

(1) Instead of having to write down the first and last terms of the product ($2x^2$ and 6) twice—as you do in the longer (vertical) method—you need write them down only once.
(2) The multiplication process requires less space, partly as a result of reason 1.
(3) The short method is mental and more direct, hence faster.

11. The two quantities that combine to make up the middle term of the binomial product (that is, the products of the inner and outer terms—$3x$ and $4x$ in the previous example) are often called *cross products.* If you will look back for a moment at the vertical method, you will see why. They are the result of multiplying the first term of one binomial by the second term of the other, and the second term of one binomial by the first term of the other (as shown by the crossed arrows).

Our next example is from frame 20, Chapter 3, where we performed the binomial multiplication by means of the long method. Now we will see how it works using the short method.

$$(-2y + 4)(y + 3)$$

$4y$

$-6y$

Multiplying the first terms gives $-2y \cdot y = -2y^2$. Adding the cross products gives $-6y + 4y = -2y$. And multiplying the last terms ($4 \cdot 3 = 12$) gives us our answer: $-2y^2 - 2y + 12$.

This method of multiplying binomials is usually referred to as finding the product *by inspection* because it is a visual or mental method that enables us to write down the answer directly.

Find the following product by inspection.

$(2x + 4)(x - 1) =$ ______________________

- - - - - - - - - - - - - - - - - - -

$2x^2 + 2x - 4$

Product of first terms:	$2x \cdot x = 2x^2$
Sum of cross products:	$4x - 2x = 2x$
Product of last terms:	$(4)(-1) = -4$

12. The practice exercises below will help you develop skill in performing mental multiplication. For (a) through (f) write in the missing terms. For (g) through (l) find the products mentally.

(a) $(k + 3)(k - 7) = k^2 - 4k -$ ________

(b) $(2a - 3)(a + 2) =$ ________ $+ a - 6$

(c) $(x - 5)(x + 4) = x^2 -$ ________ $- 20$

(d) $(3c + 2)(4c - 5) =$ ________ $- 7c - 10$

(e) $(x^2 - 2)(2x^2 + 1) = 2x^4 -$ ________ $- 2$

(f) $(5 + b)(2 - b) = 10 -$ ________ $- b^2$

(g) $(2a + 7)(a - 5) =$ ______________________

(h) $(3 - x)(5 + 2x) =$ ______________________

(i) $(k^2 - 2)(k^2 + 3) =$ ______________________

(j) $(x - 2)(x + 2) =$ ______________

(k) $(a - 2)(a - 2) =$ ______________________

(l) $(xy - 3)(xy + 4) =$ ______________________

- - - - - - - - - - - - - - - - - - -

(a) 21; (b) $2a^2$; (c) x; (d) $12c^2$; (e) $3x^2$; (f) $3b$; (g) $2a^2 - 3a - 35$; (h) $15 + x - 2x^2$; (i) $k^4 + k^2 - 6$; (j) $x^2 - 4$; (k) $a^2 - 4a + 4$; (l) $x^2y^2 + xy - 12$

Did you notice in problem (j) that the middle term disappeared? This is typical of products of binomials that differ only by their middle signs. We will discuss these further a little later on because they provide a useful method for solving a particular kind of problem.

FACTORING TRINOMIALS

13. Now let us consider the steps involved in factoring a trinomial. Factoring a trinomial is a kind of guessing game—sometimes easy and sometimes rather difficult—but usually challenging and interesting. Factoring trinomials is important because (as you can observe from your answers to the problems of frame 12) the product

of most binomials is a trinomial. Occasionally, the product is another binomial, as in problem (j). Or it could be a polynomial of four terms, as in $(x + y)(x + z) = x^2 + xy + xz + yz$. In any case, to find the two binomials whose product is a certain trinomial (which is often necessary in algebra), we must be able to factor the trinomial. The ability to do so is primarily a matter of being observant and using common sense.

Example: Factor $a^2 + 5a + 6$.

The first terms of the two binomials we are seeking are a and a since their product is a^2, the first term of our trinomial. Thus, we can write

$$a^2 + 5a + 6 = (a \qquad)(a \qquad)$$

Since the trinomial has no minus signs, there cannot be any minus signs in either of the binomials. Therefore, we can write plus signs after the first terms, giving us

$$a^2 + 5a + 6 = (a + \quad)(a + \quad)$$

All that is left is to find the two numbers which, when multiplied together, give 6 (the last term of our trinomial) and when added together give 5 (the numerical coefficient of the middle term). Fill in these last two numbers to complete the binomial factors below.

$$a^2 + 5a + 6 = (a + \_\_)(a + \_\_)$$

- - - - - - - - - - - - - - - - - -

$(a + 3)(a + 2)$

14. Here is another example of trinomial factoring.

Example: Factor $x^2 - 2x - 24$.

By examining the first term we can write at once

$$x^2 - 2x - 24 = (x \qquad)(x \qquad)$$

The fact that the middle term is minus means one of two things: either the second terms of both binomials are minus or one is minus and the other plus. If *both* terms were minus, the third term (their product) would be plus since the product of two negative numbers is positive (Chapter 2, frame 28). In this case the third term is minus, telling us that the second terms of the two binomials have opposite signs. Thus we can write

$$x^2 - 2x - 24 = (x + \quad)(x - \quad)$$

Now we must find the pair of factors of -24 with an algebraic sum of -2, the coefficient of the middle term. Since -6 and 4 are the only factors of -24 with a sum of -2, we can write

$$x^2 - 2x - 24 = (x + 4)(x - 6)$$

Could the positions of the numbers in the binomials be reversed? That is, would it be equally correct to write $(x + 6)(x - 4)$? ________________

No, because this would result in an incorrect sign for the middle term of the trinomial (plus instead of minus). This is something you must keep in mind when factoring. Remember that the sign of the larger cross product controls the sign of the middle term of the trinomial (assuming, of course, that the signs are different).

15. It is time for you to try some trinomial factoring on your own. Factor the following trinomials.

(a) $a^2 + a - 12 =$ ______________________

(b) $x^2 - 2x - 8 =$ ______________________

(c) $k^2 - k - 20 =$ ______________________

(d) $c^2 + 2c - 3 =$ ______________________ (Note: Here you must factor the third term into 3 and 1 since these are the only factors it has.)

(e) $b^2 + 5b - 14 =$ ______________________

(f) $x^2 - 5x - 6 =$ ______________________ (Tip: The obvious factors, 3 and 2, will not work.)

(g) $p^4 - p^2 - 30 =$ ______________________

(h) $q^2 - 5q + 6 =$ ______________________

(a) $(a + 4)(a - 3)$; (b) $(x - 4)(x + 2)$; (c) $(k - 5)(k + 4)$; (d) $(c + 3)(c - 1)$; (e) $(b + 7)(b - 2)$; (f) $(x - 6)(x + 1)$; (g) $(p^2 - 6)(p^2 + 5)$; (h) $(q - 3)(q - 2)$

16. You may have noticed that in the trinomials we have factored so far, the numerical coefficient of the first term has always been one (unwritten). It would be convenient if all trinomials were this way. Often, however, they are not so obliging. Let's consider how to factor trinomials whose highest-power terms have numerical coefficients greater than one.

Example: Factor $4x^2 - 8x - 21$.

Here the numerical coefficient 4 makes this trinomial different from those we previously factored. It is apparent that we cannot write at once the first term of each binomial since we do not know the correct factors of 4. However, we can start by factoring the literal coefficient of 4 (x^2) which gives us

$$4x^2 - 8x - 21 = (\ x \qquad)(\ x \qquad)$$

We know something else from our trinomial: the minus sign before the last term tells us that the signs of the last terms of the binomial factors are opposite. Therefore, we can write

$$4x^2 - 8x - 21 = (\ x + \quad)(\ x - \quad)$$

From this point on it is a matter of trial and error. The challenge is to combine the possible factors of 4 and 21 in such a way as to produce −8 as the algebraic sum of the cross products. See if you can complete the factoring. Fill in the missing binomial terms that will produce the correct factors.

$$4x^2 - 8x - 21 = (\_\_x + \_\_)(\_\_x - \_\_)$$

- - - - - - - - - - - - - - - - - - -

$(2x + 3)(2x - 7)$

17. As you gain experience in factoring you will find you have to make fewer trials. You also will make fewer errors. Bear in mind that the sign of the third term of the trinomial is your clue as to whether the signs of the last terms of the binomial factors will be like or unlike.

The following problems will give you additional practice in factoring. Supply the missing terms and signs as required. For problems (g) through (l) you are on your own!

(a) $2x^2 + x - 15 = (2x - \_\_)(x + \_\_)$

(b) $3a^2 - 2a - 5 = (3a - \_\_)(a + \_\_)$

(c) $9k^2 - 6k + 1 = (3k\_\_\_\_)(3k\_\_\_\_)$

(d) $5c^2 + 17c + 14 = (\_\_\_ + 7)(\_\_\_ + 2)$

(e) $6a^2 + a - 5 = (\_\_\_ - 5)(\_\_\_ + 1)$

(f) $9y^2 + 3y - 2 = (\_\_\_\_\_\_2)(\_\_\_\_\_\_1)$

(g) $3x^2 + 8x + 5 =$ ________________

(h) $2a^2 - 9a + 4 =$ ________________

(i) $7k^2 + 9k + 2 =$ ________________

(j) $3x^2 - 2xy - 5y^2 =$ ________________

(k) $3m^2 + 11mn - 20n^2 =$ ________________

(l) $1 - 2a - 3a^2 =$ ________________ (Do not change the order of terms.)

- - - - - - - - - - - - - - - - - - -

(a) $(2x - 5)(x + 3)$; (b) $(3a - 5)(a + 1)$; (c) $(3k - 1)(3k - 1)$;
(d) $(5c + 7)(c + 2)$; (e) $(6a - 5)(a + 1)$; (f) $(3y + 2)(3y - 1)$;
(g) $(3x + 5)(x + 1)$; (h) $(2a - 1)(a - 4)$; (i) $(7k + 2)(k + 1)$;
(j) $(3x - 5y)(x + y)$; (k) $(3m - 4n)(m + 5n)$ (l) $(1 - 3a)(1 + a)$

18. How would you factor this polynomial into prime factors?

$$5a^2b^2 - 15ab^2 + 10b^2 = \_\_\_\_\_\_\_\_\_\_\_\_\_\_\_\_$$

- - - - - - - - - - - - - - - - - - -

$5b^2(a^2 - 3a + 2) = 5b^2(a - 2)(a - 1)$

19. Did you recognize that $5b^2$ was a common monomial factor in all the terms and should, therefore, be extracted *before* trying to find the binomial factors? Learn to

look for factors that are common to all the terms of a polynomial expression. Remove the highest common factor first, then continue factoring until the polynomial can be factored no further. Practice this in the following exercises.

(a) $10x^2 + 12x + 2 =$ ______

(b) $14ah^2 + 20ah + 6a =$ ______

(c) $15x^2 - 12x - 3 =$ ______

(d) $6a^3 + 8a^2 + 2a =$ ______

(e) $5x^2y^2 + 10xy^2 + 5y^2 =$ ______

(f) $3ax^2 + 3ax - 18a =$ ______

- - - - - - - - - - - - - - - - - -

(a) $2(5x + 1)(x + 1)$; (b) $2a(7h + 3)(h + 1)$; (c) $3(5x + 1)(x - 1)$; (d) $2a(3a + 1)(a + 1)$; (e) $5y^2(x + 1)(x + 1)$; (f) $3a(x + 3)(x - 2)$

SQUARING BINOMIALS

20. Earlier in this chapter you practiced multiplying binomials. Squaring a binomial deserves some special attention because it occurs frequently, the product assumes a regular form, and it can be done easily once you understand the method.

We know, for example, that $(x + y)^2$ means $(x + y)(x + y)$ and that the trinomial product is $x^2 + 2xy + y^2$. Notice that this trinomial product consists of the square of the first term (of the binomial), twice the product of both terms, and the square of the second term. Observe this again in these two examples:

$$(x + 2)^2 = x^2 + 4x + 4$$
$$(x - 2)^2 = x^2 - 4x + 4$$

How do these last two trinomials differ from each other? ______

______

- - - - - - - - - - - - - - - - - -

They differ only by the sign of the middle term. Thus, when the sign separating the two terms of the binomial is positive, the middle term of the product is positive. When the interior binomial sign is negative, the middle term of the product is negative.

21. What we have just learned in the preceding frame gives us the following procedure for squaring a binomial.

Example: $(2x - 3)^2$

(1) Square the first term:	$(2x)^2 = 4x^2$
(2) Double the algebraic product of the two terms:	$2(2x)(-3) = -12x$
(3) Square the second term:	$(-3)^2 = +9$
(4) Arrange terms:	$(2x - 3)^2 = 4x^2 - 12x + 9$

Use this procedure to square the following binomials.

(a) $(a + 6)^2 =$ ______

(b) $(b - 3)^2 =$ ______

(c) $(2x + 1)^2 =$ ______

(d) $(x + 7)^2 =$ ______

(e) $(n - 5)^2 =$ ______

(f) $(3k + 4)^2 =$ ______

(g) $(3 - 2p)^2 =$ ______

(h) $(p + k)^2 =$ ______

- - - - - - - - - - - - - - - - - -

(a) $a^2 + 12a + 36$; (b) $b^2 - 6b + 9$; (c) $4x^2 + 4x + 1$; (d) $x^2 + 14x + 49$; (e) $n^2 - 10n + 25$; (f) $9k^2 + 24k + 16$; (g) $9 - 12p + 4p^2$; (h) $p^2 + 2pk + k^2$

THE PERFECT SQUARE TRINOMIAL

22. Reversing our analysis of the foregoing problems reveals that the factors of a perfect square trinomial are always the square of a binomial. A trinomial is a perfect square if (when it is arranged in either ascending or descending power of one letter) the first and third terms are positive perfect squares and the middle term is equal to twice the product of the square roots of the first and third terms.

Example: Factor $x^2 + 10x + 25$.

Our problem is to decide whether or not this is a perfect square trinomial. It is apparent that the first and third terms are positive, that the first term is the square of x, and that the third term is the square of 5. Also, $10x$ (the middle term) is equal to twice the product of x and 5 (the square roots of the first and third terms). Therefore, the trinomial $x^2 + 10x + 25$ is a perfect square. Since the sign of the middle term is positive, we can correctly conclude that $(x + 5)^2$ is the factored form of $x^2 + 10x + 25$.

Example: Factor $a^2 - 6a + 9$.

Once more it is evident that the first and third terms are the positive perfect squares of a and 3 and that the middle term is twice the product of a and 3. Since the sign of the middle term is negative, the factored form of $a^2 - 6a + 9$ must be $(a - 3)^2$.

Would you say that $x^2 - 2x + 1$ is a perfect square trinomial? ______

- - - - - - - - - - - - - - - - - -

Yes, because x^2 and 1 are the positive squares of x and 1 respectively, and twice the product of x and 1 is $2x$, the middle term. The negative sign of the middle term indicates that the sign of the second term of the binomial root will be minus; therefore, the factored form of $x^2 - 2x$ is $(x - 1)^2$.

23. The following exercises will help you become more adept at recognizing perfect and imperfect trinomials. Give the factored form of the trinomials below that are perfect squares. (Be careful! Some are *not* perfect squares.)

(a) $k^2 + 12k + 36 =$ ____________

(b) $a^2 - 16a + 64 =$ ____________

(c) $x^2 + x + 1 =$ ____________

(d) $9x^2 - 6x + 1 =$ ____________

(e) $z^2 + 12z - 36 =$ ____________

(f) $x^4 - 4x^2 + 4 =$ ____________

(g) $b^2 + b + \frac{1}{4} =$ ____________

(h) $81m^2 + 18m + 1 =$ ____________

- - - - - - - - - - - - - - - - - - -

(a) $(k + 6)^2$; (b) $(a - 8)^2$; (c) not a perfect square (middle term is x, not $2x$); (d) $(3x - 1)^2$; (e) not a perfect square (the term 36 is negative); (f) $(x^2 - 2)^2$; (g) $\left(b + \frac{1}{2}\right)^2$; (h) $(9m + 1)^2$

FINDING THE SUM AND DIFFERENCE OF TWO TERMS

24. So far you have learned how to quickly square binomials of the form $a + b$ and $a - b$ and how to factor perfect square trinomials. There is, however, one more binomial multiplication you should practice: the product $(a + b)(a - b)$. This is often referred to as finding the product of the sum and difference of two terms. Let's see how this works out.

$$\overset{+ab}{}$$

$$(a + b)(a - b) = a^2 - b^2$$

$$\underset{-ab}{}$$

The point to note here is that the two cross products ($+ab$ and $-ab$) add to zero, eliminating the middle term we expect to see in the product. The product of this multiplication is, therefore, a binomial rather than a trinomial. The rule for performing this kind of multiplication is quite simple:

> The product of the sum and difference of two terms is equal to the square of the first term minus the square of the second term.

Thus, $(x + 3)(x - 3) = x^2 - 9$; $(xy - 4)(xy + 4) = x^2y^2 - 16$; or $17 \times 23 = (20 - 3)(20 + 3) = 20^2 - 3^2 = 400 - 9 = 391$. The last example shows how this technique can be used to simplify some awkward multiplications of numbers where the two factors can be expressed as the sum and difference of the same two numbers (20 and 3 in the case above).

Use the rule given above to multiply these two binomials.

$(2ax + 3)(2ax - 3) =$ ______________

$4a^2x^2 - 9$

25. Although the sum-and-difference rule is generally used when multiplying algebraic expressions rather than numbers (as we saw in the final example of frame 24), it can also be used to multiply special number products in arithmetic. The only requirement is that the two numbers be expressed as the sum and difference of another pair of numbers, one of which should be a multiple of ten (such as 20, 30, 40, etc.) for ease of calculation.

For example, 35 X 25 could easily be multiplied using the sum-and-difference rule because 35 can be expressed as 30 + 5, and 25 as 30 − 5. Thus,

$$(35)(25) = (30 + 5)(30 - 5) = 30^2 - 5^2 = 900 - 25 = 875$$

A few problems of this type are included in the following exercises to aid you in recognizing multiplications that lend themselves readily to this approach. Multiply the following by inspection.

(a) $(c + 7)(c - 7) =$ ______________

(b) $(3y + 4)(3y - 4) =$ ______________

(c) $(x + ny)(x - ny) =$ ______________

(d) $(9 + 3)(9 - 3) =$ ______________ (binomial product) $=$ ______________

(Subtract mentally the square of the second term from the square of the first term.)

(e) $(2h - k)(2h + k) =$ ______________

(f) $(20 + 6)(20 - 6) =$ ______________ (binomial product) $=$ ______________

(g) $(33 \times 27) =$ (______ sum)(______ diff.) $=$ ______________ (binomial product) $=$ ______

(h) $(1 - z^2)(1 + z^2) =$ ______________

(i) $(105)(95) =$ (______ sum)(______ diff.) $=$ ______________ (binomial product) $=$ ______

(j) $\left(\frac{a}{b} + 4\right)\left(\frac{a}{b} - 4\right) =$ ______________

(a) $c^2 - 49$; (b) $9y^2 - 16$; (c) $x^2 - n^2y^2$; (d) $81 - 9 = 72$; (e) $4h^2 - k^2$; (f) $400 - 36 = 364$; (g) $(30 + 3)(30 - 3) = 900 - 9 = 891$; (h) $1 - z^4$; (i) $(100 + 5)(100 - 5) = 10{,}000 - 25 = 9975$; (j) $\frac{a^2}{b^2} - 16$

FACTORING THE DIFFERENCE OF TWO SQUARES

26. Now that you have learned how to multiply the sum and difference of two terms, we will use this knowledge to formulate a rule for factoring the difference of two squares. Consider the following procedure:

multiplying		*product*		*factoring*
$(x + y)(x - y)$	=	$x^2 - y^2$	=	$(x + y)(x - y)$

From the above we can state this rule:

To factor the difference of two squares,

(1) Take the square root of each of the squares.
(2) Write the sum of these square roots as one factor and their difference as the other.

Example: Factor $c^2 - 9y^2$.

Since $\sqrt{c^2} = c$ and $\sqrt{9y^2} = 3y$, then $c^2 - 9y^2 = (c - 3y)(c + 3y)$.

Example: Factor $1 - a^2b^2$.

Since $\sqrt{1} = 1$ and $\sqrt{a^2b^2} = ab$, then $1 - a^2b^2 = (1 - ab)(1 + ab)$.

Factor the following where possible.

(a) $a^2 - b^2 =$ ________________

(b) $16 - y^2 =$ ________________

(c) $k^2 + 25 =$ ________________

(d) $25x^2 - 64y^2 =$ ________________

(e) $36 - a^4 =$ ________________

(f) $a^2 - b^4 =$ ________________

(g) $1 - 81r^4 =$ ________________

(h) $p^4 - m^2n^2 =$ ________________

- - - - - - - - - - - - - - - - - -

(a) $(a - b)(a + b)$; (b) $(4 + y)(4 - y)$; (c) can't be factored because of the + sign; (d) $(5x + 8y)(5x - 8y)$; (e) $(6 - a^2)(6 + a^2)$; (f) $(a + b^2)(a - b^2)$; (g) $(1 - 9r^2)(1 + 9r^2) = (1 + 3r)(1 - 3r)(1 + 9r^2)$ since $(1 - 9r^2)$ can be further factored into $(1 + 3r)(1 - 3r)$; (h) $(p^2 + mn)(p^2 - mn)$

This essentially completes our discussion of special products and factoring. Before going on to the Self-Test, let's summarize the main points we have covered in this chapter.

A polynomial can be factored if:

(1) it contains a monomial factor;
(2) it is a binomial that is the difference of two squares;
(3) it is a trinomial that can be factored by trial and error or a perfect square trinomial.

Other techniques can be used to factor polynomials but they are beyond the scope of this course.

SELF-TEST

1. Multiply.

(a) $4(c + d) =$ ________________

(b) $3a(a + 4) =$ ________________

(c) $2k^2(3k - 7) =$ ________________

(d) $3xy(2x^2y - x + 4y) =$ ________________

(e) $z(a + b - cz) =$ ________________

(f) $-3m^2h(m - h + 5) =$ ________________ (frame 3)

2. Divide.

(a) $(24a - 6b)$ by $3 =$ ________________

(b) $(4x^2 + 12x)$ by $2x =$ ________________

(c) $(3m^2n + 9mn - 12mn^2)$ by $3mn =$ ________________

(d) $(2x + 2y - 2)$ by $2 =$ ________________ (frame 3)

3. Supply the missing factors.

(a) $2x^2 - 4 = 2($________________$)$

(b) $3k^3 + 6k - 9 = 3($____________$)$

(c) $4a^3b^2 - 12a^2b^2 + 8ab =$ ______$(a^2b - 3ab + 2)$

(d) $ab - bc + db =$ _____$(a - c + d)$ (frame 6)

4. Factor the following where possible.

(a) $ay + 4a =$ ________________

(b) $m + mn - 2m =$ ________________

(c) $3x + 4y =$ ________________

(d) $3a^2 - 6a + 15 =$ ________________

(e) $x^4 - 2x^3 + 3x^2 =$ ________________

(f) $2mk - 6ay + 14b + 4 =$ ________________ (frame 6)

5. Square the following monomials as indicated.
   (a) $(3ab)^2 =$ ______________
   (b) $(-2k^2m)^2 =$ ______________
   (c) $(bcd)^2 =$ ______________
   (d) $(-6x^3y^4)^2 =$ ______________
   (e) $(\frac{3}{4}k^2t)^2 =$ ______________ (frame 7)

6. Find the principal square roots of the following.
   (a) $\sqrt{81k^4} =$ ______________
   (b) $\sqrt{9a^6b^2c^8} =$ ______________
   (c) $\sqrt{\frac{25x^2}{49y^4}} =$ ______________
   (d) $\sqrt{121x^{2k}} =$ ______________ (frame 9)

7. Find the following products by inspection.
   (a) $(x + 2)(x - 7) =$ ______________
   (b) $(2a - 3)(a + 4) =$ ______________
   (c) $(3k + 4)(3k - 4) =$ ______________
   (d) $(x + y)(x + y) =$ ______________
   (e) $(x^2 - 3)(2x^2 + 6) =$ ______________
   (f) $(4g + 3t)(2g - t) =$ ______________ (frame 12)

8. Factor.
   (a) $x^2 - 3x - 18 =$ ______________
   (b) $k^2 + k - 2 =$ ______________
   (c) $a^4 - 9a^2 - 36 =$ ______________
   (d) $28 + 11k + k^2 =$ ______________ (frame 15)

9. Factor.
   (a) $4a^2 + 5a + 1 =$ ______________
   (b) $7x^2 - 15x + 2 =$ ______________
   (c) $5a^2b^2 + 7ab + 2 =$ ______________
   (d) $3q^6 + 8q^3 - 3 =$ ______________
   (e) $3 + 4a - 7a^2 =$ ______________
   (f) $2b^4 - 9b^2 + 7 =$ ______________ (frame 17)

10. Factor the following polynomials completely.

(a) $16a^2 - 4ab^2 =$ ______

(b) $3p^2 + 30p + 48 =$ ______

(c) $6x^2 - 12xy + 6y^2 =$ ______

(d) $6a^4b - 26a^3b + 28a^2b =$ ______

(e) $16r^2q^3 - 8r^2q^2 - 15r^2q =$ ______

(f) $12t^3p - 3tp - 9t^2p =$ ______ (frame 19)

11. Square the following binomials.

(a) $(n-7)^2 =$ ______

(b) $(3-4k)^2 =$ ______

(c) $(ab+cd)^2 =$ ______

(d) $(j-rt)^2 =$ ______

(e) $(8-pk)^2 =$ ______

(f) $(x^n - y^m)^2 =$ ______ (frame 21)

12. Indicate the factored form of any of the following trinomials that are perfect squares.

(a) $y^2 - 2y + 1 =$ ______

(b) $4x^2 - 12xy + 9y^2 =$ ______

(c) $64t^2 - 16t + 1 =$ ______

(d) $a^{2n} + 2a^n + 1 =$ ______

(e) $a^2b^2 - 2abc + c^2 =$ ______

(f) $1 - 2k^2 + k^4 =$ ______ (frame 23)

13. Multiply by inspection.

(a) $(a-k)(a+k) =$ ______

(b) $(4+x^2)(4-x^2) =$ ______

(c) $(10+3)(10-3) =$ ______

(d) $(a^x - b^y)(a^x + b^y) =$ ______

(e) $(54)(46) = ($______$)($______$) =$ ______

(frame 25)

14. Factor the following where possible.

(a) $k^2 - 1 =$ ______

(b) $4t^2 - p^4 =$ ______

(c) $a^2b^2 - 6c^5 =$ ______ (frame 26)

(d) $64x^6 - 81y^8 =$ ________________________________

(e) $m^4 + n^4 =$ ________________________________

(f) $16 - 9t^2 =$ ________________________________

(g) $\frac{4}{9} - 4p^4 =$ ________________________________

(h) $x^{2n} - y^{4n} =$ ________________________________ (frame 26)

Answers to Self-Test

1. (a) $4c + 4d$; (b) $3a^2 + 12a$; (c) $6k^3 - 14k^2$; (d) $6x^3y^2 - 3x^2y + 12xy^2$; (e) $az + bz - cz^2$; (f) $-3m^3h + 3m^2h^2 - 15m^2h$
2. (a) $8a - 2b$; (b) $2x + 6$; (c) $m + 3 - 4n$; (d) $x + y - 1$
3. (a) $(x^2 - 2)$; (b) $(k^3 + 2k - 3)$; (c) $4ab$; (d) b
4. (a) $a(y + 4)$; (b) $m(1 + n - 2)$; (c) not factorable; (d) $3(a^2 - 2a + 5)$; (e) $x^2(x^2 - 2x + 3)$; (f) $2(mk - 3ay + 7b + 2)$
5. (a) $9a^2b^2$; (b) $4k^4m^2$; (c) $b^2c^2d^2$; (d) $36x^6y^8$; (e) $\frac{9}{16}k^4t^2$
6. (a) $9k^2$; (b) $3a^3bc^4$; (c) $\dfrac{5x}{7y^2}$; (d) $11x^k$
7. (a) $x^2 - 5x - 14$; (b) $2a^2 + 5a - 12$; (c) $9k^2 - 16$; (d) $x^2 + 2xy + y^2$; (e) $2x^4 - 18$; (f) $8g^2 + 2gt - 3t^2$
8. (a) $(x - 6)(x + 3)$; (b) $(k + 2)(k - 1)$; (c) $(a^2 - 12)(a^2 + 3)$; (d) $(4 + k)(7 + k)$
9. (a) $(4a + 1)(a + 1)$; (b) $(7x - 1)(x - 2)$; (c) $(5ab + 2)(ab + 1)$; (d) $(3q^3 - 1)(q^3 + 3)$; (e) $(3 + 7a)(1 - a)$; (f) $(2b^2 - 7)(b^2 - 1)$
10. (a) $4a(4a - b^2)$; (b) $3(p + 8)(p + 2)$; (c) $6(x - y)(x - y)$; (d) $2a^2b(3a - 7)(a - 2)$; (e) $r^2q(4q - 5)(4q + 3)$; (f) $3tp(4t + 1)(t - 1)$
11. (a) $n^2 - 14n + 49$; (b) $9 - 24k + 16k^2$; (c) $a^2b^2 + 2abcd + c^2d^2$; (d) $j^2 - 2jrt + r^2t^2$; (e) $64 - 16pk + p^2k^2$; (f) $x^{2n} - 2x^ny^m + y^{2m}$
12. (a) $(y - 1)^2$; (b) $(2x - 3y)^2$; (c) $(8t - 1)^2$; (d) $(a^n + 1)^2$; (e) $(ab - c)^2$; (f) $(1 - k^2)^2 = (1 + k)(1 - k)(1 + k)(1 - k)$
13. (a) $a^2 - k^2$; (b) $16 - x^4$; (c) $100 - 9 = 91$; (d) $a^{2x} - b^{2y}$; (e) $(50 + 4)(50 - 4) = 2500 - 16 = 2484$
14. (a) $(k - 1)(k + 1)$; (b) $(2t - p^2)(2t + p^2)$; (c) not factorable; (d) $(8x^3 - 9y^4)(8x^3 + 9y^4)$; (e) not factorable; (f) $(4 - 3t)(4 + 3t)$; (g) $\left(\frac{2}{3} - 2p^2\right)\left(\frac{2}{3} + 2p^2\right)$; (h) $(x^n - y^{2n})(x^n + y^{2n})$

Be sure to review the frames indicated for any problems you may have missed. Then go to Chapter 5.

CHAPTER FIVE

Fractions

OBJECTIVES

The kinds of fractions we are going to be talking about in this chapter are basically no different from those you learned about in arithmetic. They may look different—primarily because in many cases they will contain letters and powers of letters—but the rules governing their addition, subtraction, multiplication, and division remain essentially the same. So techniques for handling algebraic fractions are not really as mysterious and unfamiliar as they may seem at times.

This chapter will serve as a review of fractions in general and as an introduction to their use in algebra. Thus, we will be working with both numerical fractions and fractions that contain letters. When you have completed the work in this chapter you will be able to:

- evaluate fractions;
- recognize and convert equivalent fractions,
- work with reciprocals;
- reduce fractions to lowest terms;
- multiply, divide, add, and subtract algebraic fractions;
- handle the signs of fractions;
- simplify algebraic fractions;
- combine fractions containing binomials, rearranging terms as necessary;
- change mixed expressions into single fractions in simplest form;
- change algebraic fractions into mixed expressions;
- simplify complex fractions.

WHY FRACTIONS?

1. From arithmetic you are familiar with the concept of a fraction as part of a whole. For example, we frequently picture such fractions as $\frac{1}{8}$, $\frac{1}{4}$, or $\frac{1}{2}$ as sections of a pie or parts of a circle. Fractions enable us to represent numerically the fact that we don't have exactly one, or two, or more complete objects. For the same reason we need fractions in arithmetic we need them in algebra. More importantly, however,

we need fractions to guarantee that the answer to any division will be the type of number we can use. We could not do this if we were limited to the set of integers which, as you will recall from Chapter 2 (frame 3), corresponds to the positive and negative whole numbers and zero.

For example, suppose we wish to divide 3 by 4. Can you think of some single integer that could be used to represent the result of such a division?

____

- - - - - - - - - - - - - - - - - -

Hopefully your answer is no since there is no such integer.

2. So we have the set of rational numbers, as you learned in Chapter 2. This set is considered to be all the numbers that can be formed by dividing some integer by any integer other than zero. In other words, the set of rational numbers consists of all the numbers that can be written in the form $\frac{a}{b}$ where a is any integer (or polynomial) and b is any non-zero integer (or non-zero polynomial). With this set of numbers we can show the division of any two integers or of any two rational numbers. Thus, if we wish to divide 3 by 4, we merely write the answer as $\frac{3}{4}$, and we must be able to work with this number (fraction) in any way that becomes necessary. The methods for doing so are what we will consider in this chapter.

You can see, therefore, that in algebra we think of a fraction as simply another way of writing a division. In fact, in algebra we normally write any division as a fraction.

Any rational number either has the appearance of a fraction or can be written as a fraction. How would you write the rational number 7 as a fraction?

____

- - - - - - - - - - - - - - - - - -

$\frac{7}{1}$

Any integer (other than zero) has the implied denominator 1, although it is seldom written. Don't forget that it is there!

3. With the fractions used in arithmetic it is easy to tell the size of the numbers. For example, $\frac{1}{4}$, $\frac{2}{3}$, and $\frac{3}{8}$ call up a mental picture of a certain portion of something. You can also locate these numbers on some convenient number line. However, fractions used in algebra may not have this property; in fact, they seldom do. Thus, the fractions $\frac{3}{b}$, $\frac{a}{c}$, and $\frac{2x-3}{y}$ do not call up any mental picture, nor can you locate them on a number line unless you know the values of the letters.

Let's review, very briefly, a few fundamentals about fractions. Most of the points we cover you will be familiar with from arithmetic, some others you may know as a result of our discussion here and in previous chapters, but one or two may be new to you.

The terms of a fraction are *numerator* and *denominator*. Thus, in the fraction $\frac{a}{b}$ (which we read as "a divided by b"), a is the numerator and b is the denominator.

The short horizontal line that separates the numerator from the denominator is called the *fraction bar* (or, sometimes, the *vinculum*). In addition to indicating a division the fraction bar also serves as a sign of grouping. For example $\frac{2x + 3}{5}$ means that each part of the numerator, $2x$ and 3, is to be divided by 5. Thus, if we wish to separate the two terms of the numerator, we would give each of them the denominator 5, so they would look like this: $\frac{2x}{5} + \frac{3}{5}$. In general we can say

$$\frac{a + b}{c} = \frac{a}{c} + \frac{b}{c}$$

In addition to the fraction bar, two other symbols can be used to indicate division. Of these, ÷ is the more common, although the colon (:) is sometimes used to express either division or ratio. Therefore, you can express the fraction (indicating division) $\frac{3}{5}$ as $3 \div 5$ or 3:5. A ratio (such as 3 to 5) may also be written in fractional form as $\frac{3}{5}$.

Finally, don't forget that a fraction may also represent part of a whole thing–its usual arithmetic meaning.

Test your recollection of the main points covered above by completing the following statements.

(a) The three parts of a fraction are the ______________, the ______________, and the ______________.

(b) Write the three symbols used to indicate division: ________, ________, and ________.

(c) A fraction may mean: ______________, ______________, or ______________.

- - - - - - - - - - - - - - - - - - -

(a) numerator, denominator, fraction bar; (b) ———, ÷, and : ;
(c) division, ratio, part of a whole

FRACTIONS CONTAINING ZERO

4. When studying fractions, the use of the number zero requires special attention. In Chapter 1 (frame 10) we discussed the fact that division by zero is meaningless. When discussing the division of monomials (Chapter 3), we were careful to specify that none of our denominators was zero. Now let's examine the use of zero a little more broadly.

Zero can appear in the numerator of a fraction, in the denominator, or both. The rule governing each of these cases is as follows:

Rule 1: When the numerator of a fraction is zero, the value of the fraction is zero (provided the denominator is not zero). Thus,

$$\frac{0}{4} = 0 \qquad \frac{x}{7} = 0 \text{ (if } x = 0)$$

$$\frac{x-3}{5} = 0 \text{ (if } x = 3)$$

Rule 2: $\frac{0}{0}$ is indeterminate because the quotient does not exist as a unique number.

Rule 3: Since division by zero is undefined, a fraction having a zero denominator is meaningless. These fractions are meaningless (i.e., they do not represent a real number):

$$\frac{5}{0} \cdot \frac{7}{y} \qquad \frac{a+2b}{y-3} \text{ (where } y = 3)$$

Use these rules to evaluate the following fractions.

(a) $\frac{0}{7} =$ ______________

(b) $\frac{y-7}{4}$ (when $y = 7$) = ______________

(c) $\frac{13}{z-6}$ (when $z = 6$) = ______________

(d) $\frac{x-4}{x-8}$ (when $x = 8$) = ______________

(e) For what value of x does $\frac{x-4}{2+x} = 0$? $x =$ ________

(f) For what value of x is $\frac{x-4}{2+x}$ meaningless? $x =$ ________

(g) For what value of y is $\frac{y^2+7}{y-5}$ meaningless? $y =$ ________

- - - - - - - - - - - - - - - - - -

(a) zero; (b) zero; (c) meaningless; (d) meaningless; (e) 4; (f) −2; (g) 5

EQUIVALENT FRACTIONS

5. *Equivalent fractions* are fractions that have the same value but not the same numerators and denominators. For example, the fraction $\frac{2}{4}$ can be obtained from the fraction $\frac{1}{2}$ by multiplying both terms by 2. Therefore, we say that $\frac{2}{4}$ and $\frac{1}{2}$ are equivalent fractions. Similarly, $\frac{2}{3}$ and $\frac{4}{6}$, or $\frac{3}{5}$ and $\frac{6}{10}$ are equivalent fractions. Two

important ideas are involved here. They are expressed in the following two definitions.

Definition 1: $\frac{a}{c} = \frac{b}{d}$ if, and only if, $ad = bc$.

The second idea is so important that it is called the *fundamental principle of fractions.*

Definition 2: $\frac{a}{b} = \frac{ak}{bk} = \frac{a/k}{b/k}$, provided b and k are not zero. (k stands for any number used as a common multiplier or divider for both terms of a fraction.)

These two ideas are among the most important ones dealing with fractions, so be alert to their applications.

If you were asked to determine if the two fractions $\frac{3}{4}$ and $\frac{6}{8}$ are equivalent, you could check by cross-multiplying (3 times 8 and 4 times 6), as shown in Definition 1. The result, 24 = 24, would prove their equivalence. Or, dividing the larger numerator (6) by the smaller numerator (3) and the larger denominator (8) by the smaller denominator (4) would indicate that they have the same common multiplier, 2. This would prove their equivalence according to Definition 2.

Notice that the fundamental principle (Definition 2) states that the value of a fraction is unchanged if both terms of that fraction are multiplied by the same non-zero number or divided by the same non-zero number. Use this principle to determine how the second fraction was obtained from the first in the problems below.

(a) $\frac{8}{12} = \frac{2}{3}$ (note that $8 \cdot 3 = 12 \cdot 2$) ______________________

(b) $\frac{a}{b} = \frac{3a}{3b}$ ______________________

(c) $\frac{x^2}{xy} = \frac{x}{y}$ (note that $x^2 \cdot y = xy \cdot x$) ______________________

(d) $\frac{3b - 3d}{6} = \frac{b - d}{2}$ ______________________

- - - - - - - - - - - - - - - - - - -

(a) Both terms were divided by 4.
(b) Both terms were multiplied by 3.
(c) Both terms were divided by x.
(d) Both terms were divided by 3.

Remember that *both* terms of the fraction must be multiplied or divided by the same (non-zero) number.

REDUCING FRACTIONS

6. Next we will discuss reducing fractions to lowest terms, something we need to do quite frequently in algebra. First, however, you must learn how to recognize when a fraction is in its lowest terms (that is, when it cannot be reduced further).

A fraction is in its lowest terms (simplest form) when the numerator and denominator have no common factor except 1.

A fraction is reduced to its lowest terms by dividing both numerator and denominator by their greatest common factor (GCF). Thus, in the fraction

$$\frac{8a^2y}{12a^3z}$$

4 is the largest numerical factor and a^2 is the largest literal factor. Therefore, the largest common factor of both the numerator and denominator is $4a^2$. Dividing both terms by this factor we get

$$\frac{(8a^2y) \div (4a^2)}{(12a^3z) \div (4a^2)} = \frac{2y}{3az}$$

Reduce the fraction below to its lowest terms.

$$\frac{(21x^4y^2) \div (\quad)}{(14x^2y) \div (\quad)} = \underline{\hspace{6cm}}$$

- - - - - - - - - - - - - - - - - -

$$\frac{(7x^2y)}{(7x^2y)} = \frac{3x^2y}{2}$$

7. Division of the numerator and denominator of a fraction by their greatest common factor is, of course, based on the principle stated in Definition 2 (frame 5): *Both terms of a fraction may be multiplied or divided by the same number (except zero) without changing the value of the fraction.* This process of reducing a fraction by dividing both terms by their greatest common factor is simply an application of the fundamental principle of fractions.

We have practiced reducing fractions having common monomial factors. Now let's try one having common binomial factors.

Example: Reduce $\frac{3x + 3y}{x^2 - y^2}$ to its lowest terms.

First factor both numerator and denominator into their prime factors.

$$\frac{3(x + y)}{(x + y)(x - y)}$$

Then divide both numerator and denominator by the greatest common factor, $(x + y)$.

$$\frac{3(x + y) \div (x + y)}{(x + y)(x - y) \div (x + y)} = \frac{3}{x - y}$$

Example: Reduce $\frac{y^2 - 9y + 20}{y^2 - 4y}$ to its lowest terms.

Factoring: $\frac{(y - 4)(y - 5)}{y(y - 4)}$

Dividing both terms by $(y - 4)$: $\frac{(y - 4)(y - 5) \div (y - 4)}{y(y - 4) \div (y - 4)} = \frac{y - 5}{y}$

Use these examples as a guide in reducing the following fractions to lowest terms.

(a) $\dfrac{32c^3d^3}{64c^2d} =$ ____________

(b) $\dfrac{5a - 25}{15a} =$ ____________

(c) $\dfrac{27x^2}{18x^2 - 9xy} =$ ____________

(d) $\dfrac{2x + 6}{3ax + 9a} =$ ____________

(e) $\dfrac{a^2 + a}{2 + 2a} =$ ____________

(f) $\dfrac{y^2 + 2y - 15}{2y^2 - 12y + 18} =$ ____________

- - - - - - - - - - - - - - - - - - - -

(a) $\dfrac{cd^2}{2}$; (b) $\dfrac{a - 5}{3a}$; (c) $\dfrac{{}^{1}\cancel{9} \cdot 3 \cdot {}^{1}\cancel{x} \cdot x}{{}_{1}\cancel{9} \cdot {}_{1}\cancel{x}(2x - y)} = \dfrac{3x}{2x - y}$; (d) $\dfrac{2(x + 3)}{3a(x + 3)} = \dfrac{2}{3a}$;

(e) $\dfrac{a}{2}$; (f) $\dfrac{y + 5}{2(y - 3)}$

8. Having discussed some of the correct procedures for reducing fractions to their lowest terms, we are going to devote a brief moment to considering some of the mistakes most commonly made in this area.

How *not* to Reduce Fractions to Their Lowest Terms

(1) Do *not* subtract the same number from the numerator and denominator.

For example, $\frac{3}{4}$ does *not* equal $\dfrac{3 - 1}{4 - 1}$, or $\frac{2}{3}$ because (see frame 5) $3 \cdot 3 \neq 4 \cdot 2$. Nor does subtracting x from both the numerator and denominator of the fraction $\dfrac{x + 2}{x + 3}$ result in an equivalent fraction of $\frac{2}{3}$; $\dfrac{x + 2}{x + 3}$ is already in lowest terms.

(2) Do *not* add the same number to both the numerator and denominator.

For example, $\frac{1}{3}$ does *not* equal $\dfrac{1 + 2}{3 + 2}$, or $\frac{3}{5}$, since $1 \cdot 5 \neq 3 \cdot 3$. Nor does adding 2 to both the numerator and denominator of the fraction $\dfrac{n - 2}{m - 2}$ result in an equivalent fraction of $\dfrac{n}{m}$. The fraction $\dfrac{n - 2}{m - 2}$ is in lowest terms.

(3) Do *not* try to divide the individual terms of a polynomial numerator and denominator by some common factor.

In other words, don't attempt something like the following:

$$\frac{{}^{x}\cancel{x^2} - 2x - \cancel{24}^{4}}{5\cancel{x} - \cancel{30}_{5}} \neq \frac{x - 2x - 4}{5 - 5} \neq \frac{-4 - x}{0}$$

This procedure is wrong because only parts of the numerator and denominator were divided by x and 6. The correct procedure is to factor the numerator and denominator first and then divide both by the common binomial factor $(x - 6)$.

$$\frac{\cancel{(x-6)}(x+4)}{5\cancel{(x-6)}} = \frac{x+4}{5}$$

MULTIPLYING FRACTIONS

9. Now let's recall some of the things you should do if you are to handle fractions correctly. In algebra as in arithmetic the product of two or more fractions is equal to the product of the numerators divided by the product of the denominators.

$$\frac{a}{b} \cdot \frac{c}{d} = \frac{ac}{bd}$$

The resulting fractions, of course, need to be reduced to lowest terms.

Example: $\frac{2}{3} \cdot \frac{3}{8} = \frac{6}{24} = \frac{1}{4}$

This procedure is correct but we would have saved ourselves the final reduction if we had divided by the common factors in the beginning.

$$\frac{{}^{1}\cancel{2}}{\cancel{3}_{1}} \cdot \frac{{}^{1}\cancel{3}}{\cancel{8}_{4}} = \frac{1}{4} \quad \text{or} \quad \frac{{}^{1}\cancel{2}}{\cancel{3}_{1}} \cdot \frac{{}^{1}\cancel{3}}{\cancel{4}_{2}} = \frac{1}{1} \cdot \frac{1}{2} = \frac{1}{2}$$

Always start by dividing the numerator and denominator by any common factors. For example,

$$\frac{3ab^2}{4} \cdot \frac{12}{18a^2} \quad \text{can be written} \quad \frac{3(3 \cdot 2 \cdot 2)a \cdot b \cdot b}{2 \cdot 2[3 \cdot 3 \cdot 2]a \cdot a}$$

where the numbers in parentheses are the factors of 12 and those in brackets are the factors of 18. Dividing by common factors we get

$$\frac{\cancel{3} \cdot \cancel{3} \cdot \cancel{2} \cdot \cancel{2} \cdot \cancel{a} \cdot b \cdot b}{\cancel{3} \cdot \cancel{3} \cdot \cancel{2} \cdot \cancel{2} \cdot 2 \cdot \cancel{a} \cdot a} = \frac{b^2}{2a}$$

Example:

$$\frac{x+y}{8} \cdot \frac{10}{x^2 - y^2} = \frac{{}^{1}\cancel{(x+y)}}{{}_{4}\cancel{8}} \cdot \frac{\cancel{10}^{5}}{\cancel{(x+y)}_{1}(x-y)} = \frac{5}{4(x-y)}$$

As shown in these examples, always start by factoring numerators and denominators wherever possible.

Perform the following multiplications. Be sure to factor the trinomials in problems (g) and (h) before attempting to divide out the common factors.

(a) $\frac{3}{5} \cdot \frac{5}{9} =$ ____________________

(b) $\frac{2x}{y} \cdot \frac{x}{2y} =$ ____________________

(c) $\frac{4a}{7} \cdot \frac{1}{2b^2} =$ ____________________

(d) $\frac{x^2y^2}{m^3n} \cdot \frac{m}{x^3} =$ ____________________

(e) $\frac{d}{a+b} \cdot \frac{5a+5b}{2d^2+2d} =$ ____________________

(f) $\frac{a+3}{a-5} \cdot \frac{2a-10}{3a+9} =$ ____________________

(g) $\frac{a^2-6a+5}{a+1} \cdot \frac{a+1}{a-5} =$ ____________________

(h) $\frac{a^2+6a+9}{8} \cdot \frac{4a+8}{a^2+5a+6} =$ ____________________

- - - - - - - - - - - - - - - - - -

(a) $\frac{1}{3}$; (b) $\frac{x^2}{y^2}$; (c) $\frac{2a}{7b^2}$; (d) $\frac{y^2}{m^2nx}$; (e) $\frac{5}{2(d+1)}$; (f) $\frac{2}{3}$; (g) $a-1$; (h) $\frac{a+3}{2}$

RECIPROCALS

10. The *reciprocal* of a number (in case you have forgotten) is 1 divided by the number. The reciprocal of 3, for example, is $\frac{1}{3}$; the reciprocal of b is $\frac{1}{b}$; and the reciprocal of $\frac{3}{4}$ is $\frac{4}{3}$, since $1 \div \frac{3}{4}$ is $\frac{4}{3}$. The reciprocal of a fraction is found by interchanging the numerator and the denominator. Since reciprocals are frequently used in the solution of algebraic equations, you need to be familiar with some of the rules governing them.

Rule 4: The reciprocal of a fraction is the fraction inverted. Thus, $\frac{5}{8}$ and $\frac{8}{5}$ are reciprocals.

Rule 5: The product of a number and its reciprocal is 1. Thus, $\frac{3}{4} \cdot \frac{4}{3} = 1$.

Rule 6: To divide by a number or a fraction, multiply by its reciprocal. Thus, $6 \div \frac{3}{4} = 6 \cdot \frac{4}{3} = 8$; or, $3 \div 5 = 3 \cdot \frac{1}{5} = \frac{3}{5}$.

Rule 7: To solve an equation for an unknown having a fractional coefficient, multiply both numbers by the reciprocal of the fractional coefficient. For example, if $\frac{3}{4}x = 6$, multiply both sides by $\frac{4}{3}$ (the reciprocal of $\frac{3}{4}$). This gives $\frac{4}{3} \cdot \frac{3}{4}x = \frac{4}{3} \cdot 6$, or $x = 8$.

Apply these rules to the exercises below. (Caution: Check your answers after working the problems for each rule. This will help you avoid mistakes arising from any misunderstanding of the rules.)

Rule 4: Give the reciprocals of each of the following.

(a) $\frac{6}{7}$ ________ (c) $\frac{1}{5}$ ________

(b) 8 ________ (d) $\frac{2x}{7y}$ ________

Rule 5: Supply the missing term or product.

(a) $\frac{3}{5} \cdot \frac{5}{3} =$ ________ (c) $\frac{a}{7}(\quad) = 1$

(b) $\frac{x}{5} \cdot \frac{5}{x} \cdot \frac{3}{4} =$ ________

Rule 6: Change each division to a multiplication by using the reciprocal.

(a) $6 \div \frac{2}{3} =$ ________ (c) $\frac{a}{b} \div \frac{2a}{b} =$ ________

(b) $\frac{a}{2} \div y =$ ________ (d) $\frac{x}{3} \div \frac{3}{x} =$ ________

Rule 7: Use the reciprocals to solve these equations.

(a) $\frac{2}{3}y = 4, y =$ ________ (c) $1\frac{2}{3}k = 5, k =$ ________

(b) $\frac{1}{3}a = 2, a =$ ________ (d) $\frac{7}{4}n = 14, n =$ ________

- - - - - - - - - - - - - - - - - -

Rule 4: (a) $\frac{7}{6}$; (b) $\frac{1}{8}$; (c) $\frac{5}{1}$ or 5; (d) $\frac{7y}{2x}$

Rule 5: (a) 1; (b) $\frac{3}{4}$; (c) $\frac{7}{a}$

Rule 6: (a) $6 \cdot \frac{3}{2} = 9$; (b) $\frac{a}{2} \cdot \frac{1}{y} = \frac{a}{2y}$; (c) $\frac{a}{b} \cdot \frac{b}{2a} = \frac{1}{2}$; (d) $\frac{x}{3} \cdot \frac{x}{3} = \frac{x^2}{9}$

Rule 7: (a) $\frac{3}{2} \cdot \frac{2}{3}y = \frac{3}{2} \cdot 4, y = 6$; (b) $\frac{3}{1} \cdot \frac{1}{3}a = \frac{3}{1} \cdot 14, a = 6$;
(c) $\frac{3}{5} \cdot \frac{5}{3}k = \frac{3}{5} \cdot 5, k = 3$; (d) $\frac{4}{7} \cdot \frac{7}{4}n = \frac{4}{7} \cdot 14, n = 8$

Each rule used in solving the foregoing problems is both fundamental and highly useful, hence you need to be aware of them at all times. The terms used in these practice exercises were fairly simple ones, but many of the expressions you will meet later on will appear considerably more complex. Follow these same rules, however, and you should have no trouble solving them.

DIVIDING FRACTIONS

11. The procedure for dividing fractions is the same in algebra as in arithmetic:

> To divide one fraction by another, multiply by the reciprocal of the divisor.

Does this sound very much like our Rule 6? It is.

Example: $\frac{2}{3} \div \frac{3}{4}$ (This is read "two-thirds divided by three-fourths.")

$$\frac{2}{3} \div \frac{3}{4} = \frac{2}{3} \cdot \frac{4}{3} \left(\text{since } \frac{4}{3} \text{ is the reciprocal of } \frac{3}{4}\right) = \frac{8}{9}$$

Example: $\frac{21}{y} \div 2\frac{1}{3} = \frac{21}{y} \div \frac{7}{3} = \frac{21}{y} \cdot \frac{3}{7} = \frac{\cancel{7} \cdot 3 \cdot 3}{\cancel{7} \cdot y} = \frac{9}{y}$

Notice that $2\frac{1}{3}$ was first written as $\frac{7}{3}$.

Example: $\frac{4a^2}{3ab} \div \frac{2a}{6b^2} = \frac{\overset{2}{\cancel{4}}\cancel{a^2}}{\cancel{3}\cancel{a}\cancel{b}} \cdot \frac{\overset{2}{\cancel{6}}b^{\cancel{2}}}{\cancel{2}\cancel{a}} = 4b$

Example:

$$\frac{4x^2 - 1}{2x - 6} \div \frac{2x^2 - 7x - 4}{x^2 - 7x + 12} = \frac{4x^2 - 1}{2x - 6} \cdot \frac{x^2 - 7x + 12}{2x^2 - 7x - 4} =$$

$$\frac{(2x - 1)\cancel{(2x + 1)}^1}{2\cancel{(x - 3)}_1} \cdot \frac{\cancel{(x - 4)}^1\cancel{(x - 3)}^1}{\cancel{(2x + 1)}_1\cancel{(x - 4)}_1} = \frac{2x - 1}{2}$$

Now it's your turn. Perform the indicated divisions.

(a) $\frac{7}{8} \div \frac{3}{4} =$ ____________

(b) $\frac{a}{n} \div b =$ ____________

Remember: b is the same as $\frac{b}{1}$, hence the reciprocal of b is $\frac{1}{b}$.

(c) $\frac{2a}{3b} \div \frac{6a^2}{b^2} =$ ____________

(d) $\frac{8}{x^3} \div \frac{12}{x^2} =$ ____________

(e) $\frac{ab}{(a - b)^2} \div \frac{1}{a - b} =$ ____________

(f) $\dfrac{a^2 - 9}{a^2 + 3a} \div \dfrac{a - 3}{4} =$ ____________

(g) $\dfrac{x^2 + 7x + 6}{x^2 + 6x + 5} \div \dfrac{x^3 + 6x^2}{x^2 + 5x} =$ ____________

(h) $\dfrac{x^4 - y^4}{x + y} \div \dfrac{x^2 - y^2}{4x + 4y} =$ ____________

- - - - - - - - - - - - - - - - - -

(a) $\frac{7}{6}$; (b) $\frac{a}{nb}$; (c) $\frac{b}{9a}$; (d) $\frac{2}{3x}$; (e) $\frac{ab}{a-b}$; (f) $\frac{4}{a}$; (g) $\frac{1}{x}$;
(h) $4(x^2 + y^2)$

Perhaps now you see why we spent so much time learning how to factor binomials and trinomials. You couldn't arrive at the answers to some of these problems without knowing how and when to factor the polynomials involved.

12. The process of multiplying and dividing fractions is actually quite straightforward. Of course, if you are called on to multiply and divide several fractions in the same problem, all or some of which contain binomials or trinomials, the process begins to appear a little complicated. However, the procedure is the same, so if you just take one step at a time (and try to write neatly) most problems work out fairly easily.

Most students seem to have more trouble adding and subtracting fractions than they do multiplying and dividing, mainly because they do not really understand the meaning of fractions. For example, if you are offered half an orange and (plus) an additional third of the same orange, you certainly would expect to receive more than half an orange. Yet some people calculate this as $\frac{1}{2} + \frac{1}{3} = \frac{2}{5}$, which is less than one-half. What is wrong with this kind of addition?

____

- - - - - - - - - - - - - - - - - -

The mistake lies in trying to add unlike terms. Trying to add halves and thirds is like trying to add feet and inches without first converting them into the same units, or terms. Here is the correct solution: $\frac{1}{2} + \frac{1}{3} = \frac{3}{6} + \frac{2}{6} = \frac{5}{6}$. In this solution the unlike terms (halves and thirds) have been changed into like terms (sixths) before being added. Like terms (in fractions) are those having the same denominator.

ADDING AND SUBTRACTING (COMBINING) FRACTIONS

13. From what we have considered so far we can infer a few basic rules.

To Add (or Subtract) Fractions

(1) Change each fraction (if necessary) to an equivalent fraction so all have the same denominator.

(2) Place the sum (or difference) of these new numerators over the common denominator.

(3) Reduce the resulting fraction to its lowest terms.

Using symbols we can state the foregoing as

$$\frac{a}{b} + \frac{c}{b} = \frac{a + c}{b}$$

Example: $\frac{2}{8} + \frac{3}{8} = \frac{5}{8}$

Since the denominators are the same in this case we merely had to add the numerators and place the sum over the common denominator. No reduction was necessary.

Example: $\frac{x}{4} + \frac{2x}{8} = \frac{2x}{8} + \frac{2x}{8} = \frac{4x}{8} = \frac{x}{2}$

It is apparent by inspection that the LCD (lowest common denominator) is 8. We multiply both numerator and denominator of the first fraction by 2 in order to get a common denominator (and therefore equivalent fractions); then we can combine the two fractions. Remember (from frame 5), the value of a fraction is unchanged if both terms are multiplied by the same non-zero number.

Example: Simplify $\frac{x}{2y} - \frac{4}{y^2}$; LCD = $2y^2$.

Multiplying the numerator and denominator of the first fraction by y, and the numerator and denominator of the second fraction by 2 gives a common denominator in *both* fractions. The equivalent fractions can then be combined.

$$\frac{xy}{2y^2} - \frac{2 \cdot 4}{2y^2} \text{ or } \frac{xy - 8}{2y^2}$$

Example: Combine $\frac{2a + 1}{2} + \frac{a - 3}{4} - \frac{2a}{3}$; LCD = $2 \cdot 2 \cdot 3 = 12$.

$$\frac{6(2a + 1)}{12} + \frac{3(a - 3)}{12} - \frac{4(2a)}{12} = \frac{6(2a + 1) + 3(a - 3) - 4(2a)}{12} =$$

$$\frac{12a + 6 + 3a - 9 - 8a}{12} = \frac{7a - 3}{12}$$

Practice these procedures by combining terms in the following problems.

(a) $\frac{2b}{3} - \frac{b}{2} + \frac{b}{4} =$ ______________

(b) $\frac{5a}{6} - \frac{5a}{12} - \frac{a}{3} =$ ______________

(c) $\frac{4}{n^2} + \frac{3}{n} + \frac{5}{n^3} =$ ______________

(d) $\frac{3k + 5}{5} - \frac{k + 3}{3} =$ ______________

(e) $\frac{2a - 5}{14a^2} - \frac{4 - a}{7a} + \frac{3a - 2}{2a} =$ ______________

- - - - - - - - - - - - - - - - - - -

(a) $\frac{5b}{12}$; (b) $\frac{a}{12}$; (c) $\frac{4n + 3n^2 + 5}{n^3}$;

(d) $\frac{3(3k + 5) - 5(k + 3)}{15} = \frac{9k + 15 - 5k - 15}{15} = \frac{4k}{15}$;

(e) $\frac{(2a - 5) - 2a(4 - a) + 7a(3a - 2)}{14a^2} = \frac{2a - 5 - 8a + 2a^2 + 21a^2 - 14a}{14a^2} = \frac{23a^2 - 20a - 5}{14a^2}$

14. To summarize, problems involving the combining of fractions with the same numerical denominator are worked by the rule

$$\frac{a}{b} + \frac{c}{b} = \frac{a + c}{b}$$

and then reduced if possible. Problems in which the fractions have different numerical denominators are worked by the rule

$$\frac{a}{b} + \frac{c}{d} = \frac{ad + bc}{bd}$$

and then reduced if possible. This is the method of equivalent fractions.

Now let's consider the situation in which there may be (in addition to monomial denominators) binomial or other polynomial denominators which you will have to factor before you can determine the LCD.

Example: Combine $\frac{1}{3a - 6} + \frac{1}{6a} - \frac{1}{2a + 4}$.

$$\frac{1}{3(a - 2)} + \frac{1}{6a} - \frac{1}{2(a + 2)}; \text{ LCD} = 6a(a - 2)(a + 2)$$

$$\frac{2a(a + 2) + (a - 2)(a + 2) - 3a(a - 2)}{6a(a - 2)(a + 2)} =$$

$$\frac{2a^2 + 4a + a^2 - 4 - 3a^2 + 6a}{6a(a - 2)(a + 2)} = \frac{10a - 4}{6a(a - 2)(a + 2)} =$$

$$\frac{\overset{1}{\cancel{2}}(5a - 2)}{\underset{3}{\cancel{6}}a(a - 2)(a + 2)} = \frac{5a - 2}{3a(a - 2)(a + 2)}$$

Use this procedure in the following problems.

(a) $\frac{d}{a + d} - \frac{a}{a - d} =$ ____________________

(b) $\frac{3a}{a + 3} + \frac{a^2 + 4a - 5}{a^2 - a - 12} =$ ____________________

(c) $\frac{a^2 + b^2}{a^2 - b^2} + \frac{2a}{a + b} =$ ____________________

- - - - - - - - - - - - - - - - - -

(a) $\frac{d(a-d)-a(a+d)}{(a+d)(a-d)} = \frac{ad-d^2-a^2-ad}{(a+d)(a-d)} = \frac{-a^2-d^2}{a^2-d^2}$

(b) $\frac{3a(a-4)+a^2+4a-5}{(a-4)(a+3)} = \frac{3a^2-12a+a^2+4a-5}{(a-4)(a+3)} = \frac{4a^2-8a-5}{(a-4)(a+3)}$

(c) $\frac{a^2+b^2+2a(a-b)}{(a-b)(a+b)} = \frac{a^2+b^2+2a^2-2ab}{(a-b)(a+b)} = \frac{3a^2-2ab+b^2}{(a-b)(a+b)}$

GENERAL RULE FOR FINDING THE LCD

15. In most of the problems we have encountered so far it was possible to find the LCD either by inspection or by combining denominators, factoring binomial denominators where necessary. Because it is not always so easy to find the LCD of a set of fractions, we are going to discuss a procedure you can follow that should prove helpful.

Consider, for example, the fractions $\frac{1}{12}$, $\frac{2}{15}$, $\frac{5}{9}$, and $\frac{4}{21}$. What is their lowest common denominator? We certainly don't want to multiply $12 \cdot 15 \cdot 9 \cdot 21$. Such a product would not be the lowest common denominator because 12, 15, 9, and 21 are not prime numbers. Perhaps it would help if we factored all the denominators into their prime factors. Let's try it. Factor each of the following.

(a) 12 written in terms of its prime factors would be __________________.

(b) 15 written in terms of its prime factors would be __________________.

(c) 9 written in terms of its prime factors would be __________________.

(d) 21 written in terms of its prime factors would be __________________.

- - - - - - - - - - - - - - - - - -

(a) $12 = 2^2 \cdot 3$; (b) $15 = 5 \cdot 3$; (c) $9 = 3^2$; (d) $21 = 3 \cdot 7$

16. These prime factors should furnish you a clue. Notice that the numbers 12, 15, 9, and 21 are actually composed of just the factors 2, 3, 5, and 7. Notice also that the factor 2 has a highest exponent of 2, that the factor 3 has a highest exponent of 2, and that the factors 5 and 7 have a highest exponent of 1. Thus, the LCD for the four fractions will be $2^2 \cdot 3^2 \cdot 5 \cdot 7$ or, as a product, the number 1260.

Below is the rule for finding the LCD for any set of fractions. It is given in three steps. As you read it, relate each step to the example we have just discussed.

Rule for Finding the LCD

(1) Write each denominator in prime factored form, using exponents as necessary to represent repeated factors.

(2) Write the product of all the different prime factors.

(3) Use the largest exponent required (in Step 1) for any given prime factor. The result is the LCD.

Example: Find the LCD for $\frac{1}{15}, \frac{3}{8}, \frac{5}{12}$, and $\frac{5}{18}$.

Step (1): Write the prime factors of each denominator.

$$15 = 3 \cdot 5, \quad 8 = 2^3, \quad 12 = 2^2 \cdot 3, \quad 18 = 2 \cdot 3^2$$

Step (2): Write the product of all the different prime factors.

$$2 \cdot 3 \cdot 5$$

Step (3): Use the largest exponent required for each factor.

$$2^3 \cdot 3^2 \cdot 5 = 360$$

Test your LCD of 360 to determine if it is divisible by 15, 8, 12, and 18. If it is, you have verified that 360 is the common denominator and that the rule works. *The denominator will be the lowest common denominator only if the factors used in Step 1 were prime.*

(a) $\frac{360}{15} =$ __________ (c) $\frac{360}{12} =$ __________

(b) $\frac{360}{8} =$ __________ (d) $\frac{360}{18} =$ __________

- - - - - - - - - - - - - - - - - -

(a) 24; (b) 45; (c) 30; (d) 20 (360 *is* divisible by each of the denominators.)

SIGNS OF FRACTIONS

17. In Chapter 2 (frame 33) when considering the matter of dividing signed numbers, we developed the following rule for determining the sign of the quotient:

> In dividing numbers, if the signs of the numerator and denominator are the same, the quotient will be positive; if the signs are different, the quotient will be negative.

This rule works very well where we are concerned with two signs only (that is, the sign of the numerator and the sign of the denominator). In dealing with fractions, however, we have to remain aware of three signs: the sign of the numerator, the sign of the denominator, and the sign of the fraction. When any of these signs is omitted it is understood to be plus. For example,

$$\frac{3}{4} \text{ means } +\frac{+3}{+4}$$

Now consider the four fractions below.

$+\frac{+6}{+2} = +(+3) = 3$ $-\frac{-6}{+2} = -(-3) = 3$

$+\frac{-6}{-2} = +(+3) = 3$ $-\frac{+6}{-2} = -(-3) = 3$

Notice that the value of each fraction is 3 and that each of the four fractions may be obtained from any one of the others by changing two of the three signs. This tells us that:

> Any two of the three signs of a fraction may be changed without changing the value of the fraction.

State whether each of the following is true or false.

(a) $+\frac{-6}{2} = -\frac{+6}{+2}$ ____________ (c) $-\frac{-5}{-7} = \frac{-5}{-7}$ ____________

(b) $\frac{-x}{-y} = \frac{x}{y}$ ____________ (d) $\frac{a}{b} = -\frac{-a}{b}$ ____________

- - - - - - - - - - - - - - - - - -

(a) True. The answer is still negative since two signs were changed.
(b) True. Exactly two signs were changed.
(c) False. The answer is now positive because only one sign (sign of the fraction) was changed.
(d) True. Exactly two signs were changed.

18. The fractions in the preceding frame contained monomial numerators and denominators. Now let's consider how the rule of the signs of a fraction applies where binomials, trinomials, or polynomials of more than three terms are involved.

Suppose we wish to change the fraction $\frac{a-b}{c-d}$ to an equal fraction having the denominator $d-c$. This means that the denominator $c-d$ will have signs of both its terms reversed. We accomplish this by multiplying the denominator by -1, giving the required $d-c$. But to keep the new fraction equal to the original fraction we must either change the sign of the fraction or multiply the numerator by -1. Let's elect to multiply the numerator by -1; this gives

$$\frac{a-b}{c-d} = +\frac{b-a}{d-c} \text{ or } -\frac{a-b}{d-c} \text{ (changing the sign of the fraction)}$$

Remember, the sign of a polynomial is changed by changing the sign of all its terms or by multiplying it by -1.

Supply the missing terms in the changed fractions below.

(a) $\frac{a-3}{b-2} = +\frac{?}{2-b}$ (d) $\frac{4}{a-b} = +\frac{?}{b-a}$

(b) $\frac{c-d}{x-y} = -\frac{c-d}{?}$ (e) $\frac{x-y}{b} = \frac{?}{-b}$

(c) $\frac{4-a}{x-y} = +\frac{?}{y-x}$ (f) $\frac{6}{2-y} = \frac{?}{y-2}$

- - - - - - - - - - - - - - - - - -

(a) $3-a$; (b) $y-x$; (c) $a-4$; (d) -4; (e) $y-x$; (f) -6

19. For convenience, numerators and denominators of fractions are usually arranged in descending powers of one letter (with the first term made positive), something you learned in Chapter 3 (frame 7). We will use this rule of convenience in the example below.

Example: Reduce $\dfrac{x^2 - 4x}{12 + x - x^2}$ to its lowest terms.

Arranging the denominator in descending powers of x to agree with the numerator, we get

$$\frac{x^2 - 4x}{-x^2 + x + 12}$$

To eliminate the minus sign in front of the first term of the denominator we multiply the entire denominator by -1. To keep from changing the value of the fraction, we must also change the sign of the fraction to minus. (Remember, if we change two signs out of the three we will not alter the value of a fraction.) This now gives

$$-\frac{x^2 - 4x}{x^2 - x - 12}$$

Factoring and dividing both the numerator and denominator by $(x - 4)$ we get

$$-\frac{x\cancel{(x-4)}^{1}}{{}_{1}\cancel{(x-4)}(x+3)} = -\frac{x}{x+3}$$

Follow this same general procedure in simplifying the fractions below.

(a) $\dfrac{4m - 4n}{n^2 - m^2} =$ ____________________

(b) $\dfrac{x^2 - 4x + 3}{3 - x} =$ ____________________

(c) $\dfrac{1 - k^2}{3k^2 - 6k + 3} =$ ____________________

(d) $\dfrac{x^2 + x - 20}{35 + 12x + x^2} =$ ____________________

(e) $\dfrac{(a - b)^2}{b^2 + 2ba - 3a^2} =$ ____________________

(f) $\dfrac{x^2 - 6x + 9}{9 - x^2} =$ ____________________

- - - - - - - - - - - - - - - - - -

(a) $-\dfrac{4}{m + n}$; (b) $-(x - 1) = 1 - x$; (c) $-\dfrac{k + 1}{3(k - 1)}$; (d) $\dfrac{x - 4}{x + 7}$; (e) $-\dfrac{a - b}{3a + b}$;

(f) $-\dfrac{x - 3}{x + 3}$

20. So far we have looked only at the problem of changing the sign of the numerator or denominator of a single fraction. Now we need to consider what happens when we

have a problem that involves combining several fractions, at least one of which will require that some terms be rearranged.

Example: Combine $\frac{a-3}{a^2-25}+\frac{6}{a+5}+\frac{3}{5-a}$.

It is evident that the terms of the third fraction need to be rearranged to conform to the order of the other binomials in which the letter a appears. Therefore, we multiply the denominator of the third fraction by -1, change the order of the terms by putting a first, and at the same time change the sign of the fraction from plus to minus, thus preserving the value of the fraction. This gives

$$\frac{a-3}{a^2-25}+\frac{6}{a+5}-\frac{3}{a-5}$$

By inspection we can determine that the LCD is $(a-5)(a+5)$, therefore

$$\frac{(a-3)+6(a-5)-3(a+5)}{(a-5)(a+5)}$$

$$=\frac{a-3+6a-30-3a-15}{(a-5)(a+5)}$$ (Notice that the distributive law determines this result.)

$$=\frac{4a-48}{a^2-25} \text{ or } \frac{4(a-12)}{(a-5)(a+5)}$$

Either form for the answer is acceptable; however, it is safer to factor to see if the fraction is in its lowest terms.

Apply this approach to the following problem.

Combine $\frac{4a}{a-1}+\frac{3a}{4-4a}-\frac{a^2-6}{a^2-1}$.

- - - - - - - - - - - - - - - - - -

Multiplying the denominator of the second fraction by -1 we get

$$\frac{4a}{a-1}-\frac{3a}{4a-4}-\frac{a^2-6}{a^2-1} \qquad \text{LCD} = 4(a-1)(a+1)$$

Using the LCD and applying the fundamental principle of fractions (from frame 5) gives

$$\frac{4 \cdot 4a(a+1)-3a(a+1)-4(a^2-6)}{4(a-1)(a+1)}$$

Multiplying the terms in the numerator and combining like terms gives the answer.

$$\frac{16a^2+16a-3a^2-3a-4a^2+24}{4(a-1)(a+1)} \text{ or } \frac{9a^2+13a+24}{4(a^2-1)}$$

21. To make sure you understand the procedure involved, work this additional problem.

Combine $\frac{x}{x+1}+\frac{4}{1-x}+\frac{x^2-5x-8}{x^2-1}$.

$$\frac{2(x-6)}{x-1}$$

MIXED NUMBERS

22. On the assumption that you may have forgotten some of what you once learned in arithmetic about mixed numbers, let's take a moment to review this subject. A *mixed number* consists of a whole number and a fraction. For example, $3\frac{1}{4}$, $7\frac{1}{2}$, $1\frac{3}{4}$, and $a + \frac{1}{b}$ are mixed numbers. As you will observe in the first three expressions, the plus sign is not used (or needed) in arithmetic for mixed numbers. However, it is required in algebraic mixed numbers, as shown in the fourth example. This is an important difference because $3\frac{1}{4}$ means $3 + \frac{1}{4}$, but $a\frac{1}{b}$ means a times $\frac{1}{b}$.

A mixed number can be changed into a simple fraction by adding the whole number and the fraction. However, to do so you must have a common denominator. If you are converting only one mixed number, the common denominator will be the denominator of the fractional part of the mixed number.

$$3\tfrac{2}{5} = 3 + \frac{2}{5} = \frac{3}{1} + \frac{2}{5} = \frac{15}{5} + \frac{2}{5} = \frac{17}{5}$$

Or, using an algebraic mixed number,

$$a + \frac{k}{y} = \frac{ay}{y} + \frac{k}{y} = \frac{ay + k}{y}$$

Similarly,

$$x - y - \frac{x^2}{x+y} = \frac{x-y}{1} - \frac{x^2}{x+y}$$
$$= \frac{(x-y)(x+y) - x^2}{x+y} = \frac{x^2 - y^2 - x^2}{x+y}$$
$$= -\frac{y^2}{x+y}$$

You will find yourself performing some of the above steps mentally as you develop more practice.

Change the following mixed expressions into common fractions.

(a) $4\frac{7}{8}$ = ____________

(b) $3 + \frac{4}{7}$ = ____________

(c) $\frac{a}{d} - 1$ = ____________

(d) $2k + \frac{3}{k}$ = ____________

(e) $3a + b - \frac{2}{a} =$ ____________________

(f) $a + c + d - \frac{1}{ac} =$ ____________________

(g) $2x^2 + 3x + 1 + \frac{x^2 + 1}{2x - 1} =$ ____________________

(h) $a^2 + ab - b^2 - \frac{a^3 - 2b^3}{a - 2b} =$ ____________________

(a) $\frac{39}{8}$; (b) $\frac{25}{7}$; (c) $\frac{a - d}{d}$; (d) $\frac{2k^2 + 3}{k}$; (e) $\frac{3a^2 + ab - 2}{a}$;
(f) $\frac{a^2c + ac^2 + acd - 1}{ac}$; (g) $\frac{4x^3 + 5x^2 - x}{2x - 1}$ (You could factor x out of the numerator.); (h) $-\frac{a^2b + 3ab^2 - 4b^3}{a - 2b}$ (You could factor b out of the numerator, but it's not necessary.)

23. Having considered how to change mixed numbers into fractions, let's look now at the opposite process—how to change fractions into mixed numbers. For example, to change the fraction $\frac{17}{5}$ into a mixed number (as you probably recall from arithmetic) we simply divide the numerator by the denominator, giving us an answer of $3\frac{2}{5}$. But how about an algebraic fraction such as

$$\frac{4x^2 + 8x - 3}{2x}$$

Here we proceed the same way we did in arithmetic: divide the denominator into (each term of) the numerator.

$$\frac{4x^2}{2x} + \frac{8x}{2x} - \frac{3}{2x} = 2x + 4 - \frac{3}{2x}$$

Follow this same procedure in changing the following fractions into mixed expressions.

(a) $\frac{3x^3 + x^2 - 4x + 7}{x} =$ ____________________

(b) $\frac{12x^2 + 9x + 2}{3x} =$ ____________________

(c) $\frac{4a^2b^2 - 6ab - 5}{2ab} =$ ____________________

(d) $\frac{15x^3 - 10x^2 + 5}{5x} =$ ____________________

(a) $3x^2 + x - 4 + \frac{7}{x}$; (b) $4x + 3 + \frac{2}{3x}$; (c) $2ab - 3 - \frac{5}{2ab}$;
(d) $3x^2 - 2x + \frac{1}{x}$

24. Changing a fraction with a binomial denominator into a mixed expression is slightly more involved and similar to long division in arithmetic.

Example: Change $\dfrac{x^2 + 3x + 4}{x + 1}$ into a mixed expression.

Dividing is done with the *first term* of the binomial divisor. Don't try to divide with the second term! It is simply along for the ride. First divide x^2 by x (the first term of the divisor). This gives x as the first term of the quotient (written above x^2).

$$x + 1 \overline{\smash{)}\, x^2 + 3x + 4} \quad \text{quotient: } x$$

Then multiply both terms of the divisor by the first term of the quotient (x) and subtract the resulting product from the dividend.

$$\begin{array}{r|l} & \;x \\ x + 1 & \overline{x^2 + 3x + 4} \\ & \underline{x^2 + 1x} \\ & \qquad 2x \end{array}$$

Continue (very much as in long division) until the division is complete.

$$\begin{array}{r|l} & \;x + 2 \\ x + 1 & \overline{x^2 + 3x + 4} \\ & \underline{x^2 + 1x} \\ & \qquad 2x + 4 \\ & \qquad \underline{2x + 2} \\ & \qquad\quad + 2 \end{array} \quad = x + 2 + \frac{2}{x + 1}$$

In the last term of the answer the remainder becomes the numerator and the divisor the denominator, just as in arithmetic.

Try it yourself. Change the following into a mixed expression.

$$\frac{2x^2 + 4x + 7}{2x + 2}$$

$$\begin{array}{r|l} & \;x \quad + 1 \\ 2x + 2 & \overline{2x^2 + 4x + 7} \\ & \underline{2x^2 + 2x} \\ & \qquad 2x + 7 \\ & \qquad \underline{2x + 2} \\ & \qquad\quad + 5 \end{array}$$

dividing $2x$ into $2x^2$
subtracting
dividing $2x$ into $2x$
subtracting
remainder

or $\dfrac{2x^2 + 4x + 7}{2x + 2} = x + 1 + \dfrac{5}{2x + 2}$

25. Remember, dividing is done with the *first term* of the binomial divisor. Don't try to divide with the second term; it simply trails along and is used to help form the quantity to be subtracted.

Try a few more of these to make sure you've got the idea. Change the following fractions into mixed expressions.

(a) $\dfrac{2a^2 + 3a + 2}{a + 2}$

(b) $\dfrac{k^3 - 1}{k - 1}$ (Hint: Leave space between k^3 and 1 for the missing powers of k^2 and k.)

(c) $\dfrac{5y^2 - 7y + 1}{y - 1}$

(d) $\dfrac{4x^3 - x + 1}{2x - 1}$

(a)
$$\begin{array}{r|l} & 2a \quad - 1 \\ \hline a + 2 & 2a^2 + 3a + 2 \\ & \underline{2a^2 + 4a} \\ & \quad - a + 2 \\ & \quad \underline{- a - 2} \\ & \qquad + 4 \end{array} \qquad = 2a - 1 + \frac{4}{a + 2}$$

(b)
$$\begin{array}{r|l} & k^2 + k \; + 1 \\ \hline k - 1 & k^3 \qquad\qquad + 1 \\ & \underline{k^3 - k^2} \\ & \quad k^2 \\ & \quad \underline{k^2 - k} \\ & \qquad k + 1 \\ & \qquad \underline{k - 1} \\ & \qquad + 2 \end{array} \qquad = k^2 + k + 1 + \frac{2}{k - 1}$$

(c)
$$\begin{array}{r|l} & 5y \; - 2 \\ \hline y - 1 & 5y^2 - 7y + 1 \\ & \underline{5y^2 - 5y} \\ & \quad - 2y \\ & \quad \underline{- 2y + 2} \\ & \qquad - 1 \end{array} \qquad = 5y - 2 - \frac{1}{y - 1}$$

(d)
$$\begin{array}{r|l} & 2x^2 + x \\ \hline 2x - 1 & 4x^3 \qquad -x + 1 \\ & \underline{4x^3 - 2x^2} \\ & \quad + 2x^2 \\ & \quad \underline{2x^2 - x} \\ & \qquad +1 \end{array} \quad = 2x^2 + x + \frac{1}{2x-1}$$

COMPLEX FRACTIONS

26. We have at last arrived at the final type of fraction we will consider in this chapter—*complex fractions.*

> A complex fraction has one or more fractions in the numerator, the denominator, or both.

These are examples of complex fractions:

$$\frac{\frac{3}{4}}{\frac{1}{2}} \qquad \frac{3\frac{1}{4}}{1\frac{3}{4}} \qquad \frac{1 + \frac{1}{x}}{2 - \frac{1}{x}}$$

Two methods are used to change complex fractions into simple fractions or rational numbers.

Method 1: Multiply both numerator and denominator by the LCD of all fractions appearing in the entire complex fraction and reduce, if necessary.

Example 1: Simplify $\frac{\frac{2}{3}}{\frac{3}{4}}$.

Since the LCD = 12 (3·4) we have

$$\frac{\frac{2}{3}(12)}{\frac{3}{4}(12)} = \frac{8}{9}$$

Example 2: Simplify $\dfrac{\frac{x^2 - y^2}{4}}{\frac{x + y}{2}}$.

Since the LCD = 4, we have

$$\frac{\frac{x^2 - y^2}{4}}{\frac{x + y}{2}} = \frac{\left(\frac{x^2 - y^2}{4}\right)(4)}{\left(\frac{x + y}{2}\right)(4)} = \frac{x^2 - y^2}{2(x + y)} = \frac{\overset{1}{\cancel{(x + y)}}(x - y)}{2\underset{1}{\cancel{(x + y)}}} = \frac{x - y}{2}$$

Simplify the following.

(a) $\frac{\frac{7}{2}}{\frac{2}{3}} =$ ____________________________

(b) $\dfrac{2\frac{2}{3}}{\frac{4}{5}} =$ ______________________________

(c) $\dfrac{10\frac{1}{2}}{1+\frac{1}{5}} =$ ______________________________

(d) $\dfrac{1-\dfrac{x}{y}}{1-\dfrac{x^2}{y^2}} =$ ______________________________

(a) $\dfrac{\frac{7}{2}(6)}{\frac{2}{3}(6)} = \dfrac{21}{4} = 5\frac{1}{4}$; (b) $\dfrac{\frac{8}{3}(15)}{\frac{4}{5}(15)} = \dfrac{40}{12} = \dfrac{10}{3} = 3\frac{1}{3}$; (c) $\dfrac{\frac{21}{2}(10)}{\frac{6}{5}(10)} = \dfrac{105}{12} = 8\frac{3}{4}$;

(d) $\dfrac{\left(\dfrac{y-x}{y}\right)(y^2)}{\left(\dfrac{y^2-x^2}{y^2}\right)(y^2)} = \dfrac{y(y-x)}{y^2-x^2} = \dfrac{y(y-x)}{(y-x)(y+x)} = \dfrac{y}{y+x}$

27. Now let's take a look at the second method of changing complex fractions into simple fractions or rational numbers.

Method 2: Combine the terms in the numerator, combine the terms in the denominator, then divide the numerator by the denominator.

Example 1: $\dfrac{\frac{2}{3}-\frac{1}{2}}{\frac{3}{4}-\frac{1}{2}} = \dfrac{\dfrac{4-3}{6}}{\dfrac{3-2}{4}} = \dfrac{\frac{1}{6}}{\frac{1}{4}} = \dfrac{1}{6} \div \dfrac{1}{4} = \dfrac{1}{6} \cdot \dfrac{4}{1} = \dfrac{2}{3}$

Example 2: $\dfrac{1+\dfrac{2}{x}}{1-\dfrac{4}{x^2}} = \dfrac{\dfrac{x+2}{x}}{\dfrac{x^2-4}{x^2}} = \dfrac{x+2}{x} \div \dfrac{x^2-4}{x^2}$

$= \dfrac{(x+2)}{x} \cdot \dfrac{x^2}{(x-2)(x+2)} = \dfrac{x}{x-2}$

Use the techniques above to change the following to simple fractions.

(a) $\dfrac{\dfrac{1}{z}+1}{\dfrac{1}{z^2}-1} =$

(b) $\dfrac{1-\dfrac{2}{a}-\dfrac{3}{a^2}}{1+\dfrac{1}{a}} =$

(c) $\dfrac{m + \dfrac{m}{n}}{\dfrac{1}{n} + \dfrac{1}{n^2}} =$

(d) $\dfrac{t - \dfrac{25}{t}}{t + 5} =$

(a) $\dfrac{\cancel{(1+z)}^1}{\cancel{z}} \cdot \dfrac{z^{\cancel{2}}}{(1-z)\cancel{(1+z)}_1} = \dfrac{z}{1-z}$; (b) $\dfrac{\cancel{(a+1)}^1(a-3)}{a^{\cancel{2}}} \cdot \dfrac{\cancel{a}}{\cancel{(a+1)}_1} = \dfrac{a-3}{a}$

(c) $\dfrac{m\cancel{(n+1)}^1}{\cancel{n}} \cdot \dfrac{n^{\cancel{2}}}{\cancel{(n+1)}_1} = mn$; (d) $\dfrac{(t-5)\cancel{(t+5)}^1}{t} \cdot \dfrac{1}{\cancel{(t+5)}_1} = \dfrac{t-5}{t}$

This completes our brief review of the methods of working with arithmetic fractions and the generally corresponding procedures for handling algebraic fractions. Be sure to take the Self-Test before you go to Chapter 6.

SELF-TEST

The problems that follow should afford you a general review of fractions—both arithmetic and algebraic—as we covered them in this chapter. As usual, if any of your answers do not agree with those given, check your work before turning to the frames referenced.

1. Evaluate the following fractions.

   (a) $\dfrac{0}{7} =$ ____________

   (b) $\dfrac{9-a}{7}$ (when $a = 9$) = ____________

   (c) $\dfrac{478}{x-3}$ (when $x = 3$) = ____________

   (d) $\dfrac{a-b}{b-a}$ (when $a = b$) = ____________ (frame 4)

2. Tell how the second fraction was obtained from the first in each case.

   (a) $\dfrac{9}{15} = \dfrac{3}{5}$ ____________

   (b) $\dfrac{x}{y} = \dfrac{x^2}{xy}$ ____________

   (c) $\dfrac{5a + 5d}{15} = \dfrac{a + d}{3}$ ____________

(frame 5)

3. Reduce the following fractions to lowest terms.

(a) $\frac{3xy}{9xz} =$ ____________

(c) $\frac{36c^4d^6}{48c^3d} =$ ____________

(b) $\frac{6ac}{15c^3} =$ ____________

(frame 7)

4. Perform the following multiplications after first dividing out (cancelling) common factors:

(a) $\frac{2a}{3b} \cdot \frac{b^2 - b}{2ab - 2a} =$ ____________

(b) $\frac{3a + 3b}{4a^2} \cdot \frac{2a^2}{a^2 - b^2} =$ ____________

(c) $\frac{3c - 12}{4c + 20} \cdot \frac{c^2 + 5c}{c^2 - 4c} =$ ____________

(d) $\left(\frac{2m^2 + 6}{15y^2}\right)\left(\frac{21y^4}{3m^2 + 9}\right) =$ ____________

(frame 9)

5. Give the reciprocals of each of the following.

(a) $\frac{8}{9}$ ________

(c) $\frac{2}{x}$ ________

(b) 17 ________

(frame 10)

6. Supply the missing term or product:

(a) $\frac{7}{9} \cdot \frac{9}{7} = \frac{2}{?}$

(b) $\frac{2x}{9}$ (?) = 1

(c) $\frac{a}{b} \cdot \frac{b}{c} \cdot \frac{c}{a} =$?

(d) $8 \div \frac{2}{5} =$ ____________

(e) $\frac{c}{d} \div \frac{c}{a} =$ ____________

(f) $\frac{x^2}{y^2} \div \frac{x^3}{y} =$ ____________

(frame 10)

7. Use reciprocals to solve these equations.

(a) $\frac{3}{4}y = 3$, $y =$ ____________

(b) $\frac{2}{7}z = 6$, $z =$ ____________

(c) $3\frac{2}{5}t = 17$, $t =$ ____________

(frame 10)

8. Perform the indicated divisions.

(a) $\frac{1}{x^2} \div \frac{1}{x^3} =$ ______________

(b) $\frac{a-4}{5} \div \frac{3a-12}{10} =$ ______________

(c) $\frac{b-x}{y} \div \frac{b^2-x^2}{xy} =$ ______________

(d) $\frac{x^2-9x+14}{x^2+7x+12} \div \frac{3x^2-21x}{4x^3+16x^2} =$ ______________ (frame 11)

9. Combine.

(a) $\frac{2a}{3} - \frac{a}{5} =$ ______________

(b) $\frac{1}{a} + \frac{1}{b} + \frac{1}{c} =$ ______________

(c) $\frac{x+y}{2} + \frac{x-y}{3} =$ ______________

(d) $\frac{3x^2-4x+1}{6x^2} - \frac{x+2}{3x} =$ ______________

(e) $\frac{k-t}{t} + \frac{k+s}{s} =$ ______________ (frame 14)

10. Indicate whether each of the following is true or false ($b \neq 0$ and $y \neq 0$).

(a) $-\frac{a}{b} = \frac{-a}{b}$ ______________

(b) $+\frac{3}{5} = \frac{-3}{-5}$ ______________

(c) $-\frac{-3}{+4} = -\frac{+3}{-4}$ ______________

(d) $+\frac{+x}{-y} = -\frac{-x}{-y}$ ______________ (frame 17)

11. Supply the missing terms to make the fractions equal.

(a) $\frac{b-a}{a^2-b^2} = -\frac{?}{a^2-b^2}$

(b) $\frac{-6+5x-x^2}{x-2} = -\frac{?}{x-2}$

(c) $-\frac{2a-2b}{b^2-a^2} = +\frac{2a-2b}{?}$

(d) $\frac{a^2+b^2}{a^2-4} = -\frac{?}{4-a^2}$ (frame 18)

12. Simplify the following fractions.

(a) $\frac{y^2-4}{2a-ay} =$ ______________

(b) $\frac{1-2x}{2x^2-x} =$ ______________

(c) $\dfrac{4b^2 - c^2}{4b^2 - 4bc + c^2} =$ ______

(d) $\dfrac{2x^2 - 9x - 5}{6x^2 + 7x + 2} =$ ______ (frame 19)

13. Combine.

(a) $\dfrac{x+y}{x-y} + \dfrac{1}{x+y} - \dfrac{x^2 + y^2}{y^2 - x^2} =$ ______

(b) $\dfrac{x^2 + 5x + 1}{x^2 - x - 20} + \dfrac{x+3}{5-x} =$ ______ (frame 20)

14. Change the following mixed expressions into a single fraction in simplest form.

(a) $2k - 3 + \dfrac{1}{k} =$ ______

(b) $2b + 4 - \dfrac{1}{3b} =$ ______

(c) $x^2 + x - 3 + \dfrac{3}{4x} =$ ______

(d) $y^2 - yz + z^2 + \dfrac{2}{yz} =$ ______ (frame 22)

15. Change the following fractions into mixed numbers.

(a) $\dfrac{a^3 + a^2 + 1}{a} =$ ______

(b) $\dfrac{4x^2y^2 - 6xy - 5}{2xy} =$ ______

(c) $\dfrac{10y^4 - 8y^3 + 6}{2y^3} =$ ______

(d) $\dfrac{12x^3 - 8x^2 + 4}{4x} =$ ______

(e) $\dfrac{3x^2 + 4xy - 4y^2 + 2}{x + 2y} =$ ______

(f) $\dfrac{a^3 + 1}{a - 1} =$ ______ (frames 23 and 24)

16. Simplify the following.

(a) $\dfrac{\frac{1}{x+y}}{\frac{x}{y}} =$ ______

(b) $\dfrac{\frac{x^2 + y^2}{3}}{\frac{x+y}{5}} =$ ______

(c) $\dfrac{\frac{a+b}{a-b}}{\frac{a-b}{a+b}} =$ ______

(d) $\dfrac{a - 2 + \frac{3}{a}}{1 + \frac{1}{a}} =$ ______

(frame 26)

Answers to Review Problems

1. (a) 0; (b) 0; (c) meaningless; (d) $\frac{0}{0}$ = indeterminate
2. (a) Both numerator and denominator were divided by 3.
   (b) Both numerator and denominator were multiplied by x.
   (c) Both numerator and denominator were divided by 5.
3. (a) $\frac{y}{3z}$; (b) $\frac{2a}{5c^2}$; (c) $\frac{3cd^5}{4}$
4. (a) $\frac{1}{3}$; (b) $\frac{3}{2(a-b)}$; (c) $\frac{3}{4}$; (d) $\frac{14y^2}{15}$
5. (a) $\frac{9}{8}$; (b) $\frac{1}{17}$; (c) $\frac{x}{2}$
6. (a) 2; (b) $\frac{9}{2x}$; (c) 1; (d) $8 \cdot \frac{5}{2} = 20$; (e) $\frac{c}{d} \cdot \frac{a}{c} = \frac{a}{d}$; (f) $\frac{x^2}{y^2} \cdot \frac{y}{x^3} = \frac{1}{xy}$
7. (a) $\frac{3}{4} \cdot \frac{4}{3} y = \frac{4}{3} \cdot 3$, $y = 4$; (b) $\frac{7}{2} \cdot \frac{2}{7} z = \frac{7}{2} \cdot 6$, $z = 21$; (c) $\frac{17}{5} \cdot \frac{5}{17} t = \frac{5}{17} \cdot 17$, $t = 5$
8. (a) x; (b) $\frac{2}{3}$; (c) $\frac{x}{b+x}$; (d) $\frac{4x(x-2)}{3(x+3)}$
9. (a) $\frac{7a}{15}$; (b) $\frac{bc+ac+ab}{abc}$; (c) $\frac{5x+y}{6}$; (d) $\frac{x^2-8x+1}{6x^2}$; (e) $\frac{sk+kt}{st}$
10. (a) true; (b) true; (c) true; (d) true
11. (a) $a-b$; (b) x^2-5x+6; (c) a^2-b^2; (d) a^2+b^2
12. (a) $-\frac{y+2}{a}$; (b) $-\frac{1}{x}$; (c) $\frac{2b+c}{2b-c}$; (d) $\frac{x-5}{3x+2}$
13. (a) $\frac{2x^2+2xy+2y^2+x-y}{x^2-y^2}$; (b) $-\frac{2x+11}{x^2-x-20}$
14. (a) $\frac{2k^2-3k+1}{k}$; (b) $\frac{6b^2+12b-1}{3b}$; (c) $\frac{4x^3+4x^2-12x+3}{4x}$;
    (d) $\frac{y^3z-y^2z^2+yz^3+2}{yz}$
15. (a) $a^2+a+\frac{1}{a}$; (b) $2xy-3-\frac{5}{2xy}$; (c) $5y-4+\frac{3}{y^3}$; (d) $3x^2-2x+\frac{1}{x}$;
    (e) $3x-2y+\frac{2}{x+2y}$; (f) $a^2+a+1+\frac{2}{a-1}$
16. (a) $\frac{y}{x(x+y)}$; (b) $\frac{5(x^2+y^2)}{3(x+y)}$; (c) $\frac{a^2+2ab+b^2}{a^2-2ab+b^2}$; (d) $\frac{a^2-2a+3}{a+1}$

Be sure to review the appropriate frames for any problems you may have missed. Then go to Chapter 6.

CHAPTER SIX

Exponents, Roots, and Radicals

OBJECTIVES

In previous chapters we worked with exponents, finding higher powers of numbers, and even touched briefly on the idea of square roots. In this chapter we are going to study these concepts in much greater detail. Specifically, you will learn to:

- find the squares of signed numbers and algebraic monomials;
- multiply terms having exponents and find powers of powers;
- find the square roots of numbers;
- find the square roots of fractions (both arithmetic and algebraic) whose terms are perfect squares;
- find the approximate square roots of fractions whose terms are not perfect squares;
- multiply and divide radicals;
- simplify and combine radicals;
- multiply polynomials containing radicals;
- rationalize the denominators of fractions containing radicals;
- solve radical equations.

REVIEW QUIZ

Before going ahead with our discussion of exponents, roots, and radicals, let's find out how much you remember about the terms and concepts we covered in Chapter 4.

1. The reverse of squaring a number is finding its ______________________.

2. Squaring either a positive or negative real number yields a positive result. (True / False)

3. A square root of a number is one of two e______________ f__________________ of that number.

4. The positive root of a number is called the p___________________ square root.

5. Every positive real number has _______ (how many?) square roots.

6. The symbol $\sqrt{}$ is called the r________________ sign.

7. The two square roots of a number are equal in a____________________ value but opposite in sign.

8. The number under the radical sign is called the r____________________.

9. The combination of the radical sign together with the number under it is called the r____________________.

10. The square root of a rational number is always rational. (True / False)

Answers to Review Quiz

1. square root
2. True (The square of either −2 or +2 is +4.)
3. equal factors
4. principal
5. two (one positive, one negative)
6. radical
7. absolute (that is, without regard to sign)
8. radicand
9. radical
10. False (The fact that roots are not always rational numbers is one of the principal reasons for the inclusion of this chapter.)

If you had trouble remembering any of the above answers turn back to frame 8 in Chapter 4 and review the ideas discussed there before continuing.

SOME BASIC PROPERTIES OF EXPONENTS

1. The exponent 3 in an expression such as x^3 indicates how many times the base x is to be taken as a factor. Thus, $x^3 = x \cdot x \cdot x$. Making this a bit more general, if a is a real number and m is a positive integer, we can restate some of our definitions regarding exponents and powers.

Definition 1: If m is a positive integer

$$a^m = \overbrace{a \cdot a \cdot a \cdot a \ldots a}^{m \text{ factors}}$$

Remember that a is called the *base,* m is called the *exponent,* and the result

$$a \cdot a \cdot a \cdot a \ldots a$$

is called the *power.* The value of m tells how many times the base is to be taken as a factor. In the expression 7^2, for example, the exponent, 2, indicates that the number 7 is to be used as a factor twice. Therefore, $7^2 = 7 \cdot 7 = 49$, and 49 is called the second power (or the square) of 7.

Restated below are two rules (or properties) of exponents which go hand in hand with the definition given above and which you should recognize from our work with exponents in past chapters.

Rule 1: $a^m \cdot a^n = a^{m+n}$ if m and n are positive integers.

Rule 2: $(a^m)^n = a^{m \cdot n}$ (or a^{mn}) if m and n are positive integers.

Use the definition of positive integral exponents and the two rules just given to work out the following problems.

(a) $8^2 =$ ________ (f) $(-2a^2b^3)^2 =$ ____________

(b) $(-3)^2 =$ ________ (g) $(xy^2z^3)^2 =$ ____________

(c) $12^2 =$ ________ (h) $\left(\frac{abc}{2x}\right)^2 =$ ____________

(d) $(-4a)^2 =$ ________ (e) $\left(\frac{r^2st^3}{7mn}\right)^2 =$ ________ ________

(e) $(3x^2)^2 =$ ________ (j) $\left(\frac{-3xy^2}{4x^2t^5}\right)^2 =$ ____________

- - - - - - - - - - - - - - - - - -

(a) 64; (b) 9; (c) 144; (d) $16a^2$; (e) $9x^4$; (f) $4a^4b^6$; (g) $x^2y^4z^6$; (h) $\frac{a^2b^2c^2}{4x^2}$; (i) $\frac{r^4s^2t^6}{49m^2n^2}$; (j) $\frac{9x^2y^4}{16x^4t^{10}}$.

2. So far we have used only positive integers for exponents. What if the exponent is zero? Rule 1 (of Definition 1) indicates that when multiplying like bases we add their exponents. Thus, $b^4 \cdot b^2 = b^{4+2} = b^6$. The corresponding rule for division (assuming a denominator other than zero) is to subtract the exponent of a given base in the denominator from the exponent of that same base in the numerator. Thus, $x^4 \div x^2 = x^{4-2} = x^2$. (We will consider this further in frame 4.)

Applying this last rule to the problem $x^5 \div x^5$, we get the answer x^{5-5} or x^0. However, we also know from our basic rules for division that any number (other than zero) divided by itself equals one. Therefore, $x^5 \div x^5 = 1$.

From these results and the substitution law of equality we have the following:

$$\text{If } \frac{x^5}{x^5} = 1 \text{ and } \frac{x^5}{x^5} = x^0, \text{ then } x^0 = 1.$$

Stating this in somewhat more general terms we have another definition.

Definition 2: $a^0 = 1$ if $a \neq 0$.

Hence, $7^0 = 1$, $153^0 = 1$, $357.2^0 = 1$, and so on.

As a review of what we have covered so far, work the following problems. Assume that all variables (letters) are not equal to zero.

(a) $z^5 \cdot z^4 =$ ____________

(b) $k^7 \div k^2 =$ ____________

(c) $m^9 \div m^9 =$ ____________

(d) $t^{12} \div t^5 =$ ____________

(e) $x^3(x^3) =$ ____________

(f) $7^4 \div 7^4 =$ ____________

(g) $2^4 \cdot 2^4 =$ ____________

(h) $a^3 \cdot b^2 =$ ____________

(i) $(a^3)^2 =$ ____________

(j) $(x^2y^5)^3 =$ ____________

(k) $x^0 \cdot x^4 =$ ____________

(l) $x^5 \div x^0 =$ ____________

(m) $x^6 \div x =$ ____________

(n) $x^0 \cdot x^0 =$ ____________

(o) $x^2 \div x^0 =$ ____________

- - - - - - - - - - - - - - - - - -

(a) $z^{5+4} = z^9$; (b) $k^{7-2} = k^5$; (c) $m^{9-9} = m^0 = 1$; (d) $t^{12-5} = t^7$; (e) $x^{3+3} = x^6$; (f) $7^{4-4} = 7^0 = 1$; (g) $2^{4+4} = 2^8$; (h) $a^3 \cdot b^2 = a^3b^2$ (cannot be simplified because no common base); (i) $a^{3 \cdot 2} = a^6$; (j) x^6y^{15}; (k) x^4; (l) x^5; (m) x^5; (n) 1; (o) x^2

3. Problems (i) and (j) in frame 2 involve the use of Rule 2 (frame 1). Rule 2 (frame 1) says to find the power of a power you keep the base and multiply the exponents. Let's review this concept briefly just to make sure you are clear about it.

Suppose we wish to raise the fourth power of the base b (b^4) to the third power. We write this $(b^4)^3$, which tells us to use b^4 as a factor three times ($b^4 \cdot b^4 \cdot b^4$). Following the law of multiplication we would add the exponents of the base to obtain $b^{4+4+4} = b^{12}$. We could have obtained the same result by multiplying the exponent 4 by the exponent 3. Thus, $(b^4)^3 = b^{4 \cdot 3} = b^{12}$.

Practice this property in the following problems.

(a) $(x^4)^2 =$ ____________

(b) $(a^2b^3)^3 =$ ____________

(c) $\left(\frac{a}{b}\right)^3 =$ ____________

(d) $(t^m)^n =$ ____________

(e) $(c^3)^2c^2 =$ ____________

(f) $\left(\frac{x^2y}{sg^3}\right)^3 =$ ____________

(g) $g^2(g^2)^2 =$ ____________

(h) $\left(\frac{3a^3b^2}{4mnz}\right)^3 =$ ____________

- - - - - - - - - - - - - - - - - -

(a) x^8; (b) a^6b^9; (c) $\frac{a^3}{b^3}$; (d) t^{mn}; (e) $c^{3 \cdot 2}(c^2) = c^6 \cdot c^2 = c^8$;

(f) $\frac{x^6y^3}{x^3g^9}$; (g) $g^2 \cdot g^{2 \cdot 2} = g^{2+4} = g^6$; (h) $\frac{27a^9b^6}{64m^3n^3z^3}$

4. So far we have been working with exponents that are whole numbers. Now let's consider the possibility of negative exponents. In Chapter 3 (frame 24) we mentioned the possibility of writing the fraction $\frac{1}{k^4}$ as k^{-4}, or $\frac{1}{a^n}$ as a^{-n} (where a is not equal to zero). Our justification for doing so lies in the rule for the division of like bases: *Subtract the exponent of the base in the denominator from the exponent of the base in the numerator.* Using symbols we can state this more simply as our third rule of exponents.

Rule 3: $a^m \div a^n = a^{m-n}$, if $a \neq 0$.

Let's see how this rule applies to a typical problem.

Example: $x^2 \div x^5$ $\qquad \frac{x \cdot x}{x \cdot x \cdot x \cdot x \cdot x} = \frac{x^2}{x^5} = \frac{1}{x^3}$

Applying the law of exponents for division (Rule 3) to his problem we would have

$$\frac{x^2}{x^5} = x^{2-5} = x^{-3}$$

If our law of exponents for division is to hold for all situations where the numbers m and n are whole numbers and $a \neq 0$, we must define the negative exponent as follows.

Definition 3: If $a \neq 0$ and if n is a positive integer, then $a^{-1} = \frac{1}{a}$ and $a^{-n} = \frac{1}{a^n}$. Or, conversely, $\frac{1}{a^{-n}} = a^n$.

Use this definition as necessary in working the following problems. Write your answers using positive exponents unless otherwise indicated.

(a) Write x^{-3} using a positive exponent. ____________

(b) Write $\frac{1}{b^4}$ using a negative exponent. ____________

(c) Would it be consistent with our definition of a reciprocal (Definition 3) to say that a^{-1} is the reciprocal of a? ____________

(d) Write the answer to $b^3 \div b^7$ using positive exponents. ____________

(e) Write the answer to $b^3 \div b^7$ using negative exponents. ____________

(f) Write $\frac{1}{x^{-4}}$ using a positive exponent. ____________

(a) $\frac{1}{x^3}$; (b) b^{-4}; (c) Yes, because the reciprocal of a is $\frac{1}{a}$; (d) $\frac{1}{b^4}$; (e) b^{-4}; (f) x^4

5. If you answered the problems in frame 4 correctly you are ready to try a few more applications. Write the answer to each of the following using positive exponents.

(a) $x^5 \div x^3 =$ ______________ (c) $x^{12} \div x^{13} =$ ______________

(b) $x^3 \div x^5 =$ ______________ (d) $x^6 \div x^2 =$ ______________

Write the answer to each of the above exercises using negative exponents.

(e) ______________ (g) ______________

(f) ______________ (h) ______________

- - - - - - - - - - - - - - - - - - -

(a) x^2; (b) $\frac{1}{x^2}$; (c) $\frac{1}{x}$; (d) x^4; (e) $\frac{1}{x^{-2}}$; (f) x^{-2}; (g) x^{-1}; (h) $\frac{1}{x^{-4}}$

6. Let's pause for a moment to sum up our present knowledge of exponents. First we have the definitions for all integral exponents.

Definition 1: $a^m = a \cdot a \cdot a \cdot a \ldots a$, if m is a positive, whole number.

Definition 2: $a^0 = 1$, if $a \neq 0$.

Definition 3: $a^{-1} = \frac{1}{a}$, and $a^{-n} = (a^{-1})^n$, where $a \neq 0$ and n is a positive integer.

The following rules or properties can be used with any of the exponents that have been defined.

Rule 1: $a^m \cdot a^n = a^{m+n}$

Rule 2: $(a^m)^n = a^{m \cdot n}$

Rule 3: $a^m \div a^n = a^{m-n}$ (if $a \neq 0$)

Use these definitions and rules of exponents to work the following problems.

(a) $3x^2(-5x^3) =$ ______________

(b) $(-3x^5)^2 =$ ______________

(c) Write $6a \cdot a \cdot a \cdot b \cdot b \cdot c$ using exponents. ______________

(d) Write $-12b^3c^2$ without exponents. ______________

(e) What is the value of $(15x^4)^0$? ________

(f) What is the value of $15x^0$? ________

(g) Write x^{-4} using positive exponents. ______________

(h) Write x^4 using negative exponents. ______________

(i) $x^2 \div x^7 =$ ______________

(j) $15x^7 \div (-3x^2) =$ ______________

(k) Write the reciprocal of a using positive exponents. ____________

(l) Write the reciprocal of a using negative exponents. ____________

(m) Write $\frac{x^2}{y^2}$ using negative exponents. ____________

(n) Write the product $-3x^4y^5(12x^{-5}y)$ using positive exponents. ____________

- - - - - - - - - - - - - - - - - -

(a) $-15x^5$; (b) $9x^{10}$; (c) $6a^3b^2c$; (d) $-12b\cdot b\cdot b\cdot c\cdot c$; (e) 1; (f) 15; (g) $\frac{1}{x^4}$ or $\left(\frac{1}{x}\right)^4$; (h) $\frac{1}{x^{-4}}$ or $\left(\frac{1}{x}\right)^{-4}$; (i) x^{-5} or $\frac{1}{x^5}$; (j) $-5x^5$; (k) $\frac{1}{a}$ (l) a^{-1}; (m) $\left(\frac{y}{x}\right)^{-2}$ or $\frac{y^{-2}}{x^{-2}}$; (n) $\frac{-36y^6}{x}$

If you missed problem (b) you may have forgotten that squaring -3 gives $+9$. Also, if you missed problem (f) you may have forgotten that the exponent zero applies only to the base x, not to 15 as well.

ROOTS AND RADICALS

7. You will recall that the operations of addition and multiplication have as their inverse the operations of subtraction and division, respectively. Similarly, the operation of raising a number to a power has as its inverse the operation known as *extracting the root* of a number. We touched on this briefly in Chapter 4 (frame 8).

Raising a to the power m is indicated by the expression a^m. Similarly, extracting the mth root of a is indicated by $\sqrt[m]{a}$. The symbol $\sqrt{\ }$ is called the *radical sign*, a is termed the *radicand*, and m is known as the *index*. If no index is written it is assumed to be 2, thus $\sqrt{a}$ means the positive square root of a. Defining these terms leads us to the following definition.

Definition 4: $n = \sqrt[m]{a}$, if $n^m = a$

In this case n is called the mth root of a. To see what this definition means, let's consider a few examples.

Examples:

$\sqrt{9} = 3$ because $3^2 = 9$ (Remember, $\sqrt{9}$ means $\sqrt[2]{9}$.)

$\sqrt{25} = 5$ because $5^2 = 25$

$\sqrt[3]{-8} = -2$ because $(-2)^3 = -8$

$\sqrt[3]{8} = 2$ because $2^3 = 8$

$\sqrt[4]{81} = 3$ because $3^4 = 81$

Definition 4a: A number is an nth root of a given number if it is one of the n equal factors of that number.

Examples:

A square root of a (indicated by $\sqrt{a}$) is one of two equal factors of a.

A cube root of a (indicated by $\sqrt[3]{a}$) is one of three equal factors of a.

A fourth root of a (indicated by $\sqrt[4]{a}$) is one of four equal factors of a.

Many numbers have roots (of some kind) that are exact integers or rational numbers. The following problems are with numbers of this type. Find the roots. (Give the absolute value when the index is an even number.)

(a) $\sqrt{49}$ = ________
(b) $\sqrt{64}$ = ________
(c) $\sqrt[3]{27}$ = ________
(d) $\sqrt[4]{16}$ = ________
(e) $\sqrt[3]{125}$ = ________
(f) $\sqrt{9}$ = ________
(g) $\sqrt{81}$ = ________
(h) $\sqrt[5]{32}$ = ________
(i) $\sqrt[3]{a^6}$ = ________
(j) $\sqrt[5]{a^{25}}$ = ________

- - - - - - - - - - - - - - - - - -

(a) 7; (b) 8; (c) 3; (d) 2; (e) 5; (f) 3; (g) 9; (h) 2; (i) a^2; (j) a^5

8. We have limited our discussion to extracting roots of integers. The extraction of the roots of a fraction is an easy process if the root of the fraction is a rational number. The root of such a fraction is found by extracting the root of the numerator and dividing this by the root of the denominator. Putting this into symbols we can write the following definition.

Definition 5: $\sqrt[n]{\frac{a}{b}} = \frac{\sqrt[n]{a}}{\sqrt[n]{b}}$

Example: $\sqrt{\frac{9}{16}} = \frac{\sqrt{9}}{\sqrt{16}} = \frac{3}{4}$

Work the following problems using Definition 5 and any other necessary information. Assume that all variables (letters) are positive.

(a) $\sqrt{\frac{49}{25}}$ = ________
(b) $\sqrt{\frac{4}{x^8}}$ = ________
(c) $\sqrt{\frac{36}{x^{10}}}$ = ________
(d) $\sqrt{\frac{4x^6}{9x^4}}$ = ________________
(e) $\sqrt[3]{\frac{27}{x^6}}$ = ________
(f) $\sqrt[4]{\frac{1}{16}}$ = ________
(g) $\sqrt[3]{-\frac{1}{8}}$ = ________
(h) $\sqrt{\frac{1}{25}}$ = ________
(i) $\sqrt{\frac{x^8}{y^{10}}}$ = ________

- - - - - - - - - - - - - - - - - -

(a) $\frac{7}{5}$; (b) $\frac{2}{x^4}$; (c) $\frac{6}{x^5}$; (d) $\frac{2x^3}{3x^2} = \frac{2x}{3}$; (e) $\frac{3}{x^2}$; (f) $\frac{1}{2}$; (g) $-\frac{1}{2}$; (h) $\frac{1}{5}$; (i) $\frac{x^4}{y^5}$

9. In the problems of frame 8 you worked with radicals of various orders (that is, different indices). Since our work from here on with radicals will involve only square roots, we need to discuss two important aspects of square roots.

We have defined the square root of a number as one of two equal factors of that number. By this definition the square root of 16 could be +4 (since 4•4 = 16) or −4 (since −4•−4 = 16). Because the answer to any unique combination of symbols should be a unique value (that is, there should be one and only one answer to an operation) we need the following definition.

Definition 6: $\sqrt{n}$ is considered to mean the *positive* square root of n and is called the *principal* square root of n; $-\sqrt{n}$ is considered to indicate the *negative* square root of n and should not be given unless the radical is preceded by a minus sign.

Work the following problems, being careful that you follow Definition 6. Consider all variables positive.

(a) $\sqrt{16}$ = ________

(b) $-\sqrt{16}$ = ________

(c) $-\sqrt{25x^2}$ = ________

(d) $\sqrt{36x^4}$ = ________

(e) $-\sqrt{49}$ = ________

(f) $\sqrt{64a^6}$ = ________

(g) $-\sqrt{25n^2}$ = ________

(h) $\sqrt{\frac{x^6}{25}}$ = ________

- - - - - - - - - - - - - - - - - -

(a) 4; (b) −4; (c) $-5x$; (d) $6x^2$; (e) −7; (f) $8a^3$; (g) $-5n$; (h) $\frac{x^3}{5}$

10. The second aspect of square roots that needs clarification is the matter of what to do when you see an expression such as $\sqrt{x^2}$ and don't know whether x is positive or negative. We're going to defer this question until we have developed an additional law for exponents. In the meantime, you must assume that when you are asked to take the square root of a variable raised to the second power, the variable represents a positive number.

Now let's talk a bit more about procedures for extracting roots. As we stated earlier, the roots of many numbers can be written as integers, or as non-integral rational numbers. Frequently we can find these values by inspection or by the use of tables. Extracting the roots of the numbers we have encountered so far has been fairly simple. The real problem arises when either the radicand is too large to solve by inspection or else does not give an exact root.

As an example of the second case, consider the number $\sqrt{8}$ where the radicand is not a perfect square. This kind of a number is called an *irrational number* because it cannot be represented by an integer or by a fraction composed of integers. However, we can find approximate values for irrational numbers either by the use of tables or by computation.

It would be difficult to find the square root of a radicand too large to solve by inspection (such as the number 106,929, which happens to be a perfect square). It would involve the time-consuming process of trial and error. If we are to solve such problems we need some practical and consistent way of finding the square roots of numbers. There are tables that give the approximate square roots of numbers to several decimal places. These are convenient if you happen to have such a table at hand and if it includes the number you're concerned with.

We're going to show you a procedure you can use to extract the square root of *any* number, regardless of its size. You will need to practice it a few times in order to make sure you understand it and can apply it correctly. A year from now (unless you use it in the meantime) you may have forgotten the exact procedure. Hopefully, you will at least recall that there is such a procedure and that you can return here to review it.

EXTRACTING THE SQUARE ROOT

11. In the last frame we said that the six-digit number 106,929 is a perfect square. Now we're going to verify this by finding its positive square root. From this point on we will refer to this number simply as the *radicand.* To find its square root we proceed as follows:

(1) Separate the radicand into groups of two figures each, counting from the decimal point (from right to left in this case).

10 69 29.

(2) Under the first pair, 10, place 9, the largest perfect square less than 10. Above the first pair place 3, the square root of 9. Subtract 9 from 10 to obtain the remainder 1.

```
  3
 √10 69 29.
  9
  1
```

(3) Bring down 69, the next pair of figures. The new dividend is 169.

(4) To get the new divisor, double 3 (the first figure in the root) and get 6 as the first figure of the divisor.

```
        3
       √10 69 29.
        9
6     | 1 69
```

(5) There is one more figure in the divisor, and it is also the next figure in the root. To find it, divide 6 into 16 (the first two figures of the remainder). Since it will go twice, place a 2 beside the 6 to get 62, the complete divisor, and also place the 2 above the 69, as the next figure of the root.

```
        3  2
       √10 69 29.
        9
62    | 1 69
```

(6) Now multiply this 2, the last digit of the partial root, times 62 and subtract the product, 124, from 169 to get the new remainder of 45.

(7) Bring down 29, the last pair of figures in the original number. The new dividend is now 4529. Again double the partial root (32) to get 64 and place this in front of 4529. By trial we find that 64 will go into 452 (the first three figures of the remainder) seven times. Therefore, place a 7 after 64 and also above 29 as the last figure in the root.

```
        3  2  7
       √10 69 29.
        9
62    | 1 69 29
      | 1 24
647      45 29
```

(8) Multiply 647 by 7 and place the product under 4529. Since there is no remainder, the square root of 106,929 is the whole number 327. Had there been a remainder you would place a decimal point after the final 9 (in 10 69 29) and start bringing down pairs of zeros to find the value of the root to as many decimal places as you required.

```
          3  2  7
        √10 69 29.
          9
62       1 69 29
         1 24
647        45 29
           45 29
```

12. Since this method of extracting square roots may be new to you, let's work another problem together before you try it on your own. This time we'll take a number that has some figures to the right of the decimal point (as well as to the left) and find the root to two decimal places. The radicand we will use is 23,749.98.

(1) Separate the radicand into groups of two figures, pairing left and right from the decimal point. Notice that our first "pair" is just one number and that we have to add a pair of zeros to find the root to the second decimal point.

2 37 49. 98 00

(2) The largest square smaller than 2 is 1.

(3) Doubling the first figure of the root, 1, and bringing it down gives us a trial divisor of 2, from which we get 5 as the next term in the root (since trial shows us 6 would be too large).

(4) A trial divisor of 30 (15 doubled) yields 4 as the last term of the divisor and the next figure in the root.

(5) Doubling 154 gives 308 as our next trial divisor, and since this number will go into 339 only once, we add 1 after 308 and also place it above 98 as our first decimal place in the root.

(6) Our next trial divisor, 3082, produces the figure 1 as the fifth term of the divisor and the last figure in the root, since we are seeking the square root to only two decimal places. The remainder is 879. (Note: The decimal point of the answer should be written directly above the decimal point of the radicand.)

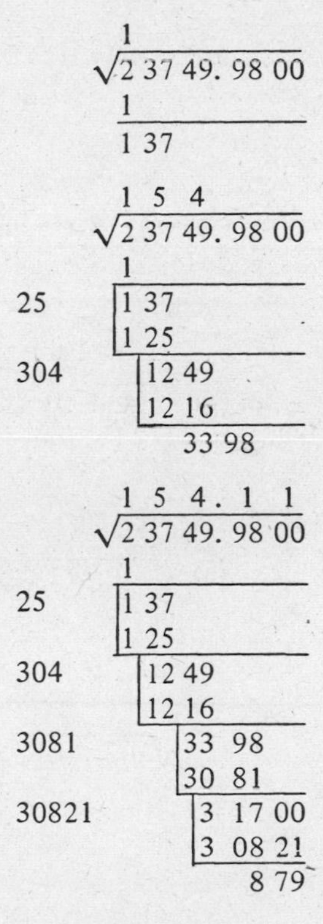

13. Now it is time for you to try a few of these problems on your own. To help you along, refer to the answer given below the dashed line.

(a) $\sqrt{529}$

(b) $\sqrt{1225}$

(c) $\sqrt{361}$

(d) $\sqrt{1764}$

(e) $\sqrt{4096}$

(f) $\sqrt{8728}$ (to one decimal place)

(g) $\sqrt{4444}$ (to two decimal places)

(h) $\sqrt{139.273}$ (to two decimal places)

(i) $\sqrt{103.8361}$ (Hint: In your first division, you must bring down two pairs of numbers before you divide.)

(j) $\sqrt{.021904}$ (Hint: Group your pairs of numbers to the right of the decimal point and proceed as usual.)

(a)

```
        2 3
      √5 29
        4
43     |1 29
       |1 29
```

(b)

```
         3  5
      √12 25
         9
65     | 3 25
       | 3 25
```

(c) 19; (d) 42; (e) 64; (f) 93.4; (g) 66.66

(h)

```
         1  1.  8  0
       √1 39. 27 30    (zero added to complete pair)
         1
21       |  39
         |  21
228      |  18 27
         |  18 24
2360           3 30    (2360 will not divide into 330)
```

(i) 10.19; (j) .148

FRACTIONAL EXPONENTS

14. Our definitions in frame 6 didn't include a definition of fractional exponents so it is time that we considered these. Reflecting upon the possible meaning of a fractional exponent such as $x^{\frac{1}{2}}$ (where x is a positive number), we might get some clue by observing that $x^{\frac{1}{2}} \cdot x^{\frac{1}{2}} = x^{\frac{1}{2}+\frac{1}{2}} = x$, by the law of multiplication of terms having the same base. Since $\sqrt{x} \cdot \sqrt{x} = x$, and $x^{\frac{1}{2}} \cdot x^{\frac{1}{2}} = x$, we might suspect that $x^{\frac{1}{2}} = \sqrt{x}$. Also, since the product of $x^{\frac{1}{3}} \cdot x^{\frac{1}{3}} \cdot x^{\frac{1}{3}}$ is $x^{(\frac{1}{3})+(\frac{1}{3})+(\frac{1}{3})}$, which is x, it would seem consistent to say that $x^{\frac{1}{3}} = \sqrt[3]{x}$.

Let's consider the quantity x^n. If we want the value of $x^{\frac{1}{n}} \cdot x^{\frac{1}{n}} \cdot x^{\frac{1}{n}} \ldots x^{\frac{1}{n}}$ (for n factors), we could write this as $\left(x^{\frac{1}{n}}\right)^n$, and our Rule 2 (frame 6) would allow us to write this as $x^{(\frac{1}{n})(n)} = x$. Although this certainly is not a proof, it seems clear that since $\left(x^{\frac{1}{n}}\right)^n = x$ and $(\sqrt[n]{x})^n = x$, we can say that $x^{\frac{1}{n}} = \sqrt[n]{x}$. If we accept the above argument, then it would be consistent to say that $x^{\frac{m}{n}} = x^{(\frac{1}{n})(m)} = (\sqrt[n]{x})^m = \sqrt[n]{x^m}$. It appears, therefore, that:

> The denominator of a fractional exponent is the index of a radical, and the numerator of that fractional exponent is the power of the radicand.

Thus, $x^{\frac{2}{3}} = \sqrt[3]{x^2}$; $x^{\frac{1}{2}} = \sqrt{x}$; $x^{\frac{3}{4}} = \sqrt[4]{x^3}$.

With the foregoing in mind, write the answers to the following problems.

(a) $4^{\frac{1}{2}} =$ ______ (e) $27^{\frac{1}{3}} =$ ______

(b) $25^{\frac{1}{2}} =$ ______ (f) $(-8)^{\frac{1}{3}} =$ ______

(c) $\left(x^{\frac{1}{2}}\right)^2 =$ ______ (g) $(16a^8)^{\frac{1}{2}} =$ ______ $(a > 0)$

(d) $8^{\frac{1}{3}} =$ ______ (h) $\left(\frac{4}{9}\right)^{\frac{1}{2}} =$ ______

- - - - - - - - - - - - - - - - - - -

(a) 2; (b) 5; (c) x; (d) 2; (e) 3; (f) -2; (g) $4a^4$; (h) $\frac{2}{3}$

15. Now that we have fractional exponents to work with, it will be easy to discuss the second aspect of square roots which we mentioned in frame 10. We stated there that $\sqrt{x^2}$ needs to be clarified. A look at the following example will illustrate why.

Example: Suppose we wish to find the value of $\sqrt{(-2)^2}$. We could write this with fractional exponents as $\left[(-2)^2\right]^{\frac{1}{2}}$. On the basis of our prior discussions there are two ways to approach this problem.

Method 1: $\left[(-2)^2\right]^{\frac{1}{2}} = \left[(-2)(-2)\right]^{\frac{1}{2}} = 4^{\frac{1}{2}} = 2$

Method 2: $\left[(-2)^2\right]^{\frac{1}{2}} = (-2)^{2\cdot\frac{1}{2}} = (-2)^1 = -2$

Since we can't have two different answers to the same problem we must agree on a method that will yield a unique (that is, one) answer. Accordingly, we establish the following definition.

Definition 7: $\sqrt{x^2} = |x|$ or $(x^2)^{\frac{1}{2}} = |x|$ (the absolute value of x).

If we follow this definition we will always achieve a unique answer. Discover this for yourself in the following problems. (Assume that the variables may be any real number.)

(a) $\sqrt{(-3)^2} =$ ______ (d) $\sqrt{(-5)^2} =$ ______

(b) $\sqrt{5^2} =$ ______ (e) $\sqrt{m^2} =$ ______

(c) $\sqrt{a^2} =$ ______ (f) $\sqrt{(-7)^2} =$ ______

- - - - - - - - - - - - - - - - - - -

(a) $|-3| = 3$; (b) $|5| = 5$; (c) $|a|$; (d) $|-5| = 5$; (e) $|m|$; (f) $|-7| = 7$

16. To summarize what we have covered thus far you will find below part of each of the definitions and rules for exponents. Your job is to complete each definition.

(a) *Definition 1:* $a^m =$ ______, if m is a positive integer.

(b) *Definition 2:* $a^0 =$ ______, if $a \neq 0$.

(c) *Definition 3:* $a^{-1} =$ ______ and $a^{-n} =$ ______, if $a \neq 0$ and n is a positive integer.

(d) *Definition 4:* If $n^m = a$, then $n =$ __________ (in terms of m and a).

(e) *Definition 5:* $\sqrt[n]{\frac{a}{b}} =$ __________

(f) *Definition 6:* $\sqrt{n}$ is considered the __________ square root of n and is called the __________ square root; $-\sqrt{n}$ is considered the __________ square root of n.

(g) *Definition 7:* $\sqrt{x^2} =$ ______, or $(x^2)^{\frac{1}{2}} =$ ______.

(h) *Rule 1:* $a^m \cdot a^n =$ __________, if m and n are positive integers.

(i) *Rule 2:* $(a^m)^n =$ __________, if m and n are positive integers.

(j) *Rule 3:* $a^m \div a^n =$ __________, if $a \neq 0$.

- - - - - - - - - - - - - - - - - -

For any that you missed, refer to the frame number given in parentheses after the answer.

(a) $\overbrace{a \cdot a \cdot a \cdot a \cdot a \ldots a}^{m \text{ factors}}$ (frame 1)
(b) 1 (frame 2)
(c) $\frac{1}{a}, \left(\frac{1}{a}\right)^n$ (frame 4)
(d) $\sqrt[m]{a}$ (frame 7)
(e) $\frac{\sqrt[n]{a}}{\sqrt[n]{b}}$ or $\frac{a}{b}^{\frac{1}{n}}$ (frame 8)
(f) positive, principal, negative (frame 9)
(g) $|x|$, $|x|$ (frame 15)
(h) a^{m+n} (frame 1)
(i) $a^{m \cdot n}$ (frame 1)
(j) a^{m-n} (frame 4)

ROOTS OF FRACTIONS

17. An understanding of and ability to use the above definitions and rules will be adequate for the problems involving exponents in this book and for most practical experiences. The statements are correct, although no attempt to prove them has been made here. Proofs of them can be found in many of the advanced algebra texts. Think of the definitions as you work problems and you will find yourself mastering them and using them more readily.

Now let's look a little further into the matter of the square roots of fractions. You will recall that the square root of a fraction may be found by dividing the square root of the numerator by the square root of the denominator (frame 8):

$$\sqrt{\frac{a}{b}} = \frac{\sqrt{a}}{\sqrt{b}}$$

Examples: $\sqrt{\dfrac{9}{25}} = \dfrac{\sqrt{9}}{\sqrt{25}} = \dfrac{3}{5}$; $\sqrt{\dfrac{25x^2}{36y^4}} = \dfrac{\sqrt{25x^2}}{\sqrt{36y^4}} = \dfrac{5x}{6y^2}$

Use this rule (from Definition 5) to find the indicated roots in the following problems. All variables (letters) are positive.

(a) $\dfrac{1}{9}$ = ____________ (e) $\dfrac{9t^4}{100s^8}$ = ____________

(b) $\dfrac{16}{49}$ = ____________ (f) $\dfrac{m^4n^8}{81p^2}$ = ____________

(c) $\dfrac{4}{64}$ = ____________ (g) $\dfrac{49x^2y^4}{144b^6}$ = ____________

(d) $\dfrac{x^4}{25}$ = ____________ (h) $\dfrac{81a^4}{121b^4}$ = ____________

- - - - - - - - - - - - - - - - - - - -

(a) $\dfrac{1}{3}$; (b) $\dfrac{4}{7}$; (c) $\dfrac{1}{4}$; (d) $\dfrac{x^2}{5}$; (e) $\dfrac{3t^2}{10s^4}$; (f) $\dfrac{m^2n^4}{9p}$; (g) $\dfrac{7xy^2}{12b^3}$;
(h) $\dfrac{9a^2}{11b^2}$

18. Our rule for finding the square root of a fraction works very well if the numerator and denominator are both perfect squares. But what if they are not? In this case the rule is as follows.

> When the denominator of a fraction is not a perfect square, multiply both the numerator and the denominator by a number that will make the denominator a perfect square.

Example: Suppose we wish to find the approximate square root of the fraction $\frac{2}{3}$. In other words, $\sqrt{\frac{2}{3}}$ = ? Following our rule we multiply the numerator and denominator by 3, giving us $\sqrt{\frac{6}{9}}$. From this we get $\dfrac{\sqrt{6}}{3}$ or (using the approximate root of 6*) $\dfrac{2.449}{3} = .816$. We could have taken the square root of 2 and divided it by the square root of 3 to begin with, but the method shown is faster and involves less chance of error.

Try the method yourself by finding the approximate square roots (to two decimal places) of the following fractions.

(a) $\sqrt{\dfrac{3}{5}}$ = ________________________________

(b) $\sqrt{\dfrac{1}{2}}$ = ________________________________

*See Table of Powers and Roots at the back of this book for approximate values.

(c) $\sqrt{\frac{4}{7}}$ = ______________________________

(d) $\sqrt{\frac{3}{4}}$ = ______________________________

- - - - - - - - - - - - - - - - - -

(a) $\frac{3 \cdot 5}{5 \cdot 5} = \frac{\sqrt{15}}{5} = \frac{3.87}{5} = .77$

(b) $\frac{1 \cdot 2}{2 \cdot 2} = \frac{\sqrt{2}}{2} = \frac{1.414}{2} = .71$

(c) $\frac{4 \cdot 7}{7 \cdot 7} = \frac{\sqrt{28}}{7} = \frac{5.29}{7} = .76$

(d) $\frac{\sqrt{3}}{2} = \frac{1.73}{2} = .86$ (or .87) (Did you notice that the denominator was already a perfect square?)

19. It isn't always necessary—or even desirable—to work out the decimal fraction equivalent of the numerator. Sometimes it's much simpler and more accurate to retain the radical in an answer. Thus, in problem (b) of frame 18 we could have left the answer as $\frac{\sqrt{2}}{2}$ or, in problem (d), $\frac{\sqrt{3}}{2}$.

Try this in the following problems. Leave the radical in the numerator instead of extracting the square root.

(a) $\sqrt{\frac{1}{3}}$ = ______________

(c) $\sqrt{\frac{4}{5}}$ = ______________

(b) $\sqrt{\frac{3}{8}}$ = ______________

(d) $\sqrt{\frac{5}{6}}$ = ______________

- - - - - - - - - - - - - - - - - -

(a) $\frac{\sqrt{3}}{3}$; (b) $\frac{\sqrt{24}}{8} = \frac{2\sqrt{6}}{8}$; (c) $\frac{2\sqrt{5}}{5}$; (d) $\frac{\sqrt{30}}{6}$

20. Before continuing with our investigation of roots and radicals we need to discuss two kinds of numbers. This will serve as a partial review of what we covered in Chapter 2 and will also extend some of the ideas we have been discussing in this chapter.

We defined a *rational number* as one that can be expressed as the ratio of integers. The following are rational numbers.

	Example
(1) All integers (that is , all positive and negative whole numbers and zero)	$3 = \frac{3}{1}$
(2) Fractions whose numerator and denominator are already integers or become integers after simplification	$\frac{2 \cdot 5}{4 \cdot 5} = \frac{1}{2}$
(3) Decimals that "terminate" after a finite number of steps (i.e., all later terms in the expansion are zero)	$\frac{687}{100} = 6.87000$
(4) Decimals that do not terminate but whose digits repeat themselves	$0.3333333\ldots\frac{1}{3}$
(5) A radical expression whose radicand is a perfect power	$\sqrt{36} = 6$; $\sqrt[3]{8} = 2$

Numbers that cannot be expressed as the ratio of two integers are known as *irrational numbers.* The square root of 2 is an example of an irrational number because it cannot be changed into a whole number or a simple fraction. Its exact value is $\sqrt{2}$ and this cannot be written as a rational number. An approximation can be made, but this approximation will be a non-terminating, non-repeating decimal.

Check the numbers that are rational.

___ (a) $\sqrt{4}$ ___ (e) $\sqrt{81}$

___ (b) $\sqrt{3}$ ___ (f) 13

___ (c) 1.748 ___ (g) .66666 . . .

___ (d) $\frac{1.5}{2}$ ___ (h) $\sqrt{8}$

- - - - - - - - - - - - - - - - - - -

You should have checked the following. (Example numbers refer to the five examples of rational numbers given in frame 20.)

(a) $\sqrt{4}$ (#5); (c) 1.748 (#3); (d) $\frac{1.5}{2}\left(=\frac{3}{4}\right)$ (#2); (e) $\sqrt{81}$ (#5);
(f) 13 (#1); (g) .66666 . . . (#4)

$\sqrt{3}$ and $\sqrt{8}$ are *irrational* numbers because they cannot be expressed as the ratio of two integers.

21. It is often necessary to simplify radicals. Not all radicals can be simplified, of course. Let's see how such simplification can take place. Can $\sqrt{36}$ be simplified? Yes, by taking its square root. Thus, $\sqrt{36} = 6$. Can $\sqrt{48}$ be simplified? Yes, if you recognize that it contains a perfect square (16) as a factor. Thus, $\sqrt{48} = \sqrt{16} \cdot \sqrt{3}$, or $4\sqrt{3}$. Can $\sqrt{\frac{2}{3}}$ be simplified? Yes, merely multiply the numerator and the denominator by a number (3 in this example) that will make the denominator a perfect square (as you learned to do in frame 18). Doing so gives $\frac{\sqrt{6}}{\sqrt{9}}$ or $\frac{\sqrt{6}}{3}$ or $\frac{1}{3}\sqrt{6}$. (Remember that $\frac{x}{3}$ and $\frac{1}{3}x$ represent the same value.)

To summarize, then, a radical can be simplified if the following statements are true.

- The radicand is a perfect square.
- The radicand contains a factor that is a perfect square.
- The radicand is a fraction.

Simplify the following radicals. (All variables are positive.)

(a) $\sqrt{81}$ = ____________

(b) $\sqrt{32}$ = ____________

(c) $\sqrt{50}$ = ____________

(d) $\sqrt{49x^6}$ = ____________

(e) $\sqrt{\frac{1}{4}}$ = ____________

(f) $\sqrt{\frac{x^2}{y^2}}$ = ____________

(g) $\sqrt{\frac{2}{7}}$ = ____________

(h) $\sqrt{b^3}$ = ____________ (Hint: $b^3 = b^2 \cdot b$)

(i) $\sqrt{\frac{1}{a}}$ = ____________

(j) $4\sqrt{27}$ = ____________

(k) $\sqrt{121}$ = ____________

(l) $3\sqrt{128}$

- - - - - - - - - - - - - - - - - -

(a) 9; (b) $4\sqrt{2}$; (c) $5\sqrt{2}$; (d) $7x^3$; (e) $\frac{1}{2}$; (f) $\frac{x}{y}$; (g) $\frac{1}{7}\sqrt{14}$;

(h) $b\sqrt{b}$; (i) $\frac{1}{a}\sqrt{a}$; (j) $12\sqrt{3}$; (k) 11; (l) $24\sqrt{2}$

ADDING AND SUBTRACTING RADICALS

22. Another useful operation is the combining of terms containing square root radicals. This can be done only when the radicals are similar. *Similar radicals* are radicals having the same index and the same radicand. Similar radicals may be combined by addition and subtraction, through application of the distributive law, as shown below.

$$a\sqrt{x} + b\sqrt{x} = (a+b)\sqrt{x}$$

Notice that $a\sqrt{x}$ and $b\sqrt{x}$ are like terms.

Examples: $4\sqrt{5} + 3\sqrt{5} - 2\sqrt{5} = (4+3-2)\sqrt{5} = 5\sqrt{5}$

$$\begin{aligned} 3\sqrt{6} + 2\sqrt{5} - \sqrt{6} + 3\sqrt{5} &= 3\sqrt{6} - \sqrt{6} + 2\sqrt{5} + 3\sqrt{5} \\ &= (3-1)\sqrt{6} + (2+3)\sqrt{5} \\ &= 2\sqrt{6} + 5\sqrt{5} \end{aligned}$$

The similarity of the radicands may not always be so obvious when the radicals are not in simplified form. For example, can we combine the following?

$3\sqrt{8} - \sqrt{50} + 6\sqrt{32}$ (Yes / No)

- - - - - - - - - - - - - - - - - -

Yes. Here's how. First simplify each radical that is not already in simplified form. Simplifying the first term gives

$$3\sqrt{8} = 3\sqrt{4 \cdot 2} = 3 \cdot 2\sqrt{2} = 6\sqrt{2}$$

Simplifying the second term gives

$$\sqrt{50} = \sqrt{25 \cdot 2} = 5\sqrt{2}$$

Finally, the third term becomes

$$6\sqrt{32} = 6\sqrt{16 \cdot 2} = 6 \cdot 4\sqrt{2} = 24\sqrt{2}$$

With the radicals in simplest form we have three similar radicals which may be combined to give

$$6\sqrt{2} - 5\sqrt{2} + 24\sqrt{2} = 25\sqrt{2}$$

23. Use the procedure demonstrated in frame 22 to simplify and combine like terms in the following problems.

(a) $\sqrt{8} - 2\sqrt{8} + 7\sqrt{8} =$ ______________________

(b) $7\sqrt{7} + 4\sqrt{7} - 3\sqrt{7} =$ ______________________

(c) $\sqrt{2} + 2\sqrt{3} + 3\sqrt{2} - \sqrt{3} =$ ______________________

(d) $2 + 6\sqrt{7} + 5 - 2\sqrt{7} =$ ______________________

(e) $4\sqrt{a} + 2\sqrt{b} - \sqrt{a} - \sqrt{b} =$ ______________________

(f) $4\sqrt{7} - \sqrt{28} - \sqrt{63} =$ ______________________

(g) $\sqrt{75} + 4\sqrt{3} + \sqrt{18} =$ ______________________

(h) $2\sqrt{9y} - \sqrt{4x} + 7\sqrt{x} - 3\sqrt{y} =$ ______________________

(i) $6\sqrt{b} + \sqrt{25a} - \sqrt{b} - 2\sqrt{a} =$ ______________________

(j) $\sqrt{2} + 3\sqrt{27} + 2\sqrt{50} - 4\sqrt{3} =$ ______________________

- - - - - - - - - - - - - - - - - -

(a) $12\sqrt{2}$; (b) $8\sqrt{7}$; (c) $4\sqrt{2} + \sqrt{3}$; (d) $7 + 4\sqrt{7}$; (e) $3\sqrt{a} + \sqrt{b}$; (f) $-\sqrt{7}$; (g) $9\sqrt{3} + 3\sqrt{2}$; (h) $5\sqrt{x} + 3\sqrt{y}$; (i) $3\sqrt{a} + 5\sqrt{b}$; (j) $11\sqrt{2} + 5\sqrt{3}$

MULTIPLYING AND DIVIDING RADICALS

24. We have experimented with combining (that is, adding and subtracting) radicals. Now let's consider the procedures for multiplying them. Here are two useful rules.

Rule 4: The product of the square roots of two or more (non-negative) real numbers is equal to the square root of their product. Thus, $\sqrt{a} \cdot \sqrt{b} = \sqrt{ab}$ or $\sqrt{a} \cdot \sqrt{b} \cdot \sqrt{c} = \sqrt{abc}$.

Rule 5: The square of the square root of a positive number equals the number being used as a radicand. Thus, $(\sqrt{8})^2 = 8$.

You will find it easiest to understand and to familiarize yourself with the application of these rules if you work the following problems.

(a) $\sqrt{3}\cdot\sqrt{4}$ = ________

(b) $\sqrt{5}\cdot\sqrt{7}$ = ________

(c) $\sqrt{2}\cdot\sqrt{18}$ = ________

(d) $\sqrt{6}\cdot\sqrt{6}$ = ________

(e) $(\sqrt{9})^2$ = ________

(f) $5\sqrt{2}\cdot 3\sqrt{2}$ = ________

(g) $\sqrt{13}\cdot\sqrt{13}$ = ________

(h) $\sqrt{12}\cdot\sqrt{3}$ = ________

(i) $3\sqrt{6}\cdot 5\sqrt{3}$ = ________

(j) $\sqrt{3}\cdot 2\sqrt{4}\cdot 3\sqrt{3}$ = ________

(a) $\sqrt{12} = 2\sqrt{3}$; (b) $\sqrt{35}$; (c) $\sqrt{36} = 6$; (d) 6; (e) 9;
(f) $15\sqrt{4} = 15\cdot 2 = 30$; (g) 13; (h) 6; (i) $15\sqrt{18} = 15\cdot 3\sqrt{2} = 45\sqrt{2}$;
(j) $6\sqrt{36} = 6\cdot 6 = 36$

25. So far we have multiplied monomials only. Now let's see how we would go about multiplying polynomials containing radicals.

Example 1: A monomial times a binomial

$$\sqrt{3}(7-\sqrt{3}) = 7\sqrt{3} - \sqrt{3}\cdot\sqrt{3} = 7\sqrt{3} - 3$$

Here we used the distributive law as well as the rules for radicals.

Example 2: A binomial times a binomial

$$\begin{aligned}(4+\sqrt{2})(2-\sqrt{2}) &= 8 + 2\sqrt{2} - 4\sqrt{2} - \sqrt{4}\\ &= 8 - 2\sqrt{2} - 2\\ &= 6 - 2\sqrt{2}\end{aligned}$$

Here we used the familiar routine for finding the product of two binomials plus the rules for radicals.

Study these examples as a guide and then perform the operations indicated below.

(a) $2(\sqrt{3}+\sqrt{7})$ = ________

(b) $3(\sqrt{2}-3)$ = ________

(c) $2\sqrt{3}(2-\sqrt{6})$ = ________

(d) $3\sqrt{5}(\sqrt{2}+3)$ = ________

(e) $\sqrt{7}(2\sqrt{7}-\sqrt{25})$ = ________

(f) $(\sqrt{2}+\sqrt{3})(\sqrt{2}-\sqrt{3})$ = ________

(g) $(3+\sqrt{5})(4-2\sqrt{5})$ = ________

(h) $(\sqrt{x}+\sqrt{y})^2$ = ________

(i) $(2\sqrt{3}-1)^2$ = ________

(j) $(a\sqrt{a}+b\sqrt{b})^2$ = ________

- - - - - - - - - - - - - - - - - - - -

(a) $2\sqrt{3} + 2\sqrt{7}$; (b) $3\sqrt{2} - 9$; (c) $4\sqrt{3} - 6\sqrt{2}$; (d) $3\sqrt{10} + 9\sqrt{5}$; (e) $14 - 5\sqrt{7}$; (f) $2 + \sqrt{6} - \sqrt{6} - 3 = -1$; (g) $12 - 2\sqrt{5} - 10 = 2 - 2\sqrt{5}$; (h) $x + 2\sqrt{xy} + y$; (i) $13 - 4\sqrt{3}$; (j) $a^3 + 2ab\sqrt{ab} + b^3$

26. Having learned something about how to multiply radicals, we now are going to see what is involved in dividing them. Actually the procedure is quite straightforward, as shown below.

	Example 1	*Example 2*
	$\dfrac{6\sqrt{24}}{2\sqrt{3}}$	$\dfrac{12\sqrt{a^6}}{3\sqrt{a^3}}$
(1) Divide the coefficients:	$3 \cdot \sqrt{\dfrac{24}{3}}$	$4 \cdot \sqrt{\dfrac{a^6}{a^3}}$
(2) Divide the radicals:	$3\sqrt{8}$	$4\sqrt{a^3}$
(3) Simplify, if possible:	$3(2\sqrt{2})$	$4(\sqrt{a^2} \cdot \sqrt{a})$
(4) Answers:	$6\sqrt{2}$	$4a\sqrt{a}$

Prove to yourself how easy this is by performing the following divisions. (Use absolute values of square root terms.)

(a) $\dfrac{6\sqrt{9}}{2\sqrt{3}} =$ ____________

(b) $\dfrac{12\sqrt{8}}{6\sqrt{2}} =$ ____________

(c) $\dfrac{15\sqrt{48}}{5\sqrt{6}} =$ ____________

(d) $\dfrac{14\sqrt{a^5}}{7\sqrt{a^3}} =$ ____________

(e) $\dfrac{6x^2\sqrt{10a}}{3x\sqrt{5a}} =$ ____________

- - - - - - - - - - - - - - - - - - - -

(a) $3\sqrt{3}$; (b) $2\sqrt{4} = 4$; (c) $3\sqrt{8} = 6\sqrt{2}$; (d) $2\sqrt{a^2} = 2a$; (e) $2x\sqrt{2}$

27. A radical is not in its simplest form if the radicand is a fraction. (We could say that a fraction is not in its simplest form if the denominator contains a radical.) In frame 18 we showed that to eliminate a radical in the denominator we multiply the denominator (and, of course, the numerator) by a number that will make the denominator a perfect square.

$$\frac{2}{\sqrt{3}} = \frac{2}{\sqrt{3}} \cdot \frac{\sqrt{3}}{\sqrt{3}} = \frac{2\sqrt{3}}{3} \text{ or } \tfrac{2}{3}\sqrt{3}$$

This procedure, known as *rationalizing the denominator,* is a useful operation to know. It is time now to learn how to rationalize some binomial denominators.

Example: Rationalize the denominator of $\dfrac{3 + \sqrt{2}}{2 - \sqrt{2}}$.

The only way we can make the radical drop out of the denominator is to multiply it by another binomial whose terms are identical but whose middle sign is opposite, that is, by $(2+\sqrt{2})$. Doing so gives

$$\frac{(3+\sqrt{2})}{(2-\sqrt{2})}\cdot\frac{(2+\sqrt{2})}{(2+\sqrt{2})} = \frac{6+5\sqrt{2}+2}{4-2} = \frac{8+5\sqrt{2}}{2} = 4 + \frac{5\sqrt{2}}{2}$$

Multiplying the irrational denominator by its conjugate binomial (one having the same terms but opposite middle sign) in order to eliminate the middle term of the resulting trinomial is an application of something we learned in Chapter 4 (frame 24): $(a+b)(a-b)=a^2-b^2$. Use this procedure whenever possible. It is a valuable tool and will work with any binomial involving square roots. You will have a chance to apply it in rationalizing the denominators of the fractions below.

(a) $\frac{2}{\sqrt{2}+3} =$ ______________________

(b) $\frac{3}{\sqrt{2}-1} =$ ______________________

(c) $\frac{1-\sqrt{3}}{2+\sqrt{3}} =$ ______________________

(d) $\frac{3\sqrt{a}-1}{\sqrt{a}+1} =$ ______________________

- - - - - - - - - - - - - - - - - - - -

(a) $\frac{2(\sqrt{2}-3)}{(\sqrt{2}+3)(\sqrt{2}-3)} = \frac{2\sqrt{2}-6}{2-9} = \frac{2\sqrt{2}-6}{-7} \left(\text{or } -\frac{2\sqrt{2}-6}{7}\right)$

(b) $\frac{3(\sqrt{2}+1)}{(\sqrt{2}-1)(\sqrt{2}+1)} = \frac{3\sqrt{2}+3}{2-1} = 3\sqrt{2}+3$

(c) $\frac{(1-\sqrt{3})(2-\sqrt{3})}{(2+\sqrt{3})(2-\sqrt{3})} = \frac{2-3\sqrt{3}+3}{4-3} = 5-3\sqrt{3}$

(d) $\frac{(3\sqrt{a}-1)(\sqrt{a}-1)}{(\sqrt{a}+1)(\sqrt{a}-1)} = \frac{3a-4\sqrt{a}+1}{a-1}$

28. Before concluding our investigation of exponents, roots, and radicals, we need to consider how to solve radical equations. Since we will be discussing equations generally in the next chapter, here we will just acquaint you with the procedure for handling an equation that contains a radical.

An equation is simply a shorthand way of saying that two expressions are equal or have the same value. We introduced the topic of equations briefly in Chapter 1 (frames 7 and 8), so you will perhaps recall such examples as $8n-3n=5n$ and $3a+2b=5b$. The part of an equation to the left of the equal sign is called the *left member*, and the part to the right of the equal sign is called (not surprisingly) the *right member*.

When we speak of *solving* an equation we mean finding the value(s) of the variable (letter) that will make the equation a true statement. Thus, in the equation $3a=6$, the solution is 2 because 2 times 3 is 6. We refer to 2 as the *root* of the equation because it satisfies the equation.

A *radical equation* is simply an equation in which the letter appears as a radicand. Thus, $\sqrt{a} = 4$ and $\sqrt{3x} - 2 = 0$ are radical equations. To solve this type of equation proceed as follows.

Example: Solve $\sqrt{2x} - 2 = 4$.

(1) Rearrange the terms of the equation so that one member contains the radical term we wish to rationalize. (Add 2 to both members.) $\sqrt{2x} = 4 + 2$

(2) Square both sides. $2x = 36$

(3) Solve for the unknown. (Divide both sides by 2.) $x = 18$

(4) Check your answer by substituting the root (18) back in the original equation.

$$\sqrt{2 \cdot 18} - 2 = 4$$
$$6 - 2 = 4$$
$$4 = 4$$

Use this procedure to solve the following equation.

$$2\sqrt{y} - 1 = 3$$

---

(1) Rearranging (adding +1 to each side): $2\sqrt{y} = 4$

(2) Squaring both sides: $4y = 16$

(3) Solving for the unknown (dividing both sides by 4): $y = 4$ (root)

(4) Checking:

$$2\sqrt{4} - 1 = 3$$
$$2 \cdot 2 - 1 = 3$$
$$4 - 1 = 3$$
$$3 = 3$$

29. In working with radical equations you need to be aware of this important fact: Since the radical always means the positive square root, equations such as $\sqrt{x} = -3$ and $\sqrt{x} + 1 = 0$ cannot be solved. Try it if you wish. You will find that your "answer" does not check. Such an answer is called an *extraneous root* since it does not satisfy the original equation.

Solve and check the following equations.

(a) $\sqrt{x} = 2$

(b) $\sqrt{x + 4} = 3$

(c) $\sqrt{x} = 1$

(d) $2\sqrt{x} - 12 = 0$

(e) $2\sqrt{a + 2} = 6$

(f) $\sqrt{2x + 6} - 6 = 0$

(g) $5 = \sqrt{4z + 1}$

(h) $3\sqrt{c - 6} = 9$

(i) $\sqrt{3y + 4} - 2 = 3$ (*Hint:* Move the 2 to the right side of the equation–by adding 2 to both members–before attempting to square both sides.)

(j) $\sqrt{\frac{x}{2}} = y$ (Solve for x.)

- - - - - - - - - - - - - - - - - -

(a) $x = 4$; check: $\sqrt{4} = 2, 2 = 2$
(b) $x = 5$; check: $\sqrt{9} = 3, 3 = 3$
(c) $x = 1$; check: $\sqrt{1} = 1$
(d) $x = 36$; check: $2\sqrt{36} - 12 = 0, 0 = 0$
(e) $a = 7$; check: $2\sqrt{7 + 2} = 6, 2 \cdot 3 = 6, 6 = 6$
(f) $x = 15$; check: $\sqrt{30 + 6} - 6 = 0, 0 = 0$
(g) $z = 6$; check: $5 = \sqrt{24 + 1}, 5 = 5$
(h) $c = 15$; check: $3\sqrt{15 - 6} = 9, 3 \cdot 3 = 9, 9 = 9$
(i) $y = 7$; check: $\sqrt{21 + 4} - 2 = 3, 5 - 2 = 3, 3 = 3$
(j) squaring: $\frac{x}{2} = y^2$; or multiplying both sides by 2, $x = 2y^2$;
check: $\sqrt{\frac{2y^2}{2}} = y, y = y$

Although considerably more can be said (and has been said) on the subject of exponents, roots, and radicals, you should, by now, feel some confidence about handling them. You will feel even more confident if you carefully work your way through the Self-Test that follows.

SELF-TEST

1. Find the indicated squares.

(a) 11^2 = ________
(b) $(-7)^2$ = ________
(c) $(2ab)^2$ = ________
(d) $(-3xy^2)^2$ = ________
(e) $(\sqrt{3}s^2t^3)^2$ = ________

(frame 1)

2. Perform the indicated operations.

(a) $x^3 \cdot x^4$ = ________
(b) $x^6 \div x^3$ = ________
(c) $3^2 \cdot 3^5$ = ________
(d) $(2a^2)^3$ = ________
(e) $(x^2y)(xy^3)$ = ________
(f) $k^3(k^4)$ = ________

(frame 2)

3. Perform the indicated operations.

(a) $a^2(a^2)^2$ = ________
(b) $(a^b)(a^c)$ = ________
(c) $(a^b)^c$ = ________
(e) $\left(\dfrac{2x^2y^3}{3abc}\right)^4$ = ________

(frame 3)

4. Perform the indicated operations (to nearest two decimal places).

(a) $\sqrt{1444}$ = ________
(b) $\sqrt{289.0}$ = ________
(c) $\sqrt{34.2}$ = ________
(d) $\sqrt{.1075}$ = ________

(frame 11)

5. Find the indicated roots (all variables are positive).

(a) $\sqrt{\dfrac{1}{4}}$ = ________
(b) $\sqrt{\dfrac{9}{25}}$ = ________
(c) $\sqrt{\dfrac{9a^2}{16b^4}}$ = ________
(d) $\sqrt{\dfrac{25x^4y^2}{49d^6}}$ = ________

(frame 17)

6. Find the approximate square roots (to two decimal places).

(a) $\sqrt{\dfrac{3}{7}}$ = ________
(b) $\sqrt{\dfrac{1}{5}}$ = ________
(c) $\sqrt{\dfrac{2}{9}}$ = ________
(d) $\sqrt{\dfrac{3}{11}}$ = ________

(frame 18)

7. Simplify.

(a) $\sqrt{\frac{5}{9}}$ = ______________ (c) $\sqrt{\frac{a}{b}}$ = ______________

(b) $\sqrt{45x^3}$ = ______________ (d) $2a\sqrt{18a^2}$ = ______________

(frame 21)

8. Simplify and combine like terms.

(a) $\sqrt{20} + 2\sqrt{5} - \sqrt{5}$ = ______________

(b) $2\sqrt{3} + 3\sqrt{12} + \sqrt{27}$ = ______________

(c) $3\sqrt{12} - \sqrt{36} + 3\sqrt{3} + 3$ = ______________

(d) $3\sqrt{2} + 2\sqrt{27} + \sqrt{12} - \sqrt{8}$ = ______________ (frame 22)

9. Perform the indicated operations.

(a) $\sqrt{3}\cdot\sqrt{7}$ = ______________ (d) $\sqrt{2}\cdot\sqrt{32}$ = ______________

(b) $\sqrt{2}\cdot\sqrt{5}\cdot\sqrt{6}$ = ______________ (e) $2\sqrt{6}\cdot 3\sqrt{8}$ = ______________

(c) $\sqrt{12}\cdot\sqrt{12}$ = ______________ (f) $\sqrt{7}\cdot\sqrt{28}$ = ______________

(frame 24)

10. Perform these multiplications.

(a) $3(\sqrt{2} + 5)$ = ______________

(b) $4\sqrt{3}(2 + \sqrt{7})$ = ______________

(c) $6(2\sqrt{6} - 4)$ = ______________

(d) $(\sqrt{a} + \sqrt{b})^2$ = ______________

(e) $(2 + 3\sqrt{2})(2 - 3\sqrt{2})$ = ______________

(f) $(\sqrt{3} - 2)^2$ = ______________ (frame 25)

11. Perform the following divisions.

(a) $\frac{8\sqrt{6}}{2\sqrt{3}}$ = ______________ (c) $\frac{8b^3\sqrt{20a^3}}{4b\sqrt{4a}}$ = ______________

(b) $\frac{25\sqrt{x^7}}{5\sqrt{x^3}}$ = ______________ (d) $\frac{z^2\sqrt{81}}{z\sqrt{27}}$ = ______________

(frame 26)

12. Rationalize the denominators.

(a) $\frac{5}{2 - \sqrt{3}}$ = ______________ (c) $\frac{2\sqrt{b} - 3}{\sqrt{b} + 2}$ = ______________

(b) $\frac{3}{3 + \sqrt{5}}$ = ______________ (frame 27)

13. Solve and check.

(a) $\sqrt{2y} = 8$

(b) $3 + \sqrt{b+4} = 6$

(c) $\sqrt{3a+4} - 2 = 3$

(d) $\sqrt{\frac{z}{3}} + 3 = 6$ (frame 28)

Answers to Self-Test

1. (a) 121; (b) 49; (c) $4a^2b^2$; (d) $9x^2y^4$; (e) $3s^4t^6$
2. (a) x^7; (b) x^3; (c) 3^7; (d) $8a^6$; (e) x^3y^4; (f) k^7
3. (a) a^6; (b) a^{b+c}; (c) a^{bc}; (d) $\frac{16x^8y^{12}}{81a^4b^4c^4}$
4. (a) 38; (b) 17; (c) 5.85; (d) .33
5. (a) $\frac{1}{2}$; (b) $\frac{3}{5}$; (c) $\frac{3a}{4b^2}$; (d) $\frac{5x^2y}{7d^3}$
6. (a) $\frac{1}{7}\sqrt{21} = .65$; (b) $\frac{1}{5}\sqrt{5} = .45$; (c) $\frac{1}{3}\sqrt{2} = .47$; (d) $\frac{1}{11}\sqrt{33} = .52$
7. (a) $\frac{1}{3}\sqrt{5}$; (b) $3x\sqrt{5x}$; (c) $\frac{1}{b}\sqrt{ab}$; (d) $6a^2\sqrt{2}$
8. (a) $3\sqrt{5}$; (b) $11\sqrt{3}$; (c) $9\sqrt{3} - 3$; (d) $\sqrt{2} + 8\sqrt{3}$
9. (a) $\sqrt{21}$; (b) $60 = 2\sqrt{15}$; (c) 12; (d) $\sqrt{64} = 8$; (e) $6\sqrt{48} = 24\sqrt{3}$; (f) 14
10. (a) $3\sqrt{2} + 15$; (b) $8\sqrt{3} + 4\sqrt{21}$; (c) $12\sqrt{6} - 24$; (d) $a + 2\sqrt{ab} + b$;
    (e) $4 - 9 \cdot 2 = -14$; (f) $7 - 4\sqrt{3}$
11. (a) $4\sqrt{2}$; (b) $5x^2$; (c) $2ab^2\sqrt{5}$; (d) $z\sqrt{3}$
12. (a) $10 + 5\sqrt{3}$; (b) $\frac{9 - 3\sqrt{5}}{4}$; (c) $\frac{2b - 7\sqrt{b} + 6}{b - 4}$
13. (a) $y = 32$; (b) $b = 5$; (c) $a = 7$; (d) $z = 27$

REVIEW TEST

Chapter 4

1. Which of the following are prime numbers? 5, 7, 9, 13, 15, 17 ________
2. Name all the factors of 18. ________
3. What is the highest common monomial factor of $3a^2bc^3 - 15ab^2c + 21ab^3c^2$?

____

4. Factor $5xy + 10x^2$. ________
5. Perform the indicated operation: $(x^3)^4 =$ ________
6. $(-4a^2bc^3)^2 =$ ________
7. Perform the indicated operation: $\left(\frac{3gt^2}{2m}\right)^2 =$ ________
8. What is the principal square root of $49x^4y^6z^2$? ________
9. $\sqrt{36t^4p^2k^{12}} =$ ________
10. $\sqrt{\frac{64m^6n^{10}}{25x^2y^4}} =$ ________
11. What is the sum of the cross products of $(2x - 3)(x + 4)$? ________
12. What is the product of $(a - 3)(3a + 5)$? ________
13. The product of two binomials is always a trinomial. (True / False)
14. Factor $k^2 - 15k + 26$. ________
15. Factor $4x^2 - 5xy - 21y^2$. ________
16. Factor $6a^3y^2 + 15a^2y^2 - 21ay^2$. ________
17. Is the trinomial $4m^2 - 10mk + 9k^2$ a perfect square? (Yes / No)
18. $(2a - 3k)(2a + 3k) =$ ________
19. Multiply 16×24 as the sum and difference of two squares.

____

20. Factor $9t^2 - 4p^2$. ________

Chapter 5

21. $\frac{0}{3} =$ ________

22. $\frac{7}{0} =$ ________________

23. $\frac{3}{2}$ and $\frac{6}{4}$ are equivalent fractions. (True / False)

24. Write the reciprocal of 7. ________

25. Write the reciprocal of $\frac{3}{8}$. ________

26. Reduce to lowest terms: $\frac{14x^3y^5}{35xy^2}$. ________________

27. Reduce to lowest terms: $\frac{a^2 + a - 12}{2a^2 + 7a - 4}$. ________________

28. Multiply: $\frac{a + 3}{a - 5} \cdot \frac{2a - 10}{3a + 9}$. ________________

29. Multiply: $\left(\frac{a^2 + 2a \div 1}{3a + 3}\right)\left(\frac{6}{a^2 - 1}\right)$. ________________

30. Combine: $\frac{5k}{6} - \frac{5k}{12} - \frac{k}{3}$. ________________

31. Combine: $\frac{3x}{x + 3} + \frac{x^2 + 4x - 5}{x^2 - x - 12}$. ________________

32. $+\frac{-a}{+b} = -\frac{+a}{+b}$ (True / False)

33. $-\frac{-a}{+b} = -\frac{+a}{+b}$ (True / False)

34. Supply the missing term: $\frac{a - 4}{b - 3} = +\frac{?}{3 - b}$

35. Simplify: $\frac{1 - a^2}{3a^2 - 6a + 3}$. ________________

36. Combine: $\frac{k}{k + 1} + \frac{4}{1 - k} + \frac{k^2 - 5k + 8}{k^2 - 1}$. ________________

37. Change this mixed expression to a common fraction: $a^2 + a + 1 - \frac{a^2 + 1}{2a - 1}$.

____________________ .

38. Change the following fraction to a mixed expression: $\frac{6m^3 + 4m^2 - 2m + 3}{2m}$.

____________________.

39. Change $\frac{2x^2 - 4x + 1}{x + 2}$ to a mixed expression. ____________________.

40. Change $\frac{\frac{1}{a} + 1}{\frac{1}{a2} - 1}$ to a simple fraction. ____________________.

Chapter 6

41. $(y^3)^5$ = ____________

42. Write $\sqrt[3]{x^2}$ using a fractional exponent. ____________

43. $\sqrt{179.56}$ = ____________

44. $\sqrt{4596}$ = ____________

45. $\sqrt{\frac{25x^6}{36y^4}}$ = ____________

46. The approximate square root of $\frac{3}{7}$ = ________.

47. $\sqrt{36} = \sqrt{4} \cdot \sqrt{9}$. (True / False)

48. $\frac{\sqrt{49}}{\sqrt{64}} = \sqrt{\frac{49}{64}}$ (True / False)

49. Write $x^{-2}y^3$ using only positive exponents. ____________

50. Simplify and combine like terms: $5\sqrt{2} - \sqrt{18} + 2\sqrt{48} - 3\sqrt{3}$. ____________

51. $x^{-3} \cdot x^4 = x$ (True / False)

52. $(\sqrt{7})^2 = 7$ (True / False)

53. $3x^{\frac{1}{2}} = \sqrt{3x}$ (True / False)

54. $2(\sqrt{3} + 2)$ = ____________________

55. $(\sqrt{a} + \sqrt{b})^2$ = ____________________

56. $\sqrt{\dfrac{21x^5}{3x^2}} =$ ______________

57. Rationalize the denominator of $\dfrac{2-\sqrt{2}}{3+\sqrt{2}}$ and simplify the fraction. ______________

58. $2\sqrt{9a} - \sqrt{4b} + 7\sqrt{a} - 3\sqrt{b} =$ ______________

59. Solve: $\sqrt{x-1} = 3$; $x =$ ______________

60. Solve and check: $\sqrt{16-3y} = 1$; $y =$ ______________

Answers to Review Test 2

Chapter 4

1. 5, 7, 13, 17
2. 1, 2, 3, 6, 9, 18
3. $3abc$
4. $5x(y + 2x)$
5. x^{12}
6. $16a^4b^2c^6$
7. $\dfrac{9g^2t^4}{4m^2}$
8. $7x^2y^3z$
9. $6t^2pk^6$
10. $\dfrac{8m^3n^5}{5xy^2}$
11. $+5x$
12. $3a^2 - 4a - 15$
13. False
14. $(k - 13)(k - 2)$
15. $(4x + 7y)(x - 3y)$
16. $3ay^2(2a + 7)(a - 1)$
17. No; to be a perfect square the middle term would have to be $-12mk$.
18. $4a^2 - 9k^2$
19. $16 \times 24 = (20 - 4)(20 + 4) = 400 - 16 = 384$
20. $(3t - 2p)(3t + 2p)$

Chapter 5

21. 0
22. Meaningless
23. True, because $3 \times 4 = 2 \times 6$
24. $\dfrac{1}{7}$
25. $\dfrac{8}{3}$
26. $\dfrac{2x^2y^3}{5}$ or $\frac{2}{5}x^2y^3$

27. $\frac{a-3}{2a-1}$
28. $\frac{2}{3}$
29. $\frac{2}{a-1}$
30. $\frac{k}{12}$
31. $\frac{4x^2-8x-5}{x^2-x-12}$
32. True
33. False
34. $4-a$
35. $\frac{a+1}{3(a-1)}$
36. $\frac{2k^2-10k+4}{k^2-1}$
37. $\frac{2a^3+a-2}{2a-1}$
38. $3m^2+2m-1+\frac{3}{2m}$
39. $2x-8+\frac{17}{x+2}$
40. $\frac{a}{1-a}$

Chapter 6

41. y^{15}
42. $x^{\frac{2}{3}}$
43. 13.4
44. 67.79
45. $\frac{5x^3}{6y^2}$
46. .65
47. True
48. True
49. $\frac{y^3}{x^2}$
50. $2\sqrt{2}+5\sqrt{3}$
51. True
52. True
53. False (It would equal $3\sqrt{x}$.)
54. $2\sqrt{3}+4$
55. $a+2\sqrt{a}\sqrt{b}+b$
56. $x\sqrt{7x}$
57. $\frac{8-5\sqrt{2}}{7}$
58. $13\sqrt{a}-5\sqrt{b}$
59. $x=10$
60. $y=5$; check:
$$\begin{aligned}\sqrt{16-3.5} &= 1\\ \sqrt{16-15} &= 1\\ \sqrt{1} &= 1\\ 1 &= 1\end{aligned}$$

CHAPTER SEVEN

Linear and Fractional Equations, and Formulas

OBJECTIVES

In Chapter 1 we discussed some of the similarities and differences between arithmetic and algebra. There are many more similarities than differences, but one of the important new concepts algebra provides is that of the *equation* or, as it often is called, the *algebraic method* of solving problems.

The idea of equations will not be entirely new to you because we introduced it briefly in the last chapter (frame 28) in connection with the solving of radical equations. But we will discuss it in much greater detail now since it is a technique that enables us to solve problems which would be very difficult or even impossible to solve using arithmetic.

When you have completed the work in this chapter you will be able to:

- solve simple equations by inspection;
- find the roots of equations of one unknown and check your answer;
- use addition, subtraction, multiplication, and division to undo their inverse operations in solving equations;
- solve equations containing fractions;
- convert solutions for negative values of the unknown to positive values;
- solve equations in which the unknown appears in both members;
- solve equations containing parentheses;
- solve equations containing binomial denominators;
- transform formulas as necessary and evaluate the unknown letter (variable) in solving problems.

SOME BASIC FACTS ABOUT EQUATIONS

1. An *equation* is simply a shorthand way of saying that two algebraic expressions are equal (that is, that they have the same value). For example, $3n = 12$ is just a quick way of saying that three times a number equals 12. Similarly, the equation $3n + 4n = 7n$ tells us that three times a number added to four times the same number is equal to seven times the number.

   We identify the part of the equation to the left of the equal sign as the *left member* and the part to the right as the *right member.* We will use these terms frequently to indicate which part of the equation we are referring to. Identify the parts of the following equation.

$$3x - 4y = 18 + 2z$$

(a) Left member ____________________

(b) Right member ____________________

(a) $3x - 4y$; (b) $18 + 2z$

2. At least one term of an equation must contain the variable (the letter whose value is unknown at the beginning of the problem, hence usually referred to as the unknown). *Solving* an equation means finding the value of the variable (unknown) that makes the equation become a true statement.

In the equation $3c = 18$, the solution is 6 because 6 times 3 equals 18. We say that the number 6 *satisfies* the equation and that 6 is the *root* of the equation. The root of an equation, then, is any number which, when substituted for the unknown (letter), makes the two sides of the equation identical.

Find the roots of the following equations.

(a) $5x = 25$

(b) $4b = 12$

(c) $x - 3 = 7$

(d) $\frac{x}{2} = 3$

(e) $t + 3 = 8$

(a) $x = 5$; (b) $b = 3$; (c) $x = 10$; (d) $x = 6$; (e) $t = 5$

3. Equations are of three types: identities, false statements, and conditional equations.

An *identity* is a statement that is true for all admissible values. Thus, $x + 3 = \frac{2x + 6}{2}$ is an identity.

A *false statement* is false for all values of the variable. Thus, $x = x + 3$ is a false statement.

A *conditional* equation is true only for a limited number of values of the variable. Thus, $2a + 1 = 7$, which is true only for the value $a = 3$, is a conditional equation.

See if you can tell whether each of the following is an identity, a false statement, or a conditional equation.

(a) $2x + 5 = 6$ ____________________

(b) $3x = 2x + 3$ ____________________

(c) $2(x + 3) = 2x + 6$ ____________________

(d) $3x = 3x + 7$ ____________________

(e) $4a - 6a = -2a$ __________________________

(f) $x - 3 = x$ __________________________

- - - - - - - - - - - - - - - - - - - -

(a) conditional equation $\left(\text{true only for } x = \frac{1}{2}\right)$; (b) conditional equation (true only for $x = 3$); (c) identity; (d) false statement ($3x$ cannot be equal to itself plus something more); (e) identity; (f) false statement

Don't be overly concerned if you didn't get all the above answers correct. The purpose of these problems was to lead you into an explanation of the methods by which we solve equations. They were simple enough to solve by inspection, which is just a mixture of arithmetic and common sense. Many equations are not this simple so it is important that you become thoroughly familiar with the rules that will assist you in solving the more difficult equations.

ROOTS OF EQUATIONS

4. All the equations presented thus far are *first degree* or *linear* equations of one variable. They are called first degree or linear because the exponent of the variable is no higher than one. (If you will look again at the problems in frames 2 and 3 you will see that this is so.) A *conditional* linear equation always has one root (that is, there is one and only one value for the variable that will make the equation a true statement). In this chapter we will use the word equation to mean a conditional linear equation.

   Answer the following questions.

   (a) What is the highest power of the variable in a linear (or first degree) equation?

   ____

   (b) How many roots has a conditional linear equation? ______________

- - - - - - - - - - - - - - - - - - - -

(a) one; (b) one

5. In order to find the solution (or root) of an equation we often have to change it to an equivalent equation which we can solve by inspection, just as we did with the problems in frame 2. Any change we make must be carefully thought out and must lead to an equation that has all (and no more) of the solutions of the given equation. Such equations are known as *equivalent equations.*

   In the example below the derived equations (2) and (3) and the original equation (1) are equivalent because 5 is the root (and the only root) of each of them.

$$
\begin{aligned}
&(1)\ 3x + 1 = x + 11\\
&(2)\ 2x + 1 = 11\\
&(3)\qquad 2x = 10\\
&\qquad\quad\ \ x = 5
\end{aligned}
$$

   When the same number is added to or subtracted from both members of an equation, the derived equation is equivalent to the original one. The same is true if both sides are multiplied or divided by any constant (except zero). If we were to multiply both members by an expression containing the unknown, we may obtain a derived equation containing a root that is not a root of the original equation. The only root of the equation $x - 1 = 2$, is 3. If we multiply both members of the equation by $x + 2$ we get $(x + 2)(x - 1) = 2(x + 2)$, which has two roots, 3 and -2.

The following operations will lead to equivalent equations. Since these operations (derived from the equality axioms we discussed in Chapter 1) are accepted here as true without proof they will be termed axioms.

Axiom 1: An equivalent equation is produced if both members of an equation are multiplied (or divided) by the same nonzero number.

Axiom 2: An equivalent equation is produced if the same number is added to (or subtracted from) both members of an equation.

There are some exceptions to these statements but their discussion is beyond the scope of this course.

Now let's look at how we might apply these axioms. Consider the equation $x - 3 = 5$. To solve it (that is, to find the numerical value of x that will satisfy the equation) we would like to get x on one side (usually the left) by itself and the other terms (known values) on the right side. We can do so by applying our second axiom (adding 3 to both sides of the equation). This removes the -3 from the left side (which we have reserved for the variable) and yields an equivalent equation that can be solved by inspection.

$$x - 3 + 3 = 5 + 3$$
$$x = 8$$

Notice that we added 3 (which is the opposite of -3) to both members (sides) of the equation. We could accomplish the same objective by means of the following rule, which we now will be able to use with other problems.

A *positive* term may be eliminated from one member of an equation by *subtracting* that term from each member of the equation. A *negative* term may be eliminated from one member of an equation by *adding* that term to both members of the equation.

Apply this rule in solving the following equation.

$$7 + y = 16$$
$$7 - \_\_\_\_ + y = 16 - \_\_\_\_$$
$$y = \_\_\_\_\_\_$$

- - - - - - - - - - - - - - - - - - -

$$7 - 7 + y = 16 - 7$$
$$y = 9$$

6. How do we solve equations for variables that are connected with other numbers that are factors (multipliers) or divisors of that variable? Consider the relationship that exists between time (T), speed (S), and distance (D). We know that distance is equal to time multiplied by speed, which gives us the relationship

$$D = T \cdot S \text{ (or } D = TS)$$

Let's imagine we have a problem in which we know the values for distance and speed but wish to solve for time. We must rewrite the equation so that T (the unknown factor) appears by itself on one side of the equal sign and the two known values, S and D, are on the other side.

Keeping in mind that our second axiom (from frame 5) allows us to multiply or divide both sides of an equation by the same number, how would you go about accomplishing the change we need? ______________________________

- - - - - - - - - - - - - - - - - - -

Divide both sides by S

$$\frac{D}{S} = \frac{T \cdot \cancel{S}}{\cancel{S}} \qquad \text{(Note: } \frac{S}{S} = 1)$$

and then interchange members to get T on the left side

$$T = \frac{D}{S}$$

7. If we had started with $T = \frac{D}{S}$ and wanted to solve for D, we would have multiplied both sides of the equation by S, thus canceling (dividing out) the S in the denominator of the right member and making it a multiplier of T in the left member.

$$T \cdot S = \frac{D \cdot \cancel{S}}{\cancel{S}} \quad \text{(or } D = TS)$$

We now have our original equation.

From the foregoing we can state the following rules.

Rule 1: A factor in any member may be changed to a value of one by dividing both members of the equation by that factor.

Rule 2: A divisor of any term can be changed to a value of one by multiplying each member of the equation by that divisor.

Now consider the equation $\frac{6}{a} = 3$. How would you solve it? Remember that when we speak of solving an equation we mean that we wish to find the value of the unknown (in this case, a) that will satisfy the equation (that is, that will make the equation a true statement). To do this we must eventually change the equation so that the letter a is on one side by itself and the remaining numbers are on the other side. Which procedure would you follow in this case? (Keep in mind the rules you have just been given.) ______________________________

- - - - - - - - - - - - - - - - - - -

Multiply both sides by a. This removes a from the denominator of the term $\frac{6}{a}$ and leads us to the equation $6 = 3a$.

$$\frac{6\cancel{a}}{\cancel{a}} = 3a$$

$$3a = 6$$

Now, what's the *next* step? ____________________

- - - - - - - - - - - - - - - - - -

Divide both sides by 3. This removes the 3 from the left member of the equation and leaves the unknown, a, by itself on the left side, which is just what we want.

$$\frac{3a}{3} = \frac{6}{3}$$
$$\frac{\cancel{3}a}{\cancel{3}} = \frac{\cancel{3}\cdot 2}{\cancel{3}}$$
$$a = 2$$

This tells us that 2 is the root of the equation and that substituting the value 2 in place of the letter a should satisfy the equation and reduce it to an identity. Let's see if it does.

$$\frac{6}{a} = 3$$
$$\frac{6}{2} = 3$$
$$3 = 3$$

8. You may feel that we belabored a rather simple problem unduly in the last frame. However, once you have learned to apply fundamental principles to simple problems you can then transfer your understanding to more difficult problems and still work correctly. To test your understanding solve this equation for k.

$$2k + 3 = 7$$

- - - - - - - - - - - - - - - - - -

$k = 2$

If you got this answer you probably correctly performed two operations involving the use of the axioms of subtraction and division: first subtracting 3 from both sides of the equation, leaving $2k = 4$; then dividing both sides by 2, giving the answer $k = 2$. This is an example of a problem involving two operations. Many algebraic problems require two or more operations in their solution; you will be seeing some of these soon.

INVERSE OPERATIONS

9. There is a term used to describe the kind of operations we have been using in the preceding examples: *inverse operations.* The phrase may be new to you but you have been performing inverse operations for some time. Addition and subtraction are inverse operations. So are multiplication and division. We simplified the equation $2k + 3 = 7$ by subtracting 3 from both members. Subtracting 3 is the procedure used because it is the inverse of adding 3; we want to eliminate the 3 that was added to the $2k$. This step yields the equivalent equation $2k = 4$, which we then solve by dividing both members by 2. To change the factor of 2 to a factor of 1 we divide by 2 which is the inverse of multiplication by 2.

For each of the examples below, indicate which inverse operation should be used in solving the problem.

(a) $z - 4 = 5$ ____________ (c) $7 + m = 12$ ____________

(b) $8x = 16$ ____________ (d) $\frac{b}{2} = 9$ ____________

- - - - - - - - - - - - - - - - - - - -

(a) addition; (b) division; (c) subtraction; (d) multiplication

10. Now it is time for you to try some problems on your own. Remember:

(1) Perform addition to undo subtraction or subtraction to undo addition. Do this when you wish to eliminate a number (that is, make it equal zero).

(2) Perform multiplication to undo division or division to undo multiplication. Do this when you wish to make a multiplier or divisor become one.

(3) Check your results in the original equation.

Solve the following equations.

(a) $3n - 5 = 7$

(b) $\frac{k}{3} + 4 = 8$

(c) $12 = \frac{3}{x}$

(d) $\frac{10}{r} = 2$

(e) $\frac{1}{4} = \frac{3}{b}$

(f) $2x + 3x - 6 = 4$ (Combine like terms first.)

(g) $\frac{2}{3}y = 8$ (Since $\frac{2}{3}$ is the multiplier we must divide both sides by $\frac{2}{3}$, which is the same as multiplying both sides by $\frac{3}{2}$.)

(h) $\frac{3}{4}t - 10 = 5$

- - - - - - - - - - - - - - - - - - - -

(a) $n = 4$; (b) $k = 12$; (c) $x = \frac{1}{4}$; (d) $r = 5$; (e) $b = 12$; (f) $x = 2$; (g) $y = 12$; (h) $t = 20$

11. Now that you are somewhat accustomed to the use of inverse operations and the axioms of equality (whatever you do to change the value of one side of an equation you must do to the other), we will extend these methods of solution to include more difficult equations and equations containing signed numbers. You are also ready for a few shortcuts.

Consider the equation $x - 3 = 4$. Using the addition/subtraction inverse operations rule, we simply add 3 to each member to get $x - 3 + 3 = 4 + 3$ or $x = 7$. Notice that the same solution could be found more quickly by merely adding +3 (mentally) to each side and writing down the answer, $x = 7$.

In the following problems you must decide whether to add a positive number or a negative number to both sides. Solve these equations by making a term become zero by adding its opposite.

	Term to be added to each side	Answer
(a) $x - 5 = 10$	______	x = ____
(b) $b - 7 = 7$	______	b = ____
(c) $2 = 1 - m$	______	m = ____
(d) $6 = 6 - y$	______	y = ____
(e) $2y + 3 = 9$	______	y = ____

- - - - - - - - - - - - - - - - - -

(a) 5, $x = 15$; (b) 7, $b = 14$; (c) -2, $m = -1$; (d) -6, $y = 0$;
(e) -3, $y = 3$

Problem (e) required that you divide by 2 as a final step. Since y is multiplied by 2 and the inverse of multiplication is division, it is necessary to divide both sides of the equation by 2.

12. Problem (e) of frame 11 gave you a chance to exercise the first rule of frame 7. Now let's take a look at a problem that requires the use of the second rule—using multiplication to undo division.

Consider the equation $\frac{x}{3} = 4$ in which 3 appears as a divisor in the left member. To solve for x we must somehow remove this denominator. We do so by making 3 a multiplier in the left member and, to preserve the equality, in the right member as well.

$$\frac{x}{3} = 4$$

$$\frac{\cancel{3}x}{\cancel{3}} = 3 \cdot 4$$

$$x = 12$$

Use the rule that both sides of an equation may be multiplied by the same non-zero quantity in solving the following equations.

(a) $\frac{y}{2} = 6$

(b) $5 = \frac{k}{5}$

(c) $\frac{z}{9} = -2$

(d) $\frac{a}{b} = c \quad (b \neq 0)$

(e) $\frac{1}{3} = \frac{x}{2}$

- - - - - - - - - - - - - - - - - - -

(a) $y = 12$; (b) $k = 25$; (c) $z = -18$; (d) $a = bc$; (e) $x = \frac{2}{3}$

13. We can modify our solution to this type of problem by using the following rule.

    If we want the coefficient of some term to become one, we multiply both sides of the equation by the reciprocal of that coefficient.

Example 1: $\frac{x}{5} = 2$

Since the coefficient of x is $\frac{1}{5}$ and the reciprocal of $\frac{1}{5}$ is 5, multiplying both sides of the equation by 5 we get

$$\cancel{5}\left(\frac{x}{\cancel{5}}\right) = 5 \cdot 2$$
$$x = 10$$

Example 2: $3x = 15$

The coefficient of x is 3 and the reciprocal of 3 is $\frac{1}{3}$, so multiplying both sides of the equation by $\frac{1}{3}$ gives

$$\frac{1}{\cancel{3}}(\cancel{3}x) = \frac{1}{3}(15)$$
$$x = 5$$

Solve the following equations by using the reciprocal rule.

	Multiply by	Solution
(a) $\frac{t}{3} = 8$	______	$t =$ ______
(b) $18 = 6m$	______	$m =$ ______
(c) $9k = 36$	______	$k =$ ______
(d) $\frac{x}{-3} = 4$	______	$x =$ ______
(e) $\frac{2k}{3} = 4$	______	$k =$ ______
(f) $2 = \frac{8}{k}$	______	$k =$ ______
(g) $\left(\frac{3}{8}\right)q = 6$	______	$q =$ ______
(h) $\frac{3m}{4} = \frac{9}{6}$	______	$m =$ ______

- - - - - - - - - - - - - - - - - - -

(a) 3; $t = 24$; (b) $\frac{1}{6}$; $m = 3$; (c) $\frac{1}{9}$; $k = 4$; (d) -3; $x = -12$;
(e) $\frac{3}{2}$; $k = 6$; (f) $k, \frac{1}{2}$; $k = 4$; (g) $\frac{8}{3}$; $q = 16$; (h) $\frac{4}{3}$; $m = 2$

Caution: Don't be confused by the different ways fractions are written. For example, problem (e) was written as $\frac{2k}{3} = 4$. It could have been written $\left(\frac{2}{3}\right)k = 4$. To help you become aware of this, problem (g) was written in the form $\left(\frac{3}{8}\right)q = 6$. Remember that the literal coefficient can be included as part of the numerator or it can appear next to the fraction. In either case it is considered part of the numerator. However, a letter that forms part of the denominator must be written under the fraction bar; thus, $\frac{5}{2q} = \frac{1}{4}$.

14. Solving an equation for x means getting the equivalent equation for positive x. If your solution is in terms of $-x$, you must rewrite it in terms of x. To do this multiply (or divide) both sides of the equation by -1. If our solution turned out to be $-x = 8$, multiplying (or dividing) both sides by -1 would give $x = -8$. If the solution were $-x = -8$, multiplication or division by -1 would give $x = 8$.

If the term containing the unknown has a negative coefficient, the reciprocal of that coefficient will be negative. In order to solve the equation, multiply both sides by the negative number which is the reciprocal of the coefficient. Thus, to solve the equation $-4x = 24$, we multiply both sides by $-\frac{1}{4}$ to obtain the solution $x = -6$.

What multiplier would you use to solve the equation $-7k = -21$? __________

- - - - - - - - - - - - - - - - - -

$-\frac{1}{7}$

15. You should be ready for a mixed set of problems that will give you many opportunities to apply the rules we have been reviewing. Solve the following equations.

(a) $x + 7 = 10$

(b) $3x - x = 6$

(c) $2 = k - 4$

(d) $2p - 6 = 4$

(e) $3 - 9 = 5r - 2r$

(f) $2y - 3y - 1 = 2$

(g) $-3 = \frac{y}{2}$

(h) $6m - 11 = 1$

(i) $\frac{-3z}{4} = -6$

(j) $4 - \frac{2x}{5} = 6$

- - - - - - - - - - - - - - - - - -

(a) $x = 3$; (b) $x = 3$; (c) $k = 6$; (d) $p = 5$; (e) $r = -2$; (f) $y = -3$; (g) $y = -6$; (h) $m = 2$; (i) $z = 8$; (j) $x = -5$

If you missed any of these you should rework them until you get the correct answers. It is important that you be able to solve the relatively simple problems easily and correctly before you try the harder ones.

EQUATIONS WITH LETTER-TERMS ON BOTH SIDES

16. In the examples and problems we have used so far the terms containing the letters have all been in one member (usually the left). Now let's consider equations in which the unknown appears in both members. You won't find this a really new or startling concept since we have used problems of this kind in the past. However, it is time we discussed in more detail the techniques for handling such problems.

> To solve equations in which the unknown appears in both members, use addition to rewrite the equation so that the terms containing the unknown are in one member and all other terms are in the other member. Combine like terms, then make the coefficient of the variable become one.

Consider the equation $6x - 2 = 8 + 4x$. Adding a $+2$ and a $-4x$ to each side gives $6x - 4x = 8 + 2$. Combining like terms we get $2x = 10$, and multiplying both sides by $\frac{1}{2}$ (to make the coefficient of the variable one), $x = 5$. To check our result we substitute 5 for x in the original equation, giving $6(5) - 2 \stackrel{?}{=} 8 + 4(5)$ or $28 \stackrel{\checkmark}{=} 28$. (Note: The equal sign with the question mark over it, $\stackrel{?}{=}$, and the equal sign with the check over it, $\stackrel{\checkmark}{=}$, are used frequently to indicate the checking process.)

Solve the equation $7 + 2y = 5y + 28$ and check your answer. Show all steps.

- - - - - - - - - - - - - - - - - -

Using addition:	$2y - 5y = 28 - 7$
Combining like terms:	$-3y = 21$
Multiplying by $-\frac{1}{3}$:	$y = -7$
Checking:	$7 + 2(-7) \stackrel{?}{=} 5(-7) + 28$
	$-7 \stackrel{\checkmark}{=} -7$

17. Mathematically speaking, there is nothing wrong with having all the terms that contain the unknown on the right side of the equation rather than on the left. It is merely conventional to have it the other way. Actually, if you have an equation in which all the terms containing the unknown are on the right side to begin with, it is easier to leave them there while working out the solution. For example, in the equation $15 = 12y - 5 - 7y$, instead of moving both y terms to the left side we leave them on the right and simply add $+5$ to both sides.

$$15 + 5 = 12y - 7y$$
$$20 = 5y$$
$$4 = y$$

As a final step you can, if you wish, switch the remaining terms to place the unknown, y, in its customary position on the left. We then have the more familiar $y = 4$.

Solve and check the following equations.

(a) $3a - 5 = 16$

(b) $3x = 7 - 4x$

(c) $6x + 21 = 84 - 3x$

(d) $7y + 28 = 5y + 6$

(e) $16 = 7 - 3x$

(f) $8x - 9 - 5x = 0$ (Don't be afraid of the zero; just add 9 to both sides.)

(g) $5r + 12 - 3r = -2 - 13 - r$

(h) $11c - 8 - 2c - 64 = 0$

(a) $a = 7, 16 \overset{\checkmark}{=} 16$; (b) $x = 1, 3 \overset{\checkmark}{=} 3$; (c) $x = 7, 63 \overset{\checkmark}{=} 63$;
(d) $y = -11, -49 \overset{\checkmark}{=} -49$; (e) $x = -3, 16 \overset{\checkmark}{=} 16$; (f) $x = 3, 0 \overset{\checkmark}{=} 0$;
(g) $r = -9, -6 \overset{\checkmark}{=} -6$; (h) $c = 8, 0 \overset{\checkmark}{=} 0$

REMOVING PARENTHESES IN EQUATIONS

18. Solving and checking the problems of frame 17 should have increased your confidence in your ability to handle equations of this type. However, be sure to work on any you missed until you can get the correct answer. Algebra, like all mathematics, requires a foundation of successful application upon which to build.

In Chapter 4 you were given some practice in removing parentheses (and other grouping symbols) that will prove helpful now. Consider the equation $(7x + 5) - (2x - 15) = 10$. The first step is to remove parentheses. (Remember to change the signs of all terms in the subtrahend since it is preceded by a minus sign.)

$$7x + 5 - 2x + 15 = 10$$

Now we can complete the solution.

$$\begin{aligned} 7x - 2x &= 10 - 5 - 15 \\ 5x &= -10 \\ x &= -2 \end{aligned}$$

Check by substituting -2 in the original equation.

$$\begin{aligned} (-14 + 5) - (-4 - 15) &\overset{?}{=} 10 \\ -9 - (-19) &\overset{?}{=} 10 \\ -9 + 19 &\overset{?}{=} 10 \\ 10 &\overset{\checkmark}{=} 10 \end{aligned}$$

Solve and check the following problem, showing all steps.

$$2a - (11 - 2a) = 37$$

Removing parentheses: $2a - 11 + 2a = 37$

Using addition: $2a + 2a = 37 + 11$

Collecting like terms: $4a = 48$

$a = 12$

Checking: $24 - (11 - 24) \stackrel{?}{=} 37$

$24 - (-13) \stackrel{?}{=} 37$

$37 \stackrel{\checkmark}{=} 37$

19. If the quantity in parentheses has a multiplier in front of it, perform the multiplication as you remove the parentheses, changing signs as required.

Example: $2(x - 5) - 3(2x + 1) = 7$

Performing the indicated multiplication: $2x - 10 - 6x - 3 = 7$

Rearranging terms: $2x - 6x = 7 + 10 + 3$

$-4x = 20$

$x = -5$

Check the above answer by substituting it in the original equation. Follow the steps indicated below.

Substitute the value of x: ______________ $\stackrel{?}{=}$ ________

Combine numbers: ______________ $\stackrel{?}{=}$ ________

Multiply: ______________ $\stackrel{?}{=}$ ________

Add terms: ______________ $\stackrel{\checkmark}{=}$ ________

$2(-5 - 5) - 3\ [2(-5) + 1] \stackrel{?}{=} 7$

$2(-10) - 3(-9) \stackrel{?}{=} 7$

$-20 + 27 \stackrel{?}{=} 7$

$7 \stackrel{\checkmark}{=} 7$

20. You should develop the habit of checking your solutions whether or not you are told to do so. This way you will catch your own errors.

Solve and check the following.

(a) $5(b+4) - 4(b+3) = 0$

(b) $3x - 4(x+2) = 5$

(c) $-(5x+4) = -6x + 3$

(d) $4(a-3) - 6(a+1) = 0$

(e) $5(2x+3) = 2(4x-1)$

(f) $7(p-5) = 14 - (p+1)$

(g) $(m-9) - (m+7) = 4m$

(h) $20 = 8 - 2(9-3x)$

Solve.

(i) $7(5t-1) - 18t = 12t - (3-t)$

(j) $5x - 4(x-6) - 2(x+6) + 12 = 0$

(k) $\frac{1}{3}(6y-9) = \frac{1}{2}(8y-4)$

(l) $24b^2 + 147 = 27 - 16b - 4b(1-6b)$

(a) $b = -8, 0 \overset{\checkmark}{=} 0$; (b) $x = -13, 5 \overset{\checkmark}{=} 5$; (c) $x = 7, -39 \overset{\checkmark}{=} -39$;
(d) $a = -9, 0 \overset{\checkmark}{=} 0$; (e) $x = -\frac{17}{2}, -70 \overset{\checkmark}{=} -70$; (f) $p = 6, 7 \overset{\checkmark}{=} 7$;
(g) $m = -4, -16 \overset{\checkmark}{=} -16$; (h) $x = 5, 20 \overset{\checkmark}{=} 20$; (i) $t = 1$; (j) $x = 24$;
(k) $y = -\frac{1}{2}$; (l) $b = -6$

FRACTIONAL EQUATIONS

21. Equations that contain fractions are usually solved more easily by changing them into equations that do not contain fractions. An equation containing fractions can be changed into an equation free of fractions by multiplying each member by a number that is exactly divisible by each denominator. The smallest number exactly divisible by each denominator is called (as you should recall from Chapter 5) the *lowest common denominator* (LCD). This process is known as *clearing fractions.* We introduced this idea in frame 13 of this chapter in connection with undoing division to solve equations of the type $\frac{x}{a} = c$. Now we will consider equations with more than one literal term in the equation and one or more of these literal terms contains a fraction with the same denominator. Let's state the governing rule first, then we will consider an example.

> If the variable has a fractional coefficient (or coefficients) involving fractions with the same denominator, clear fractions by multiplying every term of the equation by the denominator.

Example: $\frac{x}{3} + 2x = 7$

Since the common (and only) denominator of the fractional coefficient is 3, we first multiply each term by 3.

$$\frac{(3)x}{3} + (3)2x = (3)7$$

Dividing out the 3's in the first term and multiplying the others we get

$$x + 6x = 21$$
$$7x = 21$$
$$x = 3$$

Apply the rule in solving the following equation. Show all steps in your procedure.

$$\frac{3x}{4} - 2x = 5$$

Multiplying every term by 4: $\frac{(4)3x}{4} - (4)2x = (4)5$

Simplifying: $3x - 8x = 20$

Using addition: $-5x = 20$

Multiplying by $-\frac{1}{5}$: $x = -4$

22. Now let's try solving an equation with two fractions, both having the same denominator. The procedure is exactly the same as that followed in the preceding problem.

Example 1: $\frac{3y}{7} - 2 = \frac{y}{7}$

Multiplying by 7: $\frac{(7)3y}{7} - (7)2 = \frac{(7)y}{7}$

$$3y - 14 = y$$
$$y = 7$$

Example 2: $\frac{4}{x} = 5 - \frac{1}{x}$

Note that the unknown in this case is in the denominator.

Multiplying by x: $\frac{(x)4}{x} = (x)5 - \frac{(x)1}{x} \quad (x \neq 0)$

$$4 = 5x - 1$$
$$5x = 5$$
$$x = 1$$

Use our rule for clearing fractions to solve the following.

(a) $\frac{2x}{3} + x = 5$

(b) $\frac{x}{3} - x = 2$

(c) $10 - \frac{3y}{5} = y - 6$

(d) $\frac{3}{r} = 2 - \frac{7}{r}$

(e) $\frac{10}{x} - 2 = 18$

(f) $10 - \frac{z}{7} = \frac{4z}{7}$

(g) $\frac{6h}{5} + 8 = \frac{2h}{5}$

(h) $\frac{3y-1}{7} = 2y + 3$

(a) $x = 3$; (b) $x = -3$; (c) $y = 10$; (d) $r = 5$; (e) $x = \frac{1}{2}$;
(f) $z = 14$; (g) $h = -10$; (h) $y = -2$

Although problem (h) has a binomial numerator it does not change in any way the application of our rule.

$$\frac{3y-1}{7} = 2y + 3$$

Multiplying by 7: $$\frac{7(3y-1)}{7} = 7(2y) + 7(3)$$

$$3y - 1 = 14y + 21$$
$$-11y = 22$$
$$y = -2$$

23. We will consider now the case of more than one fraction in the equation, each fraction with a different denominator (or at least more than one number used as a denominator). Here we will need to use our concept of the lowest common denominator (that is, the smallest number that is divisible by each of the denominators).

What is the LCD in the following equation?

$\frac{x}{2} + \frac{x}{3} = 5$ LCD = ________

6

We arrive at this answer by multiplying the two denominators, 2 and 3, together. Since 2 and 3 are prime numbers their product is the smallest number that each will divide into an even number of times. Since 6 is the lowest denominator common to both fractions we follow our rule for clearing fractions as follows.

Multiplying all terms by 6: $$\frac{(6)x}{2} + \frac{(6)x}{3} = (6)5$$

$$3x + 2x = 30$$
$$x = 6$$

Sometimes we find both numbers and letters in the denominators.

Example: $\dfrac{10}{x} = \dfrac{25}{3x} - \dfrac{1}{3}$

Multiplying by $3x$ (the LCD): $\dfrac{(3x)10}{x} = \dfrac{(3x)25}{3x} - \dfrac{(3x)1}{3}$

The fundamental principle of fractions allows us to reduce this to:

$$3 \cdot 10 = 25 - x$$
$$x = -5$$

Solve the following equations.

(a) $\dfrac{3y}{4} - \dfrac{2y}{3} = \dfrac{3}{4}$ LCD = ________ y = ________

(b) $\dfrac{x}{4} + \dfrac{x}{3} + \dfrac{x}{2} = 26$ LCD = ________ x = ________

(c) $\dfrac{5}{x} - \dfrac{2}{x} = 3$ LCD = ________ x = ________

(d) $\dfrac{3}{4c} = \dfrac{1}{c} - \dfrac{1}{4}$ LCD = ________ c = ________

(e) $\dfrac{b-6}{b} = \dfrac{10}{7}$ LCD = ________ b = ________

- - - - - - - - - - - - - - - - - - -

(a) LCD = 12, $y = 9$; (b) LCD = 12, $x = 24$; (c) LCD = x, $x = 1$; (d) LCD = $4c$, $c = 1$; (e) LCD = $7b$, $b = -14$

HANDLING BINOMIAL DENOMINATORS

24. The last problem in the previous set included a binomial numerator. We discussed this kind of problem in frame 22 so it should not have given you any difficulty. (Review frame 22 if you need further help.)

Could you have handled a binomial denominator as easily? Let's see. Find the LCD and the value of the unknown in the following equation.

$\dfrac{3}{x-3} + \dfrac{6}{5} = \dfrac{9}{5(x-3)}$ LCD = ________ x = ________

$(x-3) \neq 0$

- - - - - - - - - - - - - - - - - - -

LCD = $5(x-3)$; $x = 2$

Here is the solution in case you need it. Multiplying by the LCD we get

$$\frac{3 \cdot 5(x-3)}{(x-3)} + \frac{6 \cdot 5(x-3)}{5} = \frac{9 \cdot 5(x-3)}{5(x-3)}$$

Simplifying each fraction by dividing the numerator and denominator by their common factors gives us

$$15 + 6(x-3) = 9$$
$$6x = 12$$
$$x = 2$$

25. In frames 15 and 16 of Chapter 5 we discussed the step-by-step procedure for finding the LCD for a group of fractions. We were concerned then with arithmetic fractions and monomial denominators. Now we are interested in finding LCD's for fractions appearing in polynomial expressions. Let's review our rule.

Rule for Finding the LCD

(1) Write each denominator in prime factored form, using exponents as necessary to represent repeated factors.

(2) Write the product of all the different prime factors.

(3) Use the largest exponent required (in Step 1) for any given prime factor. The result is the LCD.

Illustrated below is the use of this rule with a problem involving polynomials, the kind of expression encountered most frequently in algebra.

Example: Find the LCD for $\frac{2}{x^2-4}$, $\frac{3}{4x-8}$, and $\frac{4}{x^2+4x+4}$.

Step 1: Write the prime factors for each denominator.

$$x^2 - 4 = (x+2)(x-2);\ 4x - 8 = 4(x-2);\ x^2 + 4x + 4 = (x+2)^2$$

Step 2: Write the product of all factors from Step 1.

$$4(x-2)(x+2)$$

Step 3: Use the largest exponent required (in Step 1) with each factor in the product of Step 2. The result is the LCD.

$$4(x-2)(x+2)^2$$

Testing to make sure that $4(x-2)(x+2)^2$ is divisible by each of the denominators of the fractions given is an essential part of the work.

Use the rule for finding LCD for each of the following sets of fractions. Some LCD's can be found by inspection, of course, but remember that you are learning a method and you will not learn it without practice!

(a) $\frac{3}{7}$ $\frac{5}{21}$ $\frac{2}{15}$ LCD = ____________

(b) $\frac{5}{8}$ $\frac{5}{12}$ $\frac{7}{24}$ LCD = ____________

(c) $\frac{7}{45}$ $\frac{7}{18}$ $\frac{4}{5}$ $\frac{8}{9}$ LCD = ____________

(d) $\frac{3}{2x-6}$ $\frac{4}{x^2-9}$ $\frac{18}{6x+18}$ LCD = ____________

(e) $\frac{6}{x^2-1}$ $\frac{7}{x^2-2x+1}$ $\frac{9}{x+1}$ LCD = ____________

(f) $\frac{7}{x+4}$ $\frac{2}{x-3}$ $\frac{5}{x-2}$ LCD = ____________

- - - - - - - - - - - - - - - - - -

(a) 105 or $7 \cdot 5 \cdot 3$; (b) 24 or $2^3 \cdot 3$; (c) 90 or $2 \cdot 3^2 \cdot 5$;
(d) $6(x^2-9)$ or $2 \cdot 3(x-3)(x+3)$; (e) $(x-1)^2(x+1)$;
(f) $(x+4)(x-3)(x-2)$

Did you notice that the denominators in problem (f) were all prime to begin with?

26. Now that you know how to obtain the lowest common denominator when it cannot be found by inspection, it is time to return to our study of equations involving fractions. To solve such equations first multiply each term of the equation by the LCD of the fractions in that equation. The following problems will afford you some practice. (If you need help with the procedure, review frames 23 and 24.)

(a) $\frac{a-2}{3} - \frac{a+1}{4} = 1$

(b) $\frac{x+8}{x-2} = \frac{9}{4}$

(c) $\frac{4}{3(5+k)} - \frac{2}{3} = \frac{4}{(5+k)}$

(d) $\frac{8}{(x-2)} - \frac{13}{2} = \frac{3}{(2x-4)}$* (*Write this in factored form before trying to find the LCD.)

(e) $\frac{3}{(a+2)} + \frac{(a-2)}{4} = \frac{(a-3)}{4}$

(f) $\frac{4}{(x-4)} = \frac{7}{(x+2)}$

- - - - - - - - - - - - - - - - - -

(a) Multiplying by 12: $\frac{12(a-2)}{3} - \frac{12(a+1)}{4} = 12 \cdot 1$

Reducing fractions: $4(a-2) - 3(a+1) = 12$

Removing parentheses: $4a - 8 - 3a - 3 = 12$

Combining terms: $a - 11 = 12$

Adding 11 to both members: $a = 23$

(b) $x = 10$; (c) $k = -9$; (d) $x = 3$; (e) $a = -14$; (f) $x = 12$

WORKING WITH FORMULAS

27. Now let's consider a particular kind of equation, one used to express a rule in concise form. It is called a *formula.* Mathematical formulas are used to find the circumference of a circle, the height of a triangle, the volume of a cube, the speed of a moving vehicle, the gravitational pull between the earth and the moon, and the amount of energy in a unit of mass. The uses are literally endless. We worked with one such formula in frame 6 of this chapter: the time-speed-distance relationship, expressed as $D = TS$ (that is, distance is equal to speed multiplied by time, or vice versa). You may recall also that using our inverse operations rules we were able to express any one of the three quantities (T, D, or S) in terms of the other two. Thus, by using division to undo multiplication, we could divide both members by S to get $T = \frac{D}{S}$ or by T to get $S = \frac{D}{T}$.

To solve a problem by use of a formula we must know the numerical values of all the letters except the one unknown term whose value we are seeking. In the time-speed-distance relationship, for example, if we wish to find out the distance traveled by an automobile, we must know the time it traveled and the average speed at which it traveled.

Assuming an average speed of 40 miles per hour, what distance will an automobile travel in $3\frac{1}{2}$ hours?

Formula	Substitution of Known Values	Answer
______________	______________________	______________

- - - - - - - - - - - - - - - - - - -

$D = TS$; $D = 3\frac{1}{2}(40)$; $D = 140$ miles

28. Solving this type of problem is simple arithmetic once you write out the formula in terms of the unknown quantity and then substitute the known values. There are, then, three basic steps in using a formula.

(1) Write the formula in whatever form you remember it (or find it when you look it up).

(2) Change the form as necessary (according to the procedure we discussed in frame 7) to get the unknown term on the left side and all known values on the right side of the equal sign. (This is sometimes called *changing the subject of the formula.*)

(3) Substitute the known values for the letters on the right side and simplify.

In the case of the time-speed-distance formula you didn't have to perform Steps 1 and 2 since we showed you the three possible relationships of the quantities T, D, and S in the formula. Suppose, however, you were given the formula $A = bh$ (where A is the subject). This formula says that the area (A) of a rectangular surface is equal to the length of the base (b) multiplied by the height (h). Change the subject to h (that is, express the height in terms of the other two quantities).

$h =$ ____________

$$h = \frac{A}{b}$$

29. The problem in frame 28 is simply that of eliminating b in the term bh, thus isolating h. Since b is a factor of the right member this is easily accomplished by dividing both sides by b, or by multiplying by $\frac{1}{b}$ ($b \neq 0$).

There may be more than three factors in a formula. Consider the formula $V = lwh$. (Don't worry about what the letters represent just now.) To solve for w we would do what we did in the previous example. However, instead of dividing both members by one letter, we would have to divide each term by two letters–l and h.

$$\frac{V}{lh} = \frac{\cancel{l}w\cancel{h}}{\cancel{lh}} \text{ or (transposing terms) } w = \frac{V}{lh}$$

Solve the formula $F = \frac{mv^2}{2}$ for m. Hint: Write $\left(\frac{v^2}{2}\right)m = F$ and recall our discussion of multiplication by a reciprocal.

$$m = \frac{2F}{v^2}$$

Hopefully, you multiplied both members by $\frac{2}{v^2}$, the reciprocal of $\frac{v^2}{2}$.

30. Not all formulas are composed of just two monomial terms, one on the right of the equal sign and one on the left. Either or both members may include more than one term.

Consider, for example, the formula $p = 2a + b$. To solve this formula for a, we first eliminate b in the right member by adding $-b$ to both sides. Thus, $p - b = 2a$. Then we eliminate the factor 2 on the right side by dividing both members by 2, giving us $\frac{p-b}{2} = a$. The solution is now complete, but since we are used to seeing the unknown term (or subject) on the left side we can, as a final step, interchange the two members and write $a = \frac{p-b}{2}$ as our final answer.

Solve the formula $A = p + prt$ for r.

$$r = \frac{A-p}{pt}$$

31. Occasionally you will have to work with formulas in which the letter you wish to use as the subject is squared. For example, if we wish to solve our previous formula $F = \frac{mv^2}{2}$ for v, we would first multiply both sides by $\frac{2}{m}$, giving us $\frac{2F}{m} = v^2$ or, interchanging the right and left members, $v^2 = \frac{2F}{m}$. Since the subject should be v, not v^2, the final step in solving for v would be to take the square root of both sides, giving us

$$v = \sqrt{\frac{2F}{m}}$$

This tells us that to find the value of v we first have to multiply the value of F by 2, then divide the result by the value of m, and finally take the square root of the resulting number.

Evaluate the formula $v = \sqrt{\frac{2F}{m}}$ for v if $F = 54$ and $m = 3$. (Don't be concerned about the units of velocity, force, and mass which the letters represent.)

(a) Substitute the values of F and m: $v = \sqrt{\frac{2 \cdot ?}{?}}$

(b) Multiply and divide as indicated: $v = \sqrt{}$

(c) Take the square root of the resulting number: $v =$

- - - - - - - - - - - - - - - - - -

(a) $v = \sqrt{\frac{2 \cdot 54}{3}}$; (b) $v = \sqrt{36}$; (c) $v = 6$

32. In frame 28 we discussed the three steps in using a formula to solve a problem. Step (a) was simply to write down the formula in its usual form and Steps (2) and (3) were to transform the formula in such a way as to isolate the subject letter (usually as the left member), then substitute known values and evaluate. It is time now for you to perform Steps (2) and (3) successively on your own.

Find the value of m in the formula $L = 2m - w$, if $L = 40$ and $w = 20$.

(a) Transforming the formula: $m =$ ______________

(b) Substituting and solving: $m =$ ______________

- - - - - - - - - - - - - - - - - -

(a) $m = \frac{L + w}{2}$; (b) $m = 30$

Here is what your solution should look like:

$$L = 2m - w$$
$$L + w = 2m$$
$$\frac{L + w}{2} = m \quad \left(\text{or } m = \frac{L + w}{2}\right)$$
$$m = \frac{40 + 20}{2}$$
$$m = 30$$

33. Solving problems by the use of formulas is something most of us are required to do from time to time. It is common in shop work and any form of design. Even at home you sometimes use formulas to find the area of a circular flower bed or the number of square yards of carpeting you need for your living room. Transforming and evaluating formulas can be an interesting challenge. You will get more practice below.

(a) Find b if $A = bh$; $A = 48, h = 8$.

(b) Find d if $Q = 2d + k$; $Q = 9, k = 3$.

(c) Find E if $I = \dfrac{E}{R+r}$; $I = 15, R = 9, r = 3$.

(d) Find E if $C = \dfrac{nE}{R+nr}$; $C = 3, R = 6, n = 2, r = 4$.

(e) Find e if $T = 6e^2$; $T = 96$.

(a) $b = \dfrac{A}{h}$, $b = 6$; (b) $d = \dfrac{Q-k}{2}$, $d = 3$; (c) $E = I(R+r)$, $E = 180$;

(d) $E = \dfrac{C(R+nr)}{n}$, $E = 21$; (e) $e = \sqrt{\dfrac{T}{6}}$, $e = 4$

In this chapter we have discussed how to use equations and formulas to solve problems. It would not be fair to leave this subject without giving you some idea of where these formulas and equations come from. You are probably aware that formulas come from geometry (the area of a circle, the volume of a sphere, the surface of a cone, etc.) and from physics (the laws of motion, gravity, electricity, gases, hydraulics, etc.). In fact, they come from all branches of science and engineering.

Equations, however, come into being primarily as a result of a need to express number relations by the use of symbols so that problems can be solved algebraically. In Chapter 1 we worked a bit with the idea of translating word statements into mathematical expressions by the use of symbols (letters, numbers, and signs of operation). But you will need more practice translating word statements into algebraic expressions and equations before you are adept at it.

We will explore these matters in much greater depth in Chapter 12. At the moment you should concentrate on working the problems in the Self-Test. This will afford you the best possible preparation for the practical application of equations that you will find in Chapter 12.

SELF-TEST

1. Solve these equations by inspection (without paper and pencil).

(a) $z - 3 = 4$

(b) $2b = 7$

(c) $x + 3 = 7$

(d) $\frac{y}{2} = 3$

(e) $ab = c$

(frame 2)

2. Find the roots of the following equations and check your answers.

(a) $7 = b - 2$

(b) $3 = 12c$

(c) $2y + 6y = 16$

(d) $t + 6 = 2t$

(e) $7k - 4k = 10 + k$

(f) $\frac{x}{3} - 2 = 3$

(g) $b + \frac{1}{2}b = 4$

(h) $\frac{c}{2} + \frac{c}{3} = 5$

(i) $5 = \frac{b}{3} - 2$

(j) $4 + y = \frac{2}{5}y + 7$

(frame 10)

3. Solve the following by using addition.

	Term Added	Answer
(a) $y + 4 = 9$	________	$y =$ ______
(b) $c + 13 = 21$	________	$c =$ ______
(c) $5 + k = 11$	________	$k =$ ______
(d) $3 = 2 + z$	________	$z =$ ______
(e) $p - 5 = 10$	________	$p =$ ______
(f) $p + 5 = 10$	________	$p =$ ______
(g) $9 - r = 6$	________	$r =$ ______
(h) $3y = 4 + y$	________	$y =$ ______

(frame 11)

4. Solve the following equations.

(a) $\frac{r}{3} = 7$

(b) $\frac{m}{8} = -1$

(c) $\frac{t}{x} = x$

(d) $\frac{-y}{6} = -2$

(frame 12)

5. Solve the following equations.

(a) $3k = 36$

(b) $\frac{a}{9} = -3$

(c) $\frac{-x}{2} = 3$

(d) $\frac{q}{12} = \frac{5}{6}$

(e) $-2m = 1$

(f) $\frac{4}{n} = 2$ (frame 13)

6. Solve the following.

(a) $-x + 7 = 3$

(b) $4p - 18 = -2$

(c) $\frac{3z}{-2} = 9$

(d) $-3 + \frac{2a}{5} = 7$

(e) $2 + \frac{15}{a} = 7$

(f) $2y - 9y = -28$ (frame 14)

7. Solve and check the following equations.

(a) $4z - 3 = 2z + 9$

(b) $4 - 9 = 8k - 9k$

(c) $3d - 11 - d = 5$

(d) $-9c - 45 = 0$

(e) $-18 = 3x - 6x$

(f) $7b - 28 = -3b + 2$ (frame 16)

8. Solve and check the following.

(a) $3(a + 2) = -6$

(b) $5(7 - y) = 25$

(c) $12k - 3 = 5(2k + 1)$

(d) $3(x + 1) = 4(6 - x)$

Solve.

(e) $3m(2m + 4) = 2m(3m + 8) - 12$

(f) $(a + 4)(a - 3) = a(a + 4) - 24$ (frame 18)

9. Solve the following by clearing fractions.

(a) $\frac{4x}{3} - \frac{5x}{3} = -2$

(b) $\frac{2a}{5} + 6 = \frac{a}{5}$

(c) $\frac{3}{b} = 2 - \frac{7}{b}$

(d) $\frac{6y}{5} + 6 = \frac{2y}{5}$ (frame 22)

10. Solve the following.

(a) $\frac{a}{2} = 12 - \frac{a}{4}$ LCD = ________ a = ________

(b) $\frac{x}{5} + \frac{x}{6} = 11$ LCD = ________ x = ________

(c) $\frac{z}{5} + \frac{z}{3} - \frac{z}{2} = 3$ LCD = ________ z = ________

(d) $\frac{1}{x} + \frac{1}{2} = \frac{5}{x}$ LCD = ________ x = ________

(e) $\frac{n+6}{n} = \frac{7}{5}$ LCD = ________ n = ________

(f) $\frac{2c+4}{7} = \frac{c+5}{5}$ LCD = ________ c = ________ (frame 23)

11. Solve the following.

(a) $\frac{-9}{a-4} + \frac{2}{3} = \frac{-a}{a-4}$ if $(a-4) \neq 0$

(b) $\frac{3}{x-2} - \frac{4}{3} = \frac{-7}{3(x-2)}$ if $(x-2) \neq 0$

(c) $7 - \frac{9}{4-y} = \frac{-2}{4-y}$ if $(4-y) \neq 0$

(d) $\frac{3}{b+2} = \frac{4}{b+5}$ (frame 24)

12. Transform the following formulas as required and evaluate the unknown (variable). Assume the variable is positive.

(a) Find k if $Z = \frac{m^2k}{4}$; $Z = 18$, $m = 6$.

(b) Find I if $W = \frac{1}{2}LI^2$; $W = 128$, $L = 4$.

(c) Find P if $X = \frac{M}{N}P$; $X = 3$, $N = 143$, $M = 13$.

(d) Find V if $R = \frac{pVb}{M}$; $R = 12$, $M = 7$, $p = 8$, $b = 3$. (frame 32)

Answers to Review Problems

1. (a) $z = 7$; (b) $b = \frac{7}{2}$ or $3\frac{1}{2}$; (c) $x = 4$; (d) $y = 6$; (e) $a = \frac{c}{b}$

2. (a) $b = 9$; (b) $c = \frac{1}{4}$; (c) $y = 2$; (d) $t = 6$; (e) $k = 5$; (f) $x = 15$; (g) $b = \frac{8}{3}$; (h) $c = 6$; (i) $b = 21$; (j) $y = 5$

3. (a) -4, $y = 5$; (b) -13, $c = 8$; (c) -5, $k = 6$; (d) -2, $z = 1$; (e) 5, $p = 15$; (f) -5, $p = 5$; (g) -9, $r = 3$; (h) $-y$, $y = 2$
4. (a) $r = 21$; (b) $m = -8$; (c) $t = x^2$; (d) $y = 12$
5. (a) $k = 12$; (b) $a = -27$; (c) $x = -6$; (d) $q = 10$; (e) $m = -\frac{1}{2}$ (f) $n = 2$
6. (a) $x = 4$; (b) $p = 4$; (c) $z = -6$; (d) $a = 25$; (e) $a = 3$; (f) $y = 4$
7. (a) $z = 6$, $21 \overset{\checkmark}{=} 21$; (b) $k = 5$, $-5 \overset{\checkmark}{=} -5$; (c) $d = 8$, $5 \overset{\checkmark}{=} 5$; (d) $c = -5$, $0 \overset{\checkmark}{=} 0$; (e) $x = 6$, $-18 \overset{\checkmark}{=} -18$; (f) $b = 3$, $-7 \overset{\checkmark}{=} -7$
8. (a) $a = -4$, $-6 \overset{\checkmark}{=} -6$; (b) $y = 2$, $25 \overset{\checkmark}{=} 25$; (c) $k = 4$, $45 \overset{\checkmark}{=} 45$; (d) $x = 3$, $12 \overset{\checkmark}{=} 12$; (e) $m = 3$; (f) $a = 4$
9. (a) $x = 6$; (b) $a = -30$; (c) $b = 5$; (d) $y = -7\frac{1}{2}$ or $-\frac{15}{2}$
10. (a) LCD = 4, $a = 16$; (b) LCD = 30, $x = 30$; (c) LCD = 30, $z = 90$; (d) LCD = $2x$, $x = 8$; (e) LCD = $5n$, $n = 15$; (f) LCD = 35, $c = 5$
11. (a) $a = 7$; (b) $x = 6$; (c) $y = 3$; (d) $b = 7$
12. (a) $k = \frac{4Z}{m^2}$, $k = 2$; (b) $I = \sqrt{\frac{2W}{L}}$, $I = 8$; (c) $P = \frac{N}{M}X$, $P = 33$; (d) $V = \frac{RM}{pb}$, $V = 3\frac{1}{2}$

CHAPTER EIGHT

Functions and Graphs

OBJECTIVES

In this chapter we are going to investigate the use of rectangular coordinates and graphing as a means of solving linear equations. We will also consider such related concepts as ordered pairs, functions, dependent and independent variables, and various algebraic methods for simultaneous solution. This chapter begins to bring together some of the elements of geometry and algebra in the solution of algebraic equations. When you have reached the end of this chapter you will be able to:

- draw the major axes of a rectangular coordinate system, label the ends of these axes correctly, and identify the four quadrants (using Roman numerals);
- locate and plot points on the coordinate system using ordered pairs of values;
- determine the values of dependent variables for given values of the independent variable in linear equations;
- determine the x and y intercepts of linear equations;
- graph linear equations by use of their x and y intercepts plus at least one additional point;
- demonstrate that the coordinates of any point on the graph of a linear equation will, when substituted in the equation for the straight line, make the equation a true statement;
- solve pairs of linear equations simultaneously by the method of elimination;
- solve pairs of linear equations graphically;
- solve pairs of linear equations by the method of substitution.

THE PROCESS OF GRAPHING

1. You will recall from our work with equations in the last chapter that an equation is a statement that two expressions are equal. In this chapter we will be working with a particular type of equation—the first degree (or linear) equation in one or two

unknowns (variables). This means that there will be at most two variables in any given equation and that no variable will have an exponent greater than one.

In order to plot points in two dimensions or to *graph an equation* of two variables we need a coordinate system. The coordinate system we are going to use is known as the *rectangular coordinate system* because, like the corners of a rectangle, the axes are at right angles to one another. This coordinate system is also referred to as the cartesian coordinate system because it was introduced by an outstanding mathematician by the name of René Descartes.

Notice in the graph below that the rectangular coordinate system shown is formed simply by combining two number scales at right angles to each other. Their zero points (known as the *point of origin* or simply the *origin*) coincide.

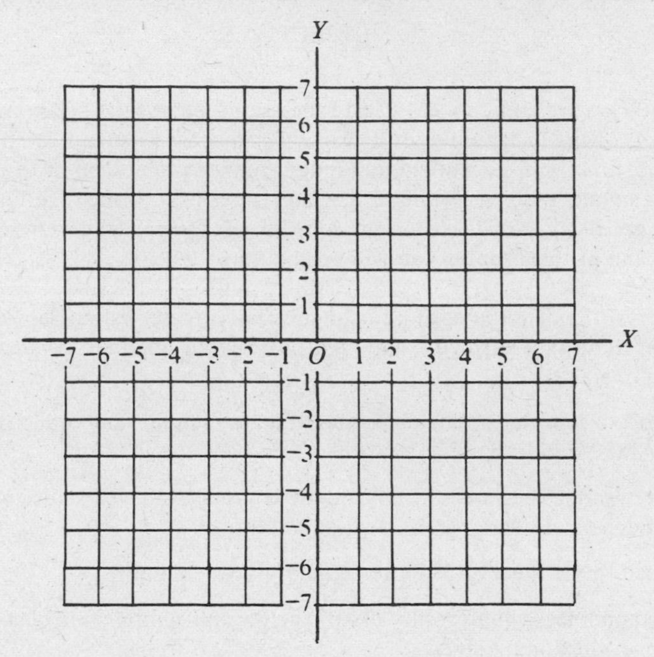

The horizontal scale is called the X axis and the vertical scale is called the Y axis. The origin, O, is the point at which the two scales intersect. This simple concept is a powerful tool that you will need to understand for many applications.

(a) The vertical scale is called the ________________ .

(b) The horizontal scale is called the ________________ .

(c) The point where the two scales intersect is known as the ________________ .

- - - - - - - - - - - - - - - - - -

(a) Y axis; (b) X axis; (c) origin

2. Here is the rectangular coordinate system again but with a few more things added.

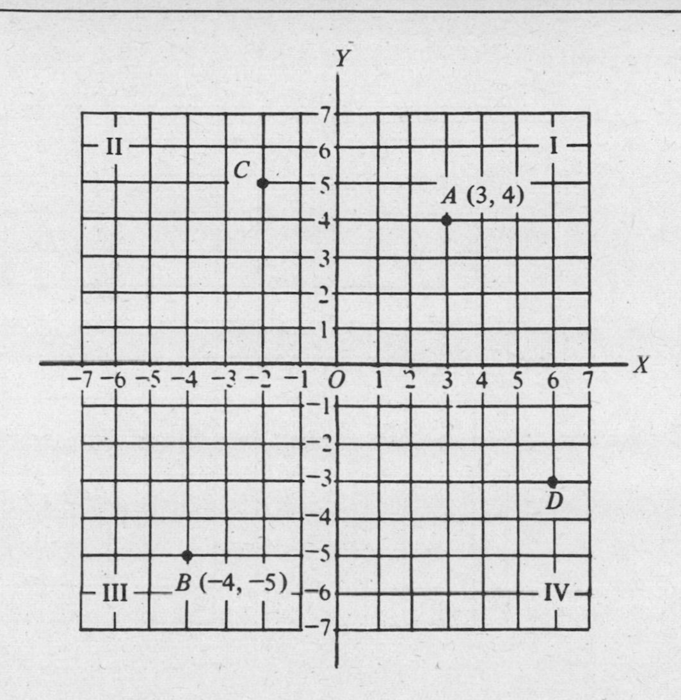

Notice that the X axis and Y axis divide the graph area into four parts. These are called *quadrants* and are numbered from I to IV starting in the upper right section and proceeding in a counterclockwise direction.

To locate a point on the graph we determine its distance along each axis. Point A, for example, is located 3 units from the origin along the X axis; we say its x coordinate is 3. Point A is also located 4 units from the origin along the Y axis; its y coordinate is 4. Points are always listed in the order (x, y), the first number representing the distance in the x direction and the second number representing the distance in the y direction (from the origin, of course). The coordinates of point A are, therefore, $(3, 4)$. Similarly, the coordinates of point B are $(-4, -5)$.

The x coordinate of a point is called the *abscissa*, or its first component. The y coordinate of a point is called the *ordinate*, or its second component.

What are the coordinates of points C and D?

- - - - - - - - - - - - - - - - - -

The coordinates of C are $(-2, 5)$ and the coordinates of D are $(6, -3)$.

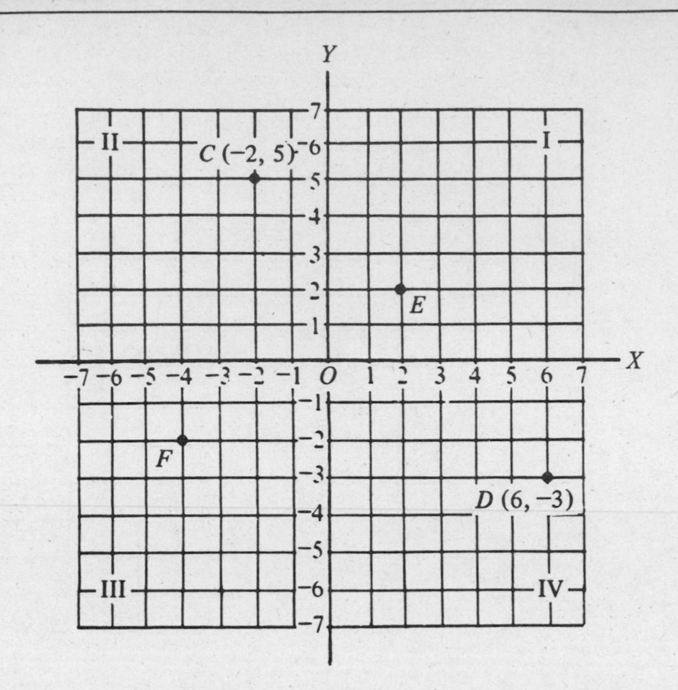

3. Let's see if you remember what we have covered so far about the terms abscissa and ordinate. Answer the following questions referring to the figures in the answer to frame 2.

(a) Which of the numbers shown at point C is the abscissa? ________

(b) Which of the numbers shown at point D is the ordinate? ________

(c) What is the abscissa of point E? ________

(d) What is the ordinate of point F? ________

(e) What is another name for the abscissa? ________________________

(f) What is another name for the ordinate? ________________________

(g) The location of points on a graph is always indicated by ordered pairs of numbers enclosed in parentheses. The first number shown is always the

____

- - - - - - - - - - - - - - - - - -

(a) −2; (b) −3; (c) 2; (d) −2; (e) first component (or x coordinate); (f) second component (or y coordinate); (g) abscissa (or first component)

If you missed any of these review frame 2 until you are clear about the answers.

4. You are already familiar with at least one rectangular coordinate system, although you may not recognize it as such. The streets of most cities intersect at right angles, resulting in a rectangular coordinate system with a main street and some major intersecting street forming the axes. These intersecting main streets would

correspond with the X and Y axes of our graphs. Street names and numbers help you locate specific buildings or geographic points, just as ordinate and abscissa enable you to specify particular points on a graph.

Latitude and longitude furnish a much larger rectangular grid system on the earth's surface, thus allowing accurate definition of geographic positions. A celestial rectangular coordinate system allows us to fix the positions of stars and other celestial bodies in space in terms of hour angle and declination.

As you probably realize by now there are many examples of rectangular coordinate systems, all of which have at least one feature in common: They contain two principal reference lines from which points anywhere in the system can be measured.

(a) The *ordinate* is the directed distance measured up or down along the Y axis. (True / False)

(b) The *abscissa* is the directed distance measured to the left or right *from* the Y axis. (True / False)

(a) True; (b) True ("*From* the Y axis" means along the X axis.)

5. Notice in the graph below the four quadrants designated as I, II, III, and IV. Notice also the signs of the number scales along the X axis and Y axis. Do you see that the numbers above the X axis and to the right of the Y axis are positive? Note too that the numbers below the X axis and to the left of the Y axis are negative.

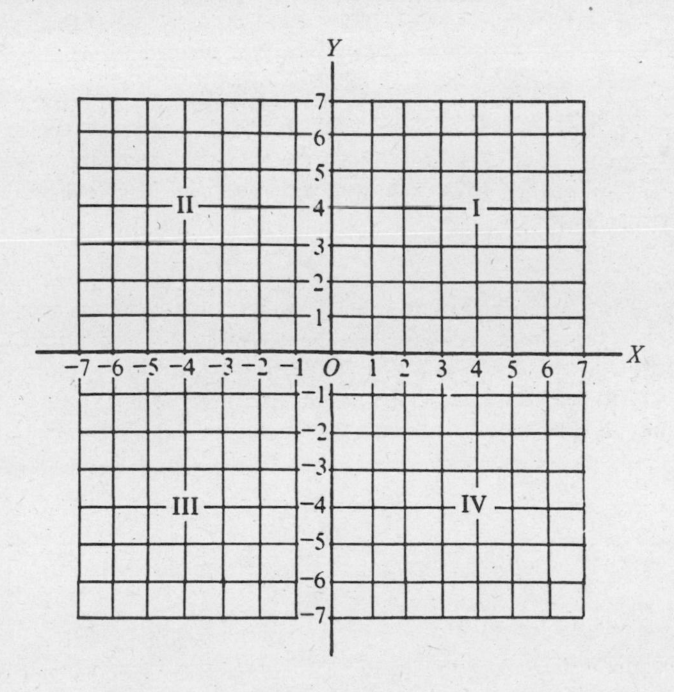

What do you think the signs of the x and y coordinates would be in each quadrant?

(a) Quadrant I ____________ (c) Quadrant III ____________

(b) Quadrant II ____________ (d) Quadrant IV ____________

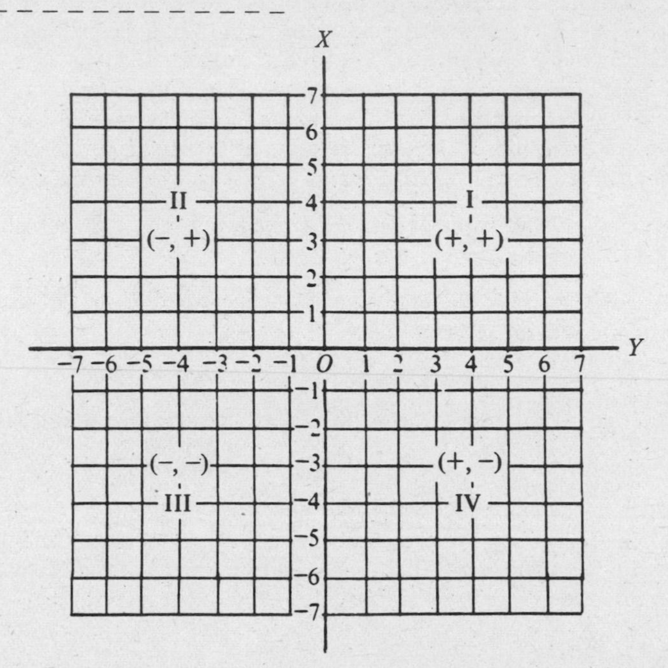

Remember that the x coordinate or abscissa (whose sign is the first sign appearing in parentheses in each quadrant) will always be positive to the right of the Y axis and negative to the left of it. The y coordinate or ordinate (whose sign appears as the second of the two in each case) will always be positive above the X axis and negative below it.

6. Now it is time for you to do some drawing. On a separate sheet of paper draw a pair of axes (like those above), establish a scale, and number positive and negative points along both axes as above. (You will find it easiest to do this if you use graph paper; otherwise use a ruler and work carefully.) When you have completed your graph, plot the following points on it.

(a) (2, 2)

(b) (5, −5)

(c) (−4, 3)

(d) (0, 4)

(e) (−4, −5)

(f) (5, 0)

(g) (0, −3)

(h) (−3, 0)

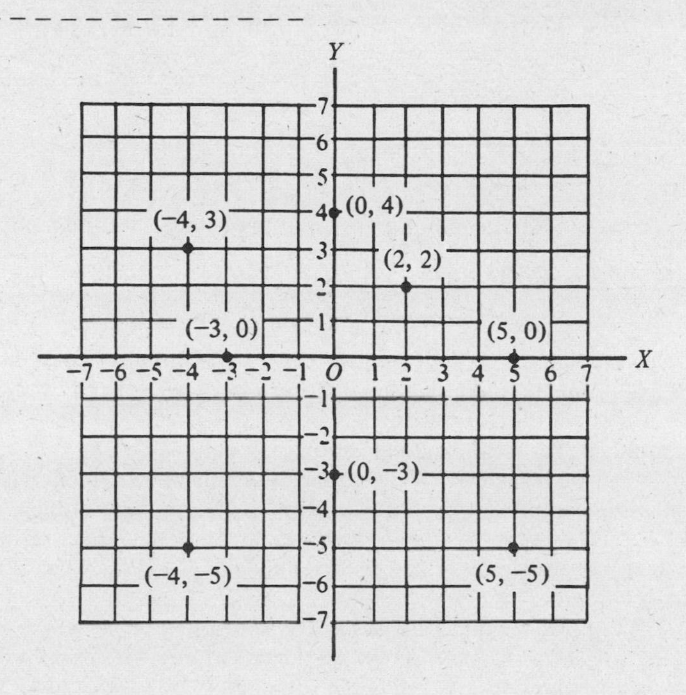

ORDERED PAIRS

7. There is a special term for the sets of coordinates that serve to locate points on a plane (that is, flat) surface. They are called *ordered pairs* because they always appear in the same order (the x value first, the y value second). You will find this idea of ordered pairs appearing frequently in mathematics. Below are some examples illustrating a few of their uses.

- What mathematicians call a "relation" between two variables (usually just an equation such as $y = x^2$) is defined as a set of ordered pairs. If we assign a whole series of values to x in the relation $y = x^2$ we wind up with a whole series of corresponding values for y. Thus, we have a "set" of ordered pairs.

- Equations and inequalities, which form much of the substance of mathematics, can be used to form ordered pairs as in the example above ($y = x^2$).

- One way (often the best) of solving equations and inequalities (which we will get into a little later on) is by graphing them on a rectangular coordinate system such as the one you have just been learning about. And graphing (sometimes called plotting) requires the use of ordered pairs.

Write the special names you have learned for each of the coordinates in an ordered pair.

(a) The first (x coordinate) is called the ______________________.

(b) The second (y coordinate) is called the ______________________.

- - - - - - - - - - - - - - - - - -

(a) abscissa (or first component); (b) ordinate (or second component)

For example, in the ordered pair $(0, -2)$ the abscissa is 0 and the ordinate is -2.

8. As we have just stated, ordered pairs can be used to plot or graph an equation. A series of points representing the equation and derived from it can be drawn on a rectangular coordinate system.

Consider, for example, the relation or equation $y = 2x$. Notice that it involves two (a pair of) variables. This is simply a shorthand statement describing the relation of the components of ordered pairs (x, y). Whatever value x represents, y will be twice as big as x; the ordered pairs $(x, 2x)$ represent this relation. Or we could say that whatever value y is, x will be only half as big since two x's are equal to one y.

The important point here is that *the value of either letter depends upon the value assigned to the other letter.* In the equation $y = 2x$ for x values of 1, 2, and 3 we get y values of 2, 4, and 6. Or for y values of 8, 10, and 12 we get x values of 4, 5, and 6.

For $y = 2x$, what are the ordered pairs in the following cases?

(a) x equals 14, 8, or 36 ______________________

(b) y equals 26, 2, or 0 ______________________

- - - - - - - - - - - - - - - - - -

(a) (14, 28), (8, 16), or (36, 72); (b) (13, 26), (1, 2), or (0, 0)

9. Because either x or y can assume different values (depending upon the values assigned to the other) x and y are called the *variables* in the relation $y = 2x$. We can also refer to the relation $y = 2x$ as a *function* and say that y is a function of x. This is true because a single value of y is created by (or determined by) each value assigned to x. A function, then, is a relation in which each first component has a single second component.

The variable used as the first component is referred to as the *independent variable* because it is more natural to assign values to it. The variable used as the second component is called the *dependent variable* because it is the variable whose value is determined by the value assigned to the first component. By assigning values to x in the relation $y = 2x + 3$ we would form ordered pairs of the form (x, y), x being the independent variable and therefore the first component, and y being the dependent variable or second component. As you can see, then, the subject of a formula (y in the case above) is the second component.

In order to make all this technically correct it should be mentioned that a function cannot be determined unless it is made clear that the first component can be *assigned* a certain set of numbers and that the second component can *assume* a certain set of numbers. If you can assume that you are free to use the set of real numbers for both components, this restriction will not be important to you.

Answer the following questions with relation to the equation (relation) $y = 3x + 5$. (Bear in mind that the subject of the formula is the second component.)

(a) You would be forming ordered pairs of the form ________________.

(b) ________________ would be the independent variable.

(c) ________________ would be the dependent variable.

(d) ________________ is a function of ________________.

- - - - - - - - - - - - - - - - - - - -

(a) (x, y); (b) x (x is the first component, therefore it would be more natural to assign values to it.); (c) y (y is the second component, therefore the values of y are determined by, or depend upon, x.); (d) y, x (The second component is a function of the first.)

PLOTTING THE LINEAR EQUATION

10. Now you are ready to try graphing the relation $y = 2x$. First you need to assign at least three values to x and see what corresponding values result for y. This way you will develop several ordered pairs which, when plotted, will represent the graph of the equation $y = 2x$.

The partially completed table below contains a series of x values. Substitute each of these values in the given equation ($y = 2x$) and write down the corresponding y values. Finally, plot each pair of coordinates and connect the resulting points with a straight line in order to find out what the graph of $y = 2x$ looks like. (You know that you will have a straight line because first degree equations are linear.)

x	3	2	1	0	−1	−3
y						

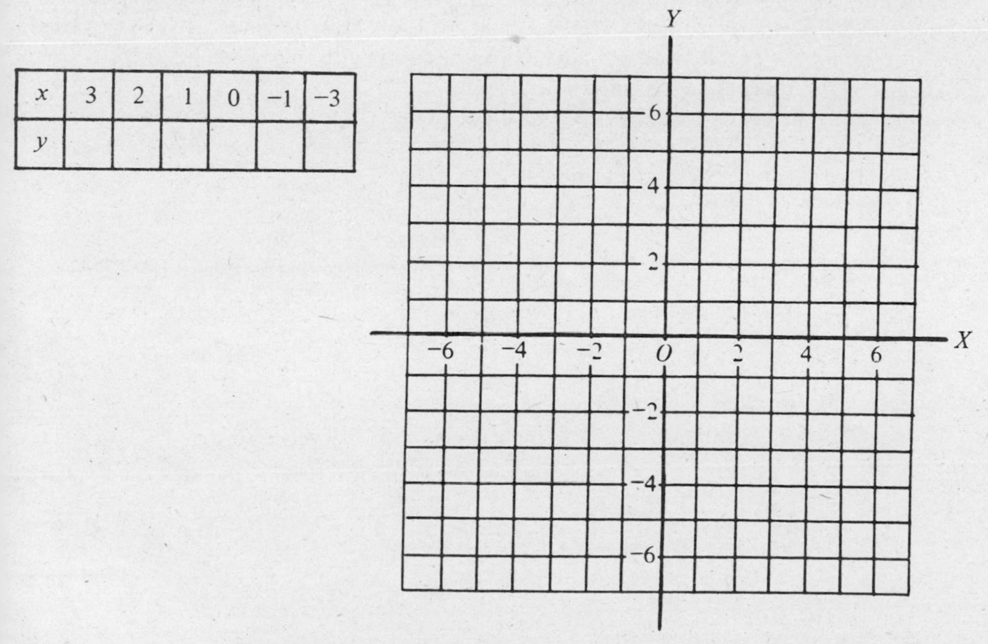

x	3	2	1	0	−1	−3
y	6	4	2	0	−2	−6

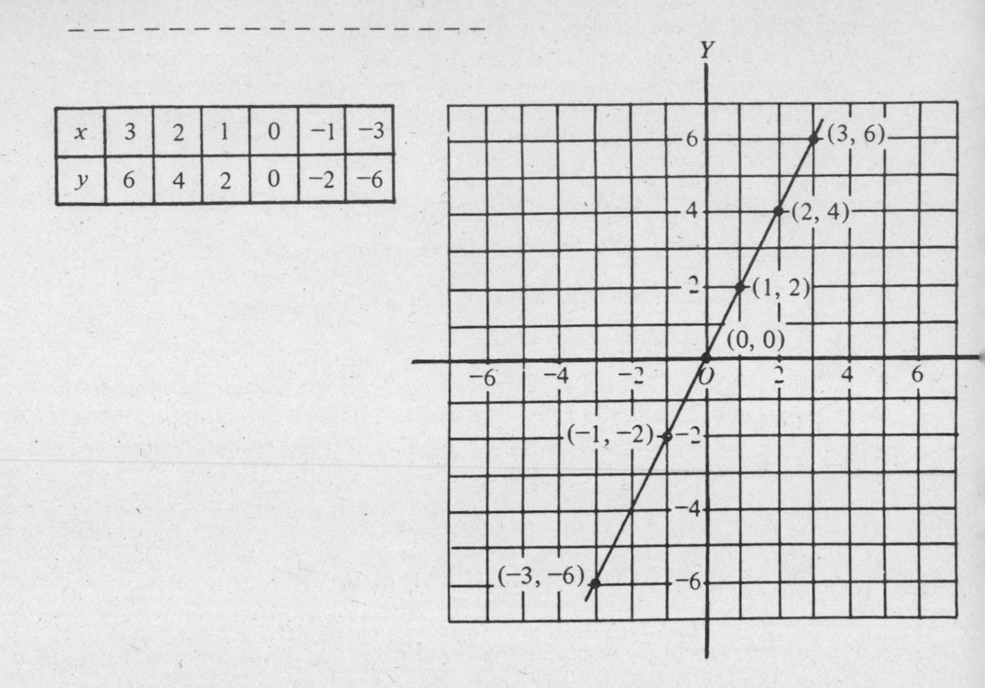

If your y values and graph don't look like this, go back and check your work using the correct graph as a guide in finding your mistake.

11. You have just graphed a linear equation; the equation $y = 2x$, when graphed, turns out to be a straight line. In Chapter 7 we described a linear equation as one in which the exponent of the variable is one. This description fits the equation $y = 2x$ since both y and x have exponents of one. Equations whose variables have exponents of exactly one will always appear as straight lines when graphed (if there are no products of two or more variables).

    Find the corresponding values of the dependent variable (y) for x values of 2, 0, and −2 in the equation $y = x + 2$. Plot the resulting pairs of coordinates and connect them with a straight line.

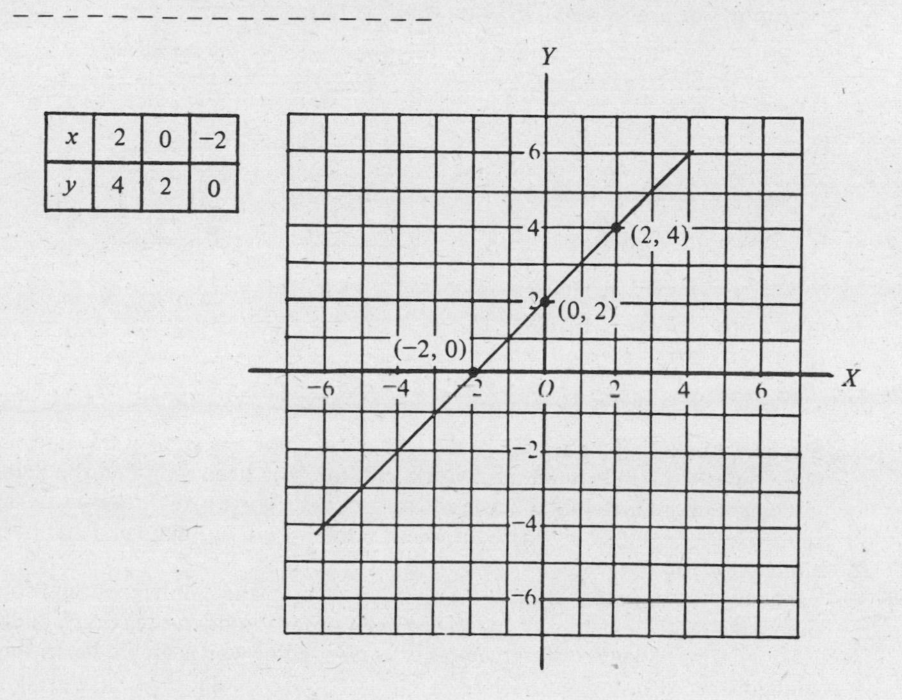

x	2	0	−2
y	4	2	0

INTERCEPTS

12. Did you notice in the problem of frame 11 that when x is zero, y is 2? And when y is zero, x is -2? Whenever x or y values are equal to zero, the point lies either on the Y axis or X axis. This means that the graph of the equation cuts (intersects) the axes at these points, as in the example of frame 11.

The special points at which a line cuts the axes are called *intercepts.* Intercepts can provide a quick way of locating at least two points for the graph. In the previous problem you found the y intercept by setting x equal to zero. You "lucked out" with the x intercept because an x value of -2 happened to produce a y value of zero. The procedure for finding the x and y intercepts (points where the graph cuts the x and y axes) is given here as a rule.

> To find the x intercept, set y equal to zero and solve the resulting equation for x. To find the y intercept, set x equal to zero and solve the resulting equation for y.

Example: Find the x and y intercepts of the equation $4x - 2y = 8$.

Setting $y = 0$: $4x - 0 = 8$

$x = 2$ x intercept: $(2, 0)$

Setting $x = 0$: $0 - 2y = 8$

$y = -4$ y intercept: $(0, -4)$

Apply this procedure in the following problems.

	x intercept	y intercept
(a) $3x + y = 9$	(____, 0)	(0, ____)
(b) $y - 5x = 10$	(____, 0)	(0, ____)
(c) $x + y = -3$	(____, 0)	(0, ____)
(d) $2y - 4x + 8 = 0$	(____, 0)	(0, ____)

- - - - - - - - - - - - - - - - - -

(a) (3, 0), (0, 9); (b) (−2, 0), (0, 10); (c) (−3, 0), (0, −3)
(d) (2, 0), (0, −4)

13. Since we know that a linear equation will always plot as a straight line and since it takes only two points to establish a straight line, you need only find the x and y intercepts in order to plot a linear equation. It is a good plan, however, to obtain a third (check) point, using any convenient value for either unknown (that is, for either x or y).

Graph the linear equation $y = 3x - 6$. Use the x and y intercepts and some other small value (such as 1, 3, or 4) for x in order to obtain the corresponding y value. This will give you three points to plot, all of which should lie on the same straight line.

- - - - - - - - - - - - - - - - - -

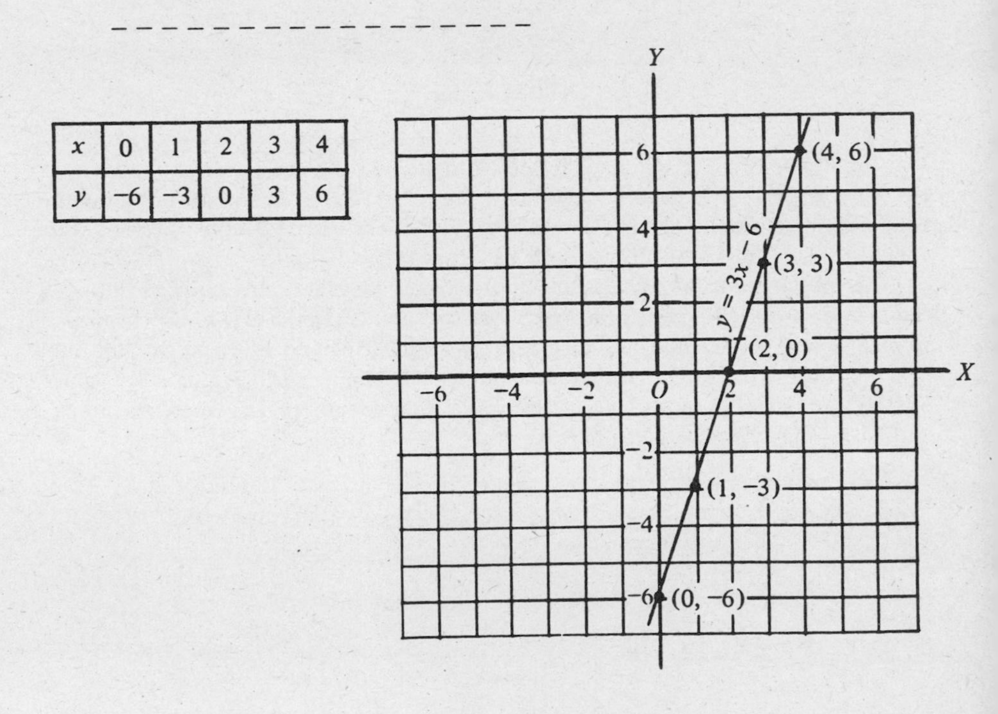

x	0	1	2	3	4
y	−6	−3	0	3	6

14. The coordinates of any point on the graph of an equation satisfy the equation. That is, the coordinates of any point on the graph, when used as the x and y values in the equation, make the equation a true statement.

Example: Consider the equation $y = 3x - 6$ (from frame 13). If we substitute the coordinate pair (1, −3) in this equation we get

$$y = 3x - 6$$
$$-3 = 3(1) - 6$$
$$-3 = -3 \text{ (a true statement)}$$

If we substitute the coordinate pair (0, −6) we get

$$-6 = 3(0) - 6$$
$$-6 = -6 \text{ (a true statement)}$$

Suppose, however, we substitute the coordinates of a point not on the graph. What do you think the result will be? Try it and see. Substitute the coordinate pair (3, −4) in the equation $y = 3x - 6$ and see what you get.

$y = 3x - 6$
$-4 = 3(3) - 6$
$-4 = 9 - 6$
$-4 = 3$ (not a true statement)

GRAPHS OF LINEAR EQUATIONS OF ONE UNKNOWN

15. We have been considering the graphing of linear equations of two unknowns. For the sake of completeness we should consider briefly what the cartesian graph of linear equations of one unknown (x or y) would look like.

Consider the equation $y = 2$. Can you picture what the graph of this equation would look like on the rectangular coordinate system? It would be a straight line composed of a series of points, all of them 2 units above the X axis and therefore parallel to the X axis. How about the graph of $x = 0$? It would simply be the Y axis. Plot the following and see for yourself.

(a) $x = -4$

(b) $y = 0$

(c) $x = 2$

(d) $y = 3\frac{1}{2}$

(e) $x = 0$

(f) $y = -5$

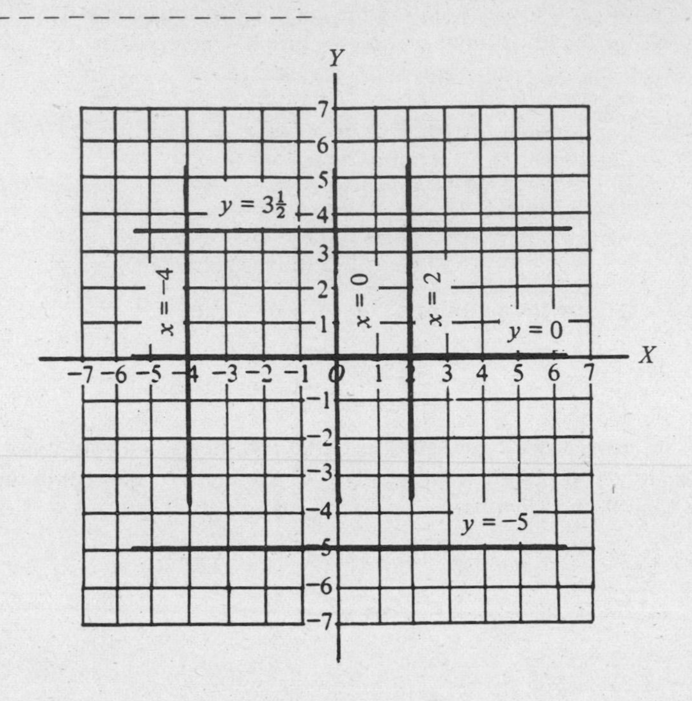

SUMMARY

It is time for us to summarize the rules we have found from our investigation of the graphing of first degree equations in one or two unknowns.

Rule 1: The graph of a first degree equation is a straight line.

Rule 2: If a point lies on the graph of an equation its coordinates satisfy the equation (that is, they reduce it to an identity or a true statement).

Rule 3: Conversely, the coordinates of points not on the graph of an equation do not satisfy the equation.

Rule 4: The graph of a first degree equation in one unknown lies on either the X axis or the Y axis or is a line parallel to one of these axes.

GRAPHIC SOLUTION OF PAIRS OF LINEAR EQUATIONS

16. In Chapter 7 you learned how to solve linear equations of one unknown. In this chapter we have considered the matter of graphing linear equations of two unknown However, the question remains: How do we find a single solution to an equation containing two unknowns? The answer is that it can't be done.

> In order to solve linear equations for unique values of the unknowns, you must have as many equations (that is, relationships between the unknowns) as there are unknown quantities.

Thus, if we have three unknowns we will need three equations containing these unknowns to find a unique solution for all three of the unknowns. Five unknowns would require five equations, nineteen unknowns nineteen equations, and so on. Now we are going to concern ourselves only with the solution of pairs of linear equations in two unknowns.

Example: Solve the following pair of linear equations graphically.

$$\begin{aligned} 2x - y &= 5 \\ x + y &= 7 \end{aligned}$$

Since these two equations will have a common solution (that is, one unique set of values for x and for y that will satisfy both equations) it is apparent that the point represented by that set of values must lie on the line graphs of both equations. The solution must represent the point at which the two lines intersect. Let's see if this is so. To plot the two equations we will arrange each in terms of y, develop at least three pairs of coordinates for each, and plot the resulting ordered pairs.

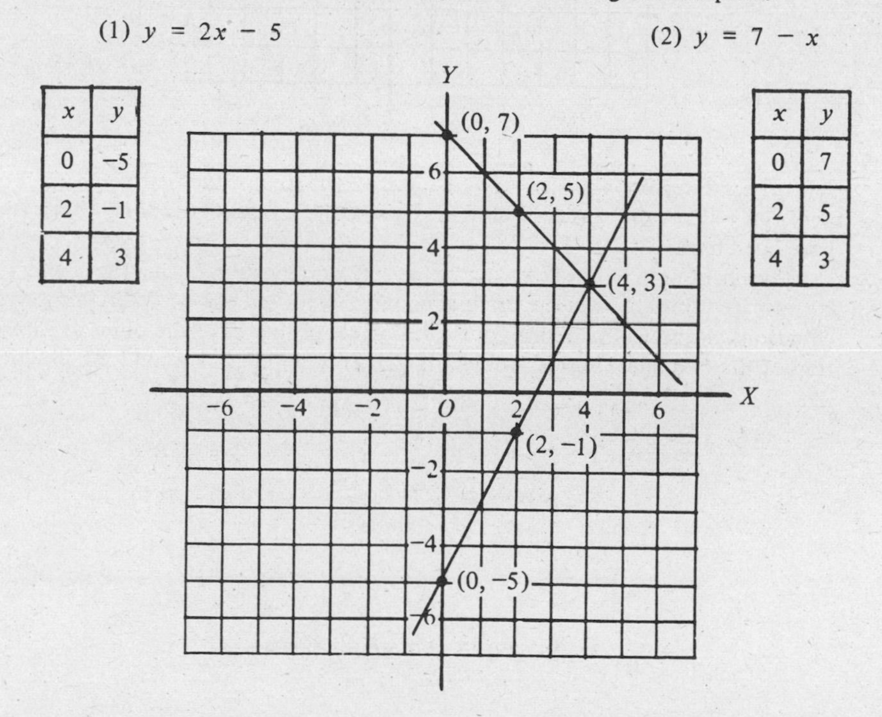

(1) $y = 2x - 5$

x	y
0	−5
2	−1
4	3

(2) $y = 7 - x$

x	y
0	7
2	5
4	3

Since the two lines intersect at the point (4, 3), $x = 4$ and $y = 3$ is the *common solution.*

Using a sheet of graph paper, follow the same procedure to solve this equation pair: (1) $2y - x = 2$; (2) $y + 2x = 6$.

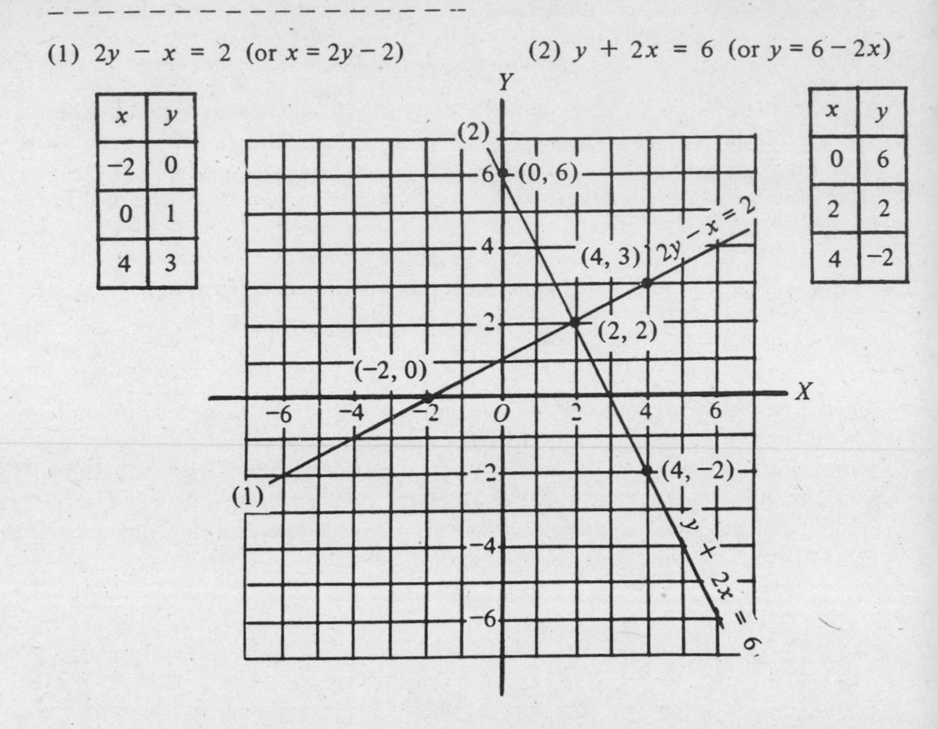

x	y
−2	0
0	1
4	3

x	y
0	6
2	2
4	−2

Both the x intercept and the y intercept are used to plot the first equation, but only the y intercept is used in plotting the second equation (although both could have been used).

Always be sure to check your graphic solution by substituting the x and y coordinate values back in the original equations, just as you do when you solve equations algebraically. Since your solution (coordinates of the point of intersection of the two lines) in this problem is (2, 2), your check should look like this:

$$\begin{array}{ll} (1)\ \ 2y - x = 2 & (2)\ \ y + 2x = 6 \\ 2\cdot 2 - 2 \overset{?}{=} 2 & 2 + 2\cdot 2 \overset{?}{=} 6 \\ 4 - 2 \overset{?}{=} 2 & 2 + 4 \overset{?}{=} 6 \\ 2 \overset{\checkmark}{=} 2 & 6 \overset{\checkmark}{=} 6 \end{array}$$

INCONSISTENT EQUATIONS

17. Equations having a common point of intersection (indicating that one pair of valu satisfies both of them) are called *consistent equations.* Equations are termed *inconsistent* if no pair of values satisfies both of them.

So that you may see what a pair of inconsistent equations looks like, graph th following system.

$$(1)\ x + y = 4$$
$$(2)\ x + y = 6$$

(1) $x + y = 4$ (or $x = 4 - y$) (2) $x + y = 6$ (or $x = 6 - y$)

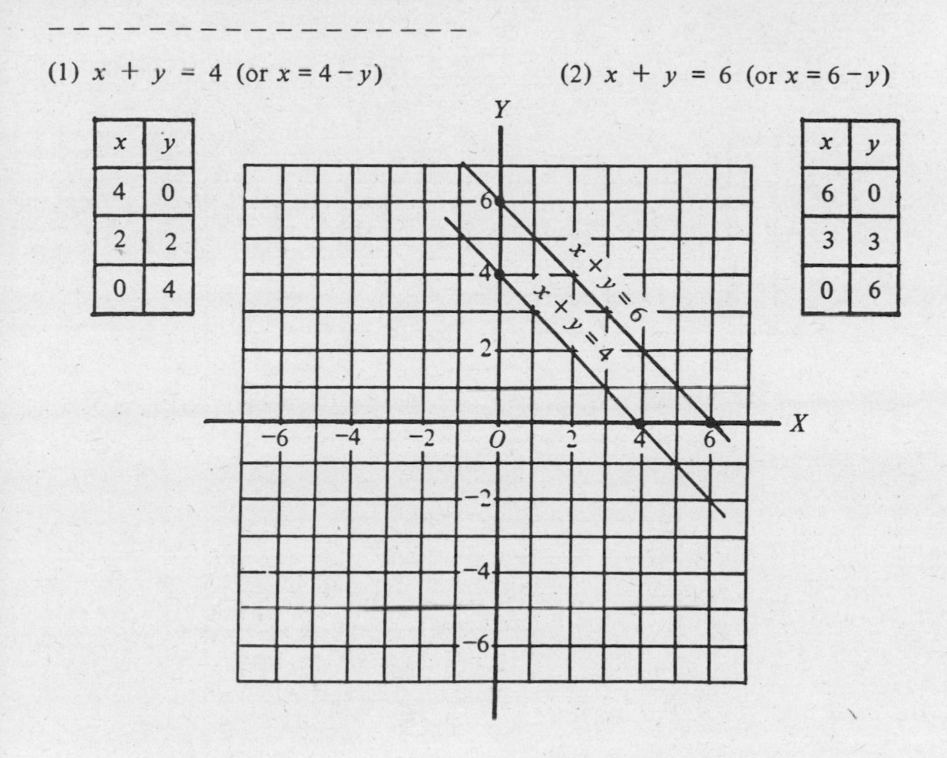

x	y
4	0
2	2
0	4

x	y
6	0
3	3
0	6

As you can see from this graph, inconsistent equations are simply parallel straight lines that cannot meet (from the standpoint of Euclidean geometry, at least), hence there is no common solution. Such equations are "inconsistent" for the obvious reason that two given numbers (represented by x and y) cannot add up to two different answers (4 and 6 in this case).

DEPENDENT EQUATIONS

8. We come, finally, to the linear equations known as *dependent equations.* Two dependent equations plot as the same line. Consider the two equations $y = x + 3$ and $2y = 2x + 6$. The second equation is simply the first equation with all terms multiplied by 2. We could, therefore, divide all the terms of the second equation by 2 and obtain the first equation. The two equations are, therefore, essentially equivalent and would produce the same graph. Prove this to yourself by graphing both equations.

(1) $y = x + 3$ (2) $2y = 2x + 6$ $\left(\text{or } y = \dfrac{2x+6}{2}\right)$

x	y
0	3
3	6
−3	0

x	y
0	3
3	6
−3	0

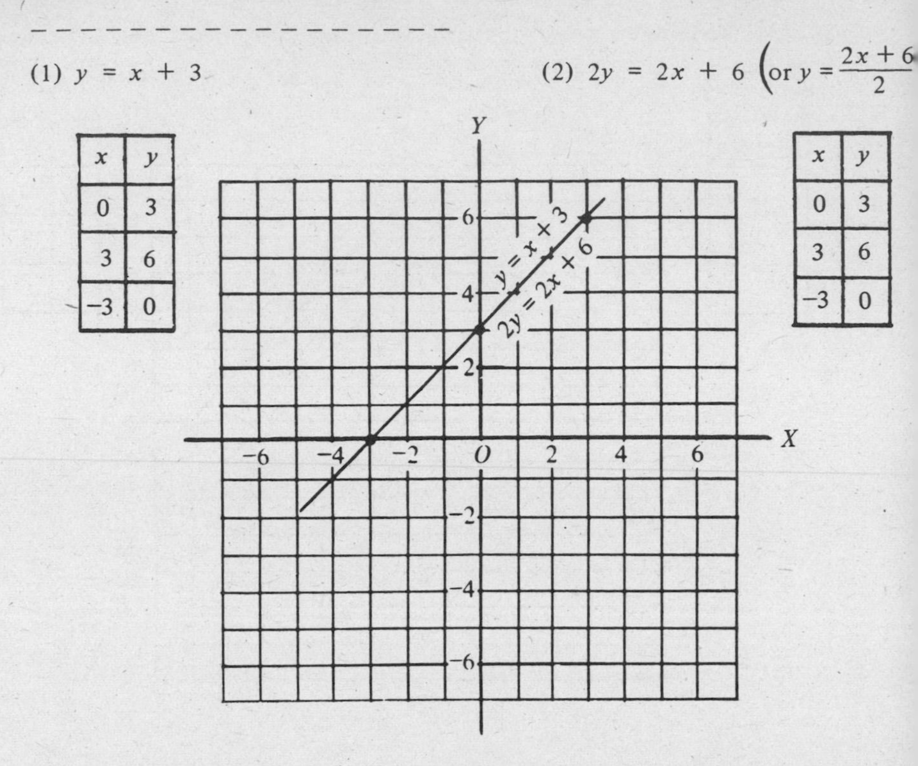

Once you see from the tables of values that the x and y values are the same for both equations, you do not have to graph them to know they represent the same line.

ALGEBRAIC SOLUTIONS OF EQUATION PAIRS

19. Let's move from the graphic solution of pairs of equations to the algebraic method of solving systems of equations in two variables. Consider, for example, this system of equations:

$$(1)\ 5x + 3y = 19$$
$$(2)\ x + 3y = 11$$

Observing that the term $3y$ appears in both equations and recalling that we can solve linear equations in one variable, it may have occurred to you that we might be able to subtract the second equation from the first and so eliminate the y term altogether. This would produce an equation in just one unknown which we could then solve.

Subtracting equation (2) from equation (1):

$$\begin{array}{lrcl} (1) & 5x + 3y & = & 19 \\ (2) & x + 3y & = & 11 \\ \hline & 4x & = & 8 \\ & x & = & 2 \end{array}$$

This yields a solution value of 2 for the variable x. But what about y? To find the value of y we merely substitute the value of x back into either of the original equations and solve for y.

$$\begin{aligned} (1)\ 5 \cdot 2 + 3y &= 19 \\ 3y &= 19 - 10 \\ 3y &= 9 \\ y &= 3 \end{aligned}$$

To check our solution we substitute the values for x and y back into both of the original equations.

$$\begin{aligned} (1)\ 5 \cdot 2 + 3 \cdot 3 &\stackrel{?}{=} 19 \\ 10 + 9 &\stackrel{?}{=} 19 \\ 19 &\stackrel{\checkmark}{=} 19 \end{aligned} \qquad \begin{aligned} (2)\ 2 + 3 \cdot 3 &\stackrel{?}{=} 11 \\ 2 + 9 &\stackrel{?}{=} 11 \\ 11 &\stackrel{\checkmark}{=} 11 \end{aligned}$$

Use this procedure to solve the following pair of equations and check your results. (Notice that the y terms are opposite in sign but have the same numerical coefficients.)

$$\begin{aligned} (1)\ 5x + 2y &= 29 \\ (2)\ 7x - 2y &= 19 \end{aligned}$$

Adding equations

$$\begin{aligned} (1)\ \ 5x + 2y &= 29 \\ (2)\ \ \underline{7x - 2y} &= \underline{19} \\ 12x &= 48 \\ x &= 4 \end{aligned}$$

Substituting 4 for x in equation (1):

$$\begin{aligned} 5 \cdot 4 + 2y &= 29 \\ 2y &= 29 - 20 \\ 2y &= 9 \\ y &= 4\tfrac{1}{2} \end{aligned}$$

Checking:

$$\begin{aligned} (1)\ 5 \cdot 4 + 2\left(4\tfrac{1}{2}\right) &\stackrel{?}{=} 29 \\ 29 &\stackrel{\checkmark}{=} 29 \end{aligned} \qquad \begin{aligned} (2)\ 7 \cdot 4 - 2\left(4\tfrac{1}{2}\right) &\stackrel{?}{=} 19 \\ 28 - 9 &\stackrel{?}{=} 19 \\ 19 &\stackrel{\checkmark}{=} 19 \end{aligned}$$

The only difference between this problem and the previous one is the fact that the y terms in the two equations have opposite signs rather than the same sign. To eliminate the y terms, therefore, we add them, allowing them to cancel one another. Addition served the same purpose here that subtraction did in the previous problem.

20. So far we have had numerical coefficients of the same absolute value for one of the unknowns in both equations. Now we need to think about what we would do if the absolute values were not the same. Consider, for example, these two equations:

$$\begin{aligned} (1)\ \ 2x - y &= 5 \\ (2)\ 3x + 3y &= 21 \end{aligned}$$

There are two ways to proceed in eliminating one of the variables in both equations. We could make the numerical coefficients of x alike if we multiplied every term in the first equation by 3 and every term in the second equation by 2. Then x would have the coefficient 6 in both equations. However, this would require two multiplications. Although you could divide the second equation by 3, a simpler way is to match up the absolute value of the coefficients of y by multiplying the first equation by 3. Using this latter method complete the solution and check your results.

Multiplying the first equation by 3: $6x - 3y = 15$

Adding the second equation: $3x + 3y = 21$

$$9x = 36$$
$$x = 4$$

Replacing x in the second equation (just for variety): $3 \cdot 4 + 3y = 21$

$$y = 3$$

Check: (1) $2 \cdot 4 - 3 \stackrel{?}{=} 5$ (2) $3 \cdot 4 + 3 \cdot 3 \stackrel{?}{=} 21$

$5 \stackrel{\checkmark}{=} 5$ $21 \stackrel{\checkmark}{=} 21$

SOLUTION BY SUBSTITUTION

21. The strategy of obtaining matching absolute values for the numerical coefficients of the same unknown in two equations and then using addition or subtraction to eliminate one of the unknowns is often referred to as *solution by elimination.* However, it also is possible to obtain a solution by the method known as *substitution.*

To illustrate the procedure of substitution, consider the same pair of equations you worked with in frame 20:

$$(1)\ 2x - y = 5$$
$$(2)\ 3x + 3y = 21$$

Start by solving the first equation for y (writing y as a function of x).

$$-y = 5 - 2x$$
$$y = 2x - 5$$

Next, substitute this expression for y (that is, $2x - 5$) in the second equation and solve for x.

$$3x + 3(2x - 5) = 21$$
$$3x + 6x - 15 = 21$$
$$9x = 36$$
$$x = 4$$

Substituting 4 for x in either of the original equations gives $y = 3$. Thus, we have the solution pair (4, 3).

Use the substitution method to solve this pair of equations: (1) $x - y = 10$; (2) $2x = 3y + 14$.

Solving in (1) for x: $x = y + 10$

Substituting this for x in (2):

$$\begin{aligned} 2(y + 10) &= 3y + 14 \\ 2y + 20 &= 3y + 14 \\ 2y - 3y &= 14 - 20 \\ -y &= -6 \\ y &= 6 \end{aligned}$$

Substituting this value for y in either of the original equations gives the answer $x = 16$.

By now you should see that solving a pair of linear equations algebraically is not really difficult. However, you may need more practice to become proficient, so be sure to do the Self-Test before going on to the next chapter.

Although solution of pairs of linear equations by algebraic methods (elimination or substitution) is usually simpler than graphing, it is important that you appreciate the power of graphs in gaining insight into mathematical situations. You are urged to practice both techniques in order to add to your total of mathematical skills.

SELF-TEST

1. Use graph paper to plot the following.

   (a) Draw the major axes of a rectangular coordinate system and label these axes.

   (b) Identify the four quadrants using Roman numerals.

   (c) Plot these points: (2, 4), (−3, 3), (5, −3), and (−4, −4).

   (d) The x coordinate of a point is called the ______________.

   (e) The y coordinate of a point is called the ______________.

   (frames 1 and 6)

2. In the equations below find the values of the dependent variable for each value of the independent variable given.

   (a) In the equation $y = x^2$, if $x = 2, 3,$ and 4, y will equal ______, ______, and ______.

   (b) In the equation $x = 2y + 3$, if $y = -1, 2,$ and -2, $x =$ ______, ______, and ______.

   (c) In the equation $2m = 3k - 5$, if $k = 1, 3,$ and 5, $m =$ ______, ______, and ______.

   (frames 9 and 10)

3. Find the x and y intercepts of the following linear equations.

	x intercept	y intercept
(a) $3x + y = 4$	(____, 0)	(0, ____)
(b) $x = y + 2$	(____, 0)	(0, ____)
(c) $2x + 3y - 4 = 0$	(____, 0)	(0, ____)
(d) $2y - x + 7 = 0$	(____, 0)	(0, ____)

(frame 12)

4. Using graph paper, graph the four equations given in problem 3. In each case you will need the intercepts plus at least one additional point to establish your straight line. (frame 13)

5. Select one convenient point on each of the straight lines you plotted in problem 4 (other than the points you used for graphing) and substitute its coordinates in the corresponding equation to prove that it makes the equation a true statement (that is, reduces it to an identity). (frame 14

6. Referring to problems 4 and 5 select one point *not* on the plotted straight line in each case and substitute its coordinates for x and y in the corresponding equation to prove that they do not satisfy the equation. (frame 14

7. Solve the following pairs of equations graphically.

(a) (1) $3x + 4y = -6$
(2) $y = 3 - 3x$

(b) (1) $2y - x = 2$
(2) $y + 2x = 6$

(c) (1) $2y = x - 4$
(2) $x + 4y = 4$

(frame 16

8. Solve the following pairs of linear equations simultaneously by use of the elimination procedure.

(a) (1) $3x - y = 21$
(2) $2x + y = 4$

(b) (1) $3k + 5p = 9$
(2) $3k - p = -9$

(c) (1) $7a + t = 42$
(2) $3a - t = 8$

(d) (1) $7r = 5s + 15$
(2) $2r = s + 9$

(e) (1) $2y - 7x = 2$
(2) $3x = 14 - y$

(frame 19

9. Solve the following pairs of linear equations by the substitution procedure.

(a) (1) $y = 2x$
(2) $7x - y = 35$

(b) (1) $r = 4t - 1$
(2) $6t + r = 79$

(c) (1) $a = b + 2$
(2) $3a + 4b = 20$

(d) (1) $3p = 27 - q$
(2) $2q = 3p$

(frame 21

Answers to Self-Test

1. (a) through (c)

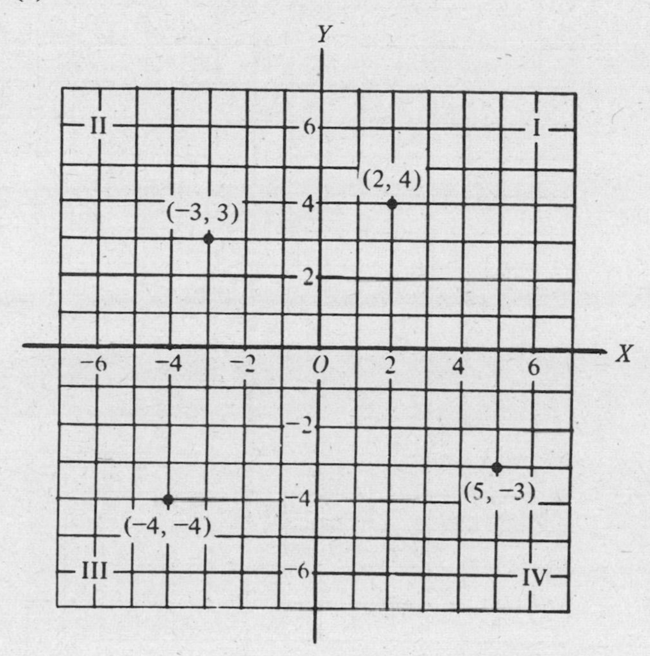

(d) abscissa; (e) ordinate

2. (a) 4, 9, 16; (b) 1, 7, −1; (c) −1, 2, 5
3. (a) $(\frac{4}{3}, 0)$, $(0, 4)$; (b) $(2, 0)$, $(0, -2)$; (c) $(2, 0)$, $(0, \frac{4}{3})$ (d) $(7, 0)$, $(0, \frac{7}{2})$
   (Remember: These are intercepts, not coordinate pairs.)

4.

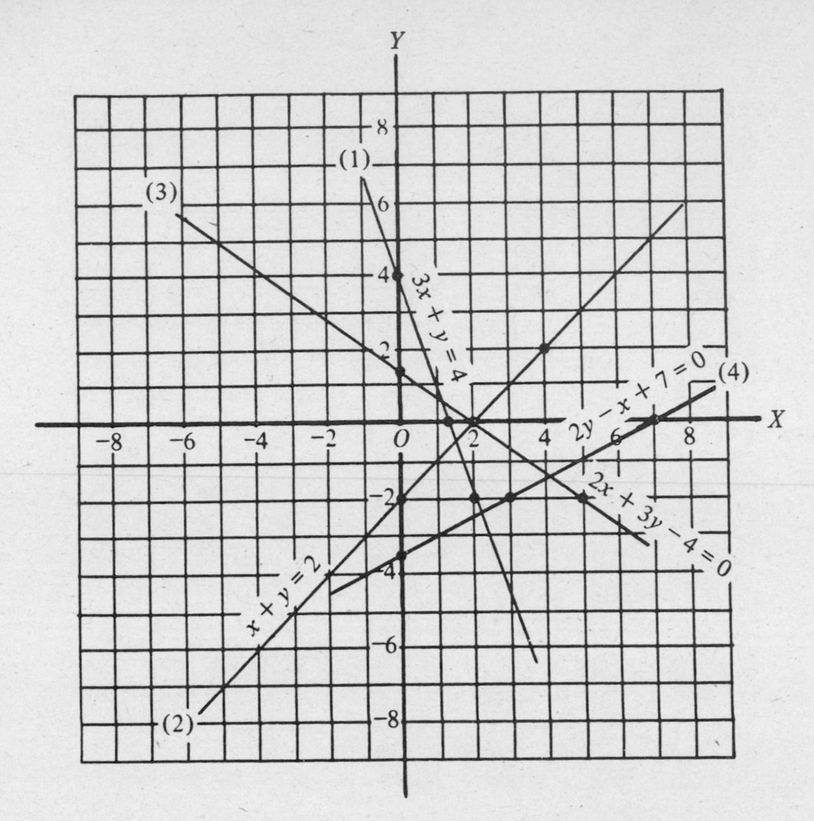

5. You can use the third point shown on each of the lines above to prove it will satisfy the equation. However, any other point on the line will serve equally well.
6. Any random point not on the graph of the equation being checked will serve to prove this fact.

7. (a) (1) $3x + 4y = -6$

$$y = \frac{-3x - 6}{4}$$

(2) $y = 3 - 3x$

x	y
0	$-\frac{3}{2}$
−2	0
2	−3

x	y
0	3
1	0
2	−3

(2, −3) solution

(b) (1) $2y - x = 2$

$x = 2y - 2$

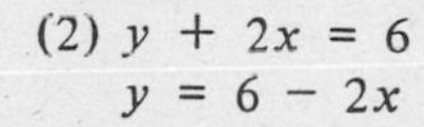

(2) $y + 2x = 6$

$y = 6 - 2x$

x	y
−2	0
0	1
2	2

x	y
0	6
2	2
3	0

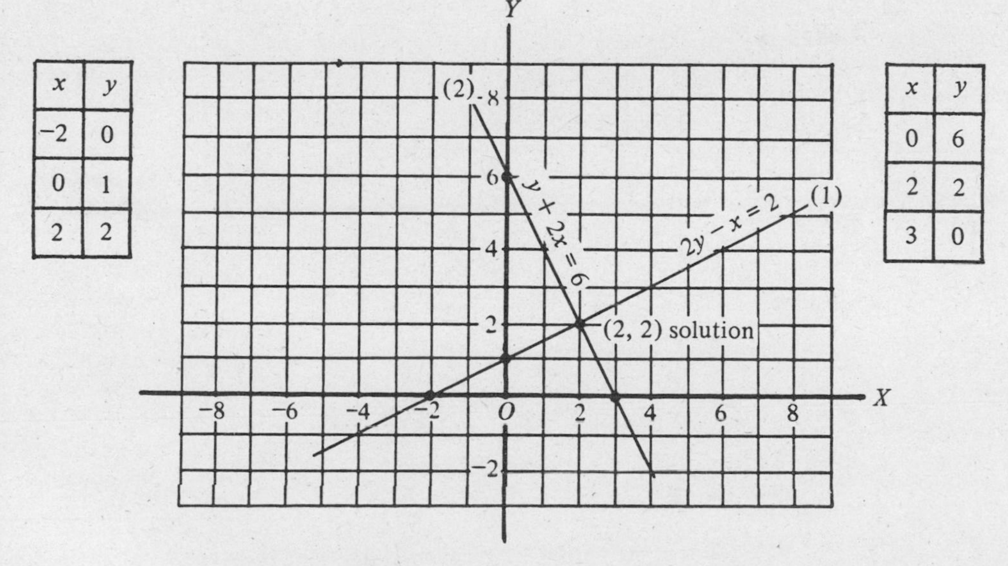

7. (continued)

(c) (1) $2y = x - 4$
$x = 2y + 4$

(2) $x + 4y = 4$
$x = 4 - 4y$

x	y
4	0
0	−2
2	−1

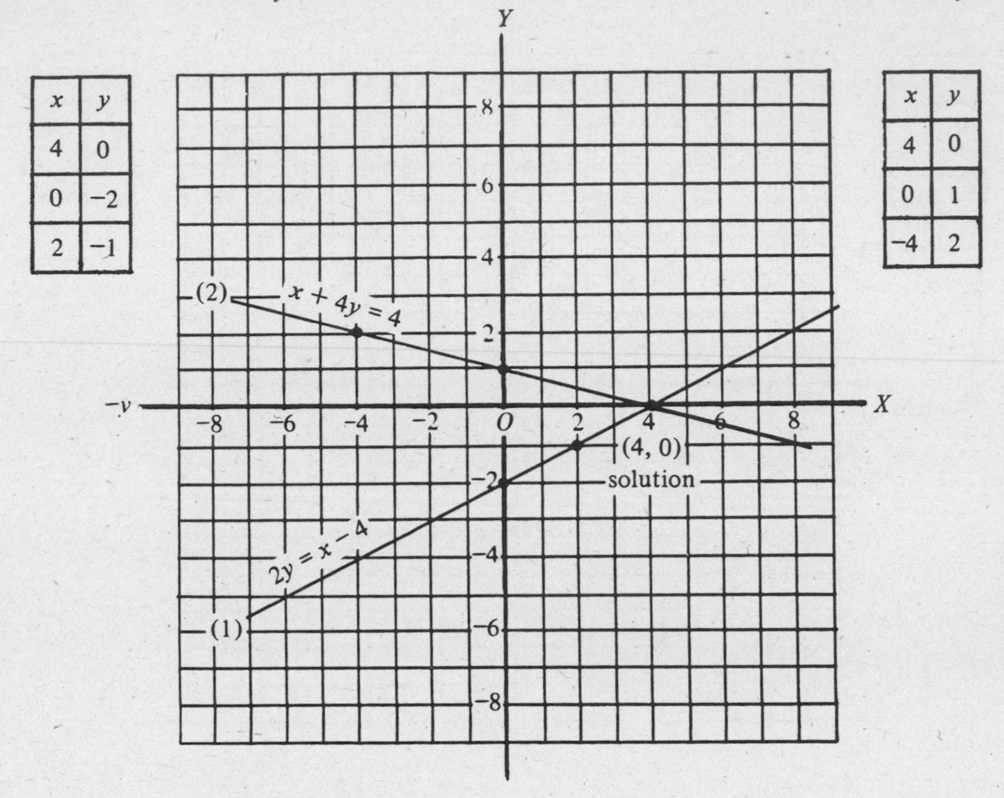

x	y
4	0
0	1
−4	2

8. (a)

Adding:

$$
\begin{aligned}
(1)\quad 3x - y &= 21 \\
(2)\quad 2x + y &= 4 \\
\hline
5x &= 25 \\
x &= 5
\end{aligned}
$$

Substituting $x = 5$ back in either of the original equations: $y = -6$

(b)

Subtracting:

$$
\begin{aligned}
(1)\quad 3k + 5p &= 9 \\
(2)\quad 3k - p &= -9 \\
\hline
6p &= 18 \\
p &= 3
\end{aligned}
$$

Substituting 3 for p: $k = -2$

(c)

Adding:

$$
\begin{aligned}
(1)\quad 7a + t &= 42 \\
(2)\quad 3a - t &= 8 \\
\hline
10a &= 50 \\
a &= 5
\end{aligned}
$$

Substituting 5 for a: $t = 7$

(d) Multiplying (2) by 5:

Subtracting (1):

$$
\begin{aligned}
10r &= 5s + 45 \\
7r &= 5s + 15 \\
\hline
3r &= 30 \\
r &= 10
\end{aligned}
$$

Substituting 10 for r: $s = 11$

(e) Rearranging (2): $y + 3x = 14$

Multiplying by 2: $2y + 6x = 28$

Subtracting (1): $2y - 7x = 2$

$$13x = 26$$

$$x = 2$$

Substituting 2 for x: $y = 8$

9. (a) Substituting (1) in (2):

$$7x - 2x = 35$$
$$5x = 35$$
$$x = 7$$
$$y = 14$$

(b) Substituting (1) in (2):

$$6t + (4t - 1) = 79$$
$$10t = 80$$
$$t = 8$$
$$r = 31$$

(c) Substituting (1) in (2):

$$3(b + 2) + 4b = 20$$
$$7b = 14$$
$$b = 2$$
$$a = 4$$

(d) Rearranging (2): $q = \frac{3p}{2}$

Substituting q in (1): $3p = 27 - \frac{3p}{2}$

Clearing fractions: $6p = 54 - 3p$

$$p = 6$$
$$q = 9$$

CHAPTER NINE

Quadratic Equations

OBJECTIVES

In the last chapter we investigated various methods of solving linear (first order or first degree) equations of one or more unknowns. Quadratic (second order or second degree) equations in one unknown are simply a logical progression from what you have already learned. This type of equation is not completely new to you since in Chapter 4 we discussed procedures for factoring quadratic expressions. In this chapter we will consider various ways of solving equations involving quadratic expressions. Specifically, when you reach the end of the chapter you will be able to:

- write quadratic equations in standard form;
- factor quadratic equations, solve for roots, and check your solutions;
- square binomials;
- find the principal square roots of trinomials;
- solve quadratic equations by completing the square;
- use the general quadratic formula to solve quadratic equations;
- solve quadratic equations graphically.

THE MEANING OF QUADRATIC EQUATIONS

1. A quadratic equation (in one unknown) is simply an equation in which the highest power of the unknown is the second power. Thus, $x^2 - 3x + 2 = 0$ is a quadratic equation because the unknown x is raised to a second (and no higher) power.

   The general (or standard) form of a quadratic equation is $ax^2 + bx + c = 0$; a represents the numerical coefficient of x^2 (the second degree term of the unknown), b is the numerical coefficient of x (the first power of the unknown), and c represents the constant. In other words, a, b, and c represent the real number coefficients and x represents the variable. In the equation $x^2 - 3x + 2 = 0$, $a = 1$, $b = -3$, and $c = 2$.

   What are the values of a, b, and c in the equation $x^2 - 49 = 0$?

   (a) a = ________ (c) c = ________

   (b) b = ________

(a) 1; (b) 0; (c) −49

Since there is no first degree term in x, the coefficient b has a value of zero.

2. Quadratic equations are divided into two classes: *complete quadratic equations* (such as $x^2 - 3x + 2 = 0$) and *incomplete quadratic equations* (such as $x^2 - 49 = 0$ or $x^2 - 6x = 0$).

   An *incomplete quadratic equation* is of the form $ax^2 + c = 0$ (which lacks the term containing the first power of the unknown) or of the form $ax^2 + bx = 0$ (where the constant term is zero).

   Since one way of solving many quadratic equations involves factoring, we will practice first the general procedure discussed in Chapter 4 for factoring expressions of the form $ax^2 + bx + c$. Factor $x^2 + 2x - 15$ into its two binomial factors.

$x^2 + 2x - 15 = (x - 3)(x + 5)$

3. How did you find the factors in the above problem? Factoring some types of trinomials is largely a matter of educated guessing. Since the numerical coefficient of x^2 is 1 in the trinomial $x^2 + 2x - 15$, it is apparent that the first term in each binomial is simply x. The next step is to examine the factors of 15 to discover if any two of them differ by 2 (the numerical coefficient of the middle term of the trinomial). Three times 5 and 1 times 15 are the only possible (integral) factors of 15. The correct pair of factors is 3 and 5. It is then necessary to make the sign of the larger factor plus (+5) and the sign of the smaller factor minus (−3) in order to arrive at the correct sign for the middle term. Since the coefficient of x^2 is 1, this trinomial is somewhat easier to factor than if x had some other coefficient. However, if a (in $ax^2 + bx + c$) is not 1, the general procedure for factoring is much the same as that used above.

   Factor the trinomial $4x^2 + 2x - 6$.

$2(x - 1)(2x + 3)$

4. In the problem of frame 3 it is first necessary to consider the use of the distributive law to remove the common factor 2. Next, you have to factor the trinomial $2x^2 + x - 3$. Since the coefficients are small the range of factors is limited and it should require only a small amount of experimenting with the various combinations of factors to arrive at the correct solution.

   Factor the following quadratic polynomials to make sure you understand the process.

(a) $x^2 + 4x - 21 =$ ______________________

(b) $x^2 + 6x + 8 =$ ______________________

(c) $4x^2 - 7x - 2 =$ ______________________

(d) $x^2 - 9 =$ ______________________

(e) $9x^2 - 6x + 1 =$ ______________________

(f) $3x^2 - x =$ ______________________

(g) $2y^2 - 5y - 25 =$ ______________________

(h) $k^2 - 4k =$ ______________________

- - - - - - - - - - - - - - - - -

(a) $(x + 7)(x - 3)$; (b) $(x + 2)(x + 4)$; (c) $(4x + 1)(x - 2)$; (d) $(x - 3)(x + 3)$; (e) $(3x - 1)(3x - 1)$; (f) $x(3x - 1)$; (g) $(y - 5)(2y + 5)$; (h) $k(k - 4)$

Did you remember to classify problem (d) as the product of the sum and difference of the same two terms? Keep in mind also that problem (e) is the square of a binomial. You should try to recognize and mentally identify such quadratic expressions; they have special properties that can be used to advantage when factoring.

SOLUTION BY FACTORING

5. Having reviewed the procedure for factoring quadratic expressions, our next step is to consider how to solve equations involving quadratic expressions. There are several available methods for solving quadratic equations. The first method we will look at is that of *solution by factoring.* Here are the steps essential to this method.

Step 1: Write the quadratic equation in standard form (that is, $ax^2 + bx + c = 0$).

Step 2: Factor the left member into first degree factors.

Step 3: Set each factor containing the variable equal to zero.

Step 4: Solve the resulting first degree equations.

Step 5: Check the roots of these first degree equations in the original quadratic equation.

Example: Solve (find the roots of) $x^2 + 4x = 21$.

Step 1: Since the equation is not in standard form our first step is to add -21 to both members, giving us $x^2 + 4x - 21 = 0$.

Step 2: Factoring the left member we get $(x + 7)(x - 3) = 0$.

Step 3: Since both factors contain the variable we will write the two first degree equations $x + 7 = 0$ and $x - 3 = 0$. This step uses the concept called the *zero factor law* which states that if $a \cdot b = 0$, then $a = 0$, $b = 0$, or both a and $b = 0$. This concept can be used only when one member of an equation is zero.

Step 4: Solve the equation $x + 7 = 0$ to get the root $x = -7$, and solve the equation $x - 3 = 0$ to get the root $x = 3$.

Step 5: Check the roots $x = -7$ and $x = 3$ in the original equation.

$$(-7)^2 + 4(-7) = 21;\ 49 - 28 = 21;\ 21 = 21$$
$$3^2 + 4(3) = 21;\ 9 + 12 = 21;\ 21 = 21$$

Notice that we have a conditional second degree equation that has two solutions (roots): -7 and 3. This is one of the major differences between first degree and second degree equations; first degree equations have only one solution.

6. Quadratic equations will have either two unequal roots or one root that occurs with a multiplicity of two (that is, the same value will occur for both roots).

Review the factoring procedure by completing the blanks below to solve the equation $3x^2 + 13x - 10 = 0$.

(a) It is not necessary to change the form of this equation prior to factoring because it is already in what form? ______________________.

(b) ______________ the left member to obtain $(3x - 2)(x + 5) = 0$.

(c) Write the two linear equations: ______________________ and ______________________.

(d) Solve the first linear equation to obtain $x =$ ________.

(e) Solve the second to obtain $x =$ ________.

(f) Check your solution values in the original equation. Show your work.

- - - - - - - - - - - - - - - - - -

(a) standard form; (b) Factor; (c) $3x - 2 = 0$ and $x + 5 = 0$;

(d) $\frac{2}{3}$; (e) -5

(f) $3\left(\frac{4}{9}\right) + 13\left(\frac{2}{3}\right) - 10 = 0;\quad 4 + 26 - 30 \stackrel{?}{=} 0;\quad 0 \stackrel{\checkmark}{=} 0$

$3(25) + 13(-5) - 10 = 0;\quad 75 - 65 - 10 \stackrel{?}{=} 0;\quad 0 \stackrel{\checkmark}{=} 0$

7. Solve the following quadratic equations by the factoring method. Check each of the roots in the original quadratic equation from which it was derived. Remember that a number is a root of an equation if it makes the equation a true statement. You will not know for certain whether or not you made a mistake in deriving a numerical value as a root unless you test it in the original equation.

(a) $x^2 + 6x + 8 = 0$

(b) $4x^2 - 7x - 2 = 0$

(c) $x^2 - 9 = 0$

(d) $9x^2 - 6x = -1$

(e) $3x^2 = x$

(f) $2y^2 - 5y = 25$

(g) $k^2 - 4k = 0$

- - - - - - - - - - - - - - - - - -

(a) $x = -2$; check: $4 - 12 + 8 \stackrel{?}{=} 0$, $0 \stackrel{\checkmark}{=} 0$

$x = -4$; check: $16 - 24 + 8 \stackrel{?}{=} 0$, $0 \stackrel{\checkmark}{=} 0$

(b) $x = -\frac{1}{4}$, $x = 2$

(c) $x = 3$; check: $9 - 9 \stackrel{?}{=} 0$, $0 \stackrel{\checkmark}{=} 0$

$x = -3$; check: $9 - 9 \stackrel{?}{=} 0$, $0 \stackrel{\checkmark}{=} 0$

(d) $x = \frac{1}{3}$, $x = \frac{1}{3}$ (This is an example of a multiple root.)

(e) $x = 0$; check: $3(0)^2 \stackrel{?}{=} 0$, $0 \stackrel{\checkmark}{=} 0$

$x = \frac{1}{3}$; check: $3(\frac{1}{3})^2 \stackrel{?}{=} \frac{1}{3}$, $\frac{1}{3} \stackrel{\checkmark}{=} \frac{1}{3}$

(f) $y = 5$, $y = -2\frac{1}{2}$

(g) $k = 0$; check: $0 - 0 \stackrel{?}{=} 0$, $0 \stackrel{\checkmark}{=} 0$

$k = 4$; check: $16 - 16 \stackrel{?}{=} 0$, $0 \stackrel{\checkmark}{=} 0$

SOLVING INCOMPLETE QUADRATIC EQUATIONS

8. It is apparent that the method of solution by factoring will not always work since not all quadratic trinomials are factorable. Later in this chapter we will discuss some methods that can be used to solve any quadratic equation. Now, however, let us turn our attention to the two types of incomplete quadratic equations and some of their solutions.

An incomplete quadratic equation with the constant term equal to zero can be represented by the general form $ax^2 + bx = 0$. This type does not present any new requirements in the way of solution because it can always be solved by factoring. However, if we examine the equation $ax^2 + bx = 0$ we can learn some interesting facts about the roots of such an equation.

(1) If we factor the left member we obtain $x(ax + b) = 0$ by the distributive law.

(2) Since both factors contain the variable x we have two linear equations without further factoring: $x = 0$ and $ax + b = 0$.

(3) If we solve the equation $x = 0$, the root is 0. If we solve the equation $ax + b = 0$, the root is $-\frac{b}{a}$.

From this we learn that one solution to an equation of the type $ax^2 + bx = 0$ is $x = 0$ and the other solution is $x = -\frac{b}{a}$.

Solve the following equations using either the factoring method or the method discussed above.

(a) $y^2 - 2y = 0$

(b) $4x^2 = 28x$ (Remember to change to standard form $ax^2 + bx = 0$.)

(c) $7k^2 - 35k = 0$

(d) $3a^2 - 18a = 0$

(e) $2x^2 + 3x = 0$

(a) $y = 0, y = 2$; (b) $x = 0, x = 7$; (c) $k = 0, k = 5$;

(d) $a = 0, a = 6$

	$3a^2 - 18a = 0$	
Factoring:	$3a(a - 6) = 0$	
Two linear equations:	$3a = 0$	$a - 6 = 0$
	$a = 0$	$a = 6$

(e) $2x^2 + 3x = 0$

$x(2x + 3) = 0$

$x = 0 \quad 2x + 3 = 0$

$$x = -\frac{3}{2} \left(\text{i.e., } -\frac{b}{a}\right)$$

These problems should point out to you that this type of incomplete quadratic equation is easy to solve by factoring. Just remember to write the equation in the standard form $ax^2 + bx = 0$ before you do anything else.

SOLUTION BY EXTRACTION OF ROOTS

9. If an incomplete quadratic equation is of the type $ax^2 = c$ (where the coefficient of the first degree term is zero) we have a slightly different situation because factoring may or may not work. Let's discuss first the procedure to follow if factoring is not possible.

If an incomplete quadratic of the form $ax^2 = c$ cannot be solved by factoring, we can use the method of extraction of roots, as given below.

Step 1: Solve for the square of the variable. This will yield an equation of the form $x^2 = \frac{c}{a}$.

Step 2: The roots of $x^2 = \frac{c}{a}$ will be the roots of the two equations $x = \sqrt{\frac{c}{a}}$ and $x = -\sqrt{\frac{c}{a}}$ (if $\frac{c}{a}$ is positive). There will be no real number solution if $\frac{c}{a}$ is negative.

Step 3: Check results back in the original equation. It is most important that the work in Step 2 be thought of properly. It is the result of the property of real numbers which states that $a^2 = b^2$ if and only if $a = b$ or $a = -b$. We need to remember that when we solve an equation we are seeking *all* of the numbers that make the statement true.

Therefore, the solution to $x^2 = \frac{c}{a}$ must include both the principal square root of $\frac{c}{a}$ and the negative square root of $\frac{c}{a}$.

From what we have just discussed would you say that the following statement is true or false? Incomplete quadratic equations of the form $ax^2 = c$ can only be solved by the method of extraction of roots. (True / False)

False

Often they can be solved by factoring; do so if you can. But if you can't, then use the extraction of roots procedure.

10. Now let's look at more examples of the method of extraction of roots in solving incomplete quadratic equations of the type $ax^2 = c$.

Example 1: Solve the equation $x^2 - 25 = 0$. (Although this could be solved by factoring we will use the method of extraction of roots.)

Step 1: $x^2 = 25$

Step 2: $x = \sqrt{25}$ and $x = -\sqrt{25}$, hence $x = 5$ and $x = -5$.

Step 3: Check results (roots) in the original equation.
Does $5^2 - 25 = 0$? Yes. Does $(-5)^2 - 25 = 0$? Yes.

Example 2: Solve $9x^2 - 25 = 0$ using the method of extraction of roots

Step 1: $x^2 = \frac{25}{9}$

Step 2: $x = \sqrt{\frac{25}{9}}$ and $x = -\sqrt{\frac{25}{9}}$, hence $x = \frac{5}{3}$ and $-\frac{5}{3}$.

Step 3: Check. Does $9(\frac{5}{3})^2 - 25 = 0$? Yes. Does $9(-\frac{5}{3})^2 - 25 = 0$? Yes.

You will frequently see the above answers written in the form $x = \pm\frac{5}{3}$. This form is quite acceptable. We have not used it because we are stressing the fact that we need both of the possible square roots in order to have a complete solution to the equation. Also, these two examples could have been worked out by factoring, although it might have taken longer.

Solve the following equations by the method of extraction of roots. In (c) and (d) start by dividing both terms by the coefficient of x^2.

(a) $x^2 - 49 = 0$

(b) $x^2 = 36$

(c) $4x^2 - 36 = 0$

(d) $3x^2 = 48$

(a) $x = \pm 7$; (b) $x = \pm 6$; (c) $x = \pm 3$; (d) $x = \pm 4$

11. Now let us consider a third example of solution by the method of extraction of roots that is slightly more difficult.

Example: Solve $7x^2 - 5 = 0$. If we were to attempt to use the method of factoring here we would have to think of the correct set of factors involving irrational fractions. We will find it easier, therefore, to use the method of extraction of roots.

Step 1: $7x^2 = 5,\ x^2 = \frac{5}{7}$

Step 2: $x = \sqrt{\frac{5}{7}}$ and $x = -\sqrt{\frac{5}{7}}$

We should now write the radicals in their simplest form. To simplify $\sqrt{\frac{5}{7}}$ we multiply (as you will recall from Chapter 6) both the numerator and denominator by $\sqrt{7}$ in order to rationalize the denominator. Therefore, $\sqrt{\frac{5}{7}}$ becomes $\sqrt{\frac{35}{49}}$ or $\frac{\sqrt{35}}{7}$.

Thus, the roots are $x = \frac{\sqrt{35}}{7}$ and $x = -\frac{\sqrt{35}}{7}$.

Step 3: Check. Does $7\left(\frac{\sqrt{35}}{7}\right)^2 - 5 = 0$? Yes. Does $7\left(-\frac{\sqrt{35}}{7}\right)^2 - 5 = 0$? Yes.

Let's complete our discussion of the method of extraction of roots by examining the equation $x^2 + 4 = 0$. If we attempt a solution by the method of extraction of roots we get:

$$x^2 = \sqrt{-4}$$
$$x = \sqrt{-4} \text{ and } x = -\sqrt{-4}$$

Neither of these last two equations has solutions in the real number system because there is no real number x such that $x \cdot x = -4$. If you run into this type of problem you will have to indicate an answer of "no real roots." The solution to this kind of problem is discussed in algebra texts where the complex number system is introduced.

Solve the following quadratic equations by the method of extraction of roots.

(a) $x^2 - 1 = 0$

(b) $\frac{x^2}{2} = 8$

(c) $x^2 + 36 = 0$

(d) $3x^2 - 2 = 0$

(a) $x = \pm 1$; (b) $x = \pm 4$; (c) no real roots; (d) $x = \pm\frac{\sqrt{6}}{3}$

12. A quadratic equation is not always in the standard form $ax^2 + bx + c = 0$ which we referred to at the beginning of this chapter. In fact, there will be times when you will have to change the form of an equation just to see whether it *is* in fact a quadratic equation. For example, $x = 4 - \frac{3}{x}$ may not at first appear to be a quadratic equation. However, clearing fractions and transforming the resulting equation gives us $x^2 - 4x + 3 = 0$, which is somewhat more recognizable.

If a quadratic equation is not in standard form, perform whatever operations are necessary to transform it to standard form. Possible operations are any of the axioms we have developed for transforming to equivalent equations. A partial list of these would include the following:

(1) Clearing of fractions by multiplying all terms by the LCD. (Note: If the LCD includes the variable you will need to check all apparent solutions to be sure that no denominators become zero. If some value of the variable makes the denominator become 0, that value cannot be a root.)

(2) Removing parentheses and other grouping symbols.

(3) Removing radical signs by squaring both members of the equation.

(4) Collecting like terms.

Use whichever of these procedures are necessary to express the following quadratic equations in standard form. Write the equation with a positive coefficient for the second degree term.

(a) $\sqrt{y^2 - 5y} = 3y$ (e) $3b^2 = -5b$

(b) $20 + 6k = 2k^2$ (f) $7(x^2 - 9) = x(x - 5)$

(c) $\frac{10}{c} + 1 = 4c$ (g) $18 = 2x^2$

(d) $p^2 = 5p - 4$ (h) $y(8 - 2y) = 6$

(a) $8y^2 + 5y = 0$; (b) $2k^2 - 6k - 20 = 0$; (c) $4c^2 - c - 10 = 0$; (d) $p^2 - 5p + 4 = 0$; (e) $3b^2 + 5b = 0$; (f) $6x^2 + 5x - 63 = 0$; (g) $2x^2 - 18 = 0$; (h) $2y^2 - 8y + 6 = 0$

SOLUTION BY COMPLETING THE SQUARE

13. We have mentioned that some quadratic equations cannot be solved by factoring. One method that can be used to solve such quadratic equations is called *completing the square.* To use this method you must be able to square binomials quickly, find the square root of perfect square trinomials, and form a perfect square trinomial when two of its terms are given. You practiced the first two of these procedures in Chapter 4. The following exercises will test your recall and help sharpen your skills. Square these binomials.

(a) $(y + 1)^2 =$ __________

(b) $(k + 3)^2 =$ __________

(c) $(x - 1)^2 =$ __________

(d) $(y - 7)^2 =$ __________

(e) $(a - 5)^2 =$ __________

(f) $(x + 4)^2 =$ __________

(g) $(p + 10)^2 =$ __________

(h) $\left(a - \frac{1}{2}\right)^2 =$ __________

(a) $y^2 + 2y + 1$; (b) $k^2 + 6k + 9$; (c) $x^2 - 2x + 1$;
(d) $y^2 - 14y + 49$; (e) $a^2 - 10a + 25$; (f) $x^2 + 8x + 16$;
(g) $p^2 + 20p + 100$; (h) $a^2 - a + \frac{1}{4}$

Some of the following trinomials are perfect squares. Find the principal (positive) square roots of those that are perfect squares.

(i) $a^2 + 2a + 1$ __________

(j) $x^2 + 6x - 9$ __________

(k) $y^2 + y + 1$ __________

(l) $b^2 - 10b + 25$ __________

(m) $x^2 - 2x + 1$ __________

(n) $z^2 - 16z + 64$ __________

(o) $a^2 + a + \frac{1}{4}$ __________

(p) $x^2 - 4x + 8$ __________

(i) $(a + 1)$; (j) not a perfect square; (k) not a perfect square; (l) $(b - 5)$;
(m) $(x - 1)$; (n) $(z - 8)$; (o) $\left(a + \frac{1}{2}\right)$; (p) not a perfect square

14. Do you remember that the answer to problems (a) through (h) in frame 13 are called *perfect square trinomials*? It is important to keep in mind that for a trinomial to be a perfect square trinomial both the first and third terms must be perfect squares and the middle term must equal twice the product of the square roots of the first and third terms. Thus, $(x + a)^2 = x^2 + 2ax + a^2$.

To solve a quadratic equation by completing the square it is necessary to make one member of the equation a perfect square. For example, the left member of the equation $x^2 + 6x - 7 = 0$ is not a perfect square trinomial because the third term, -7, is not the correct value. However, the binomial $x^2 + 6x$ could be converted to a perfect square trinomial by adding a third term that is equal to the square of one-half the coefficient of x (that is, by adding 3^2 or 9). The procedure for completing the square is as follows:

(1)	Eliminate the constant from the left member of the equation.	$x^2 + 6x = 7$
(2)	Divide both members by the numerical coefficient of the first (squared) term, if the coefficient is not 1.	Step 2 is not necessary in this case.
(3)	Square one-half the coefficient of x and add this amount to both sides: $\left(\frac{6}{2}\right)^2 = 9$.	$x^2 + 6x + 9 = 7 + 9$

(4) Write the trinomial in its factored form (a binomial squared). $(x + 3)^2 = 16$

(5) Use the method for extraction of roots. $x + 3 = \pm 4$

(6) Solve the resulting two equations. $x + 3 = 4$, $x = 1$ $\quad$ $x + 3 = -4$, $x = -7$

(7) Check both roots in the original equation. This step left to you.

Apply this procedure (completing the square) to solve the following quadratic equation: $x^2 - 4x - 5 = 0$.

---

(1) $x^2 - 4x - 5 = 0;$
(2) $x^2 - 4x = 5$
(3) $x^2 - 4x + 4 = 5 + 4$
(4) $(x - 2)^2 = 9$
(5) $x - 2 = \pm 3$
(6) $x - 2 = 3$ $\quad$ $x - 2 = -3$
$x = 5$ $\quad$ $x = -1$

(7) $5^2 - 4 \cdot 5 - 5 \stackrel{?}{=} 0$ $\quad$ $(-1)^2 - 4(-1) - 5 \stackrel{?}{=} 0$

$0 \stackrel{\checkmark}{=} 0$ $\quad$ $0 \stackrel{\checkmark}{=} 0$

This equation could have been solved by factoring. It is more important, however, that you concentrate on the method rather than the numbers.

15. Now for some practice on your own. Solve the following equations by completing the square. Don't forget to check your answers. Remember to apply Step 2 in problem (h).

(a) $x^2 - 12x = 13$ $\quad$ (e) $y^2 - 20y = 96$

(b) $a^2 - 6a = 7$ $\quad$ (f) $k^2 + 18k + 17 = 0$

(c) $x^2 + 2x = 24$ $\quad$ (g) $a^2 - 3a - 18 = 0$

(d) $m^2 - 12m + 11 = 0$ $\quad$ (h) $2x^2 + 10x - 12 = 0$

(a) $x = 13, x = -1$; (b) $a = 7, a = -1$; (c) $x = 4, x = -6$;
(d) $m = 11, m = 1$; (e) $y = 24, y = -4$; (f) $k = -1, k = -17$

(g) $a^2 - 3a - 18 = 0$

$$a^2 - 3a + \frac{9}{4} = 18 + \frac{9}{4}$$
$$\left(a - \frac{3}{2}\right)^2 = \frac{81}{4}$$
$$a - \frac{3}{2} = \pm\frac{9}{2}$$
$$a = 6, -3$$

(h) $2x^2 + 10x - 12 = 0$

$$x^2 + 5x + \frac{25}{4} = 6 + \frac{25}{4}$$
$$\left(x + \frac{5}{2}\right)^2 = \frac{49}{4}$$
$$x = 1, -6$$

16. All of the problems in frame 15 could have been solved by factoring; however, they lent themselves to useful practice in completing the square. You also probably noticed that certain coefficients of the x term result in our having a fraction for our third term. Add this fraction to the right term of the equation, just as you would any whole number. The fact that it is a fraction makes no difference in the procedure; just be careful to use correct arithmetic and algebraic operations in arriving at your final answers.

The following problem is slightly harder. Don't be alarmed by the number of fractional and binomial terms. Simply clear fractions by finding the LCD and multiply as you learned to do in Chapter 5. The equation can then be written in the standard form for quadratic equations. Solve by completing the square.

$$\frac{a-2}{a+2} - \frac{a+2}{a-1} = \frac{a^2-32}{a^2+a-2}$$

- - - - - - - - - - - - - - - - - -

Simplified form of the equation: $a^2 + 7a - 30 = 0$
Solutions: $a = 3, -10$

17. Now let's apply the method of completing the square to solve a quadratic equation that cannot be solved by factoring.

Example: Solve the equation: $x^2 + 2x - 5 = 0$.

Clearly this trinomial is not factorable. Using the method of completing the square we proceed as follows:

$$x^2 + 2x = 5$$
$$x^2 + 2x + 1 = 5 + 1$$
$$(x + 1)^2 = 6$$
$$x + 1 = \sqrt{6} \text{ or } x + 1 = -\sqrt{6}$$
$$x = -1 + \sqrt{6} \text{ or } x = -1 - \sqrt{6}$$

The solutions obtained are irrational real numbers and are perfectly valid, as we will see by checking. Does $(-1 + \sqrt{6})^2 + 2(-1 + \sqrt{6}) - 5 = 0$?

$$(1 - 2\sqrt{6} + 6) - 2 + 2\sqrt{6} - 5 = 7 - 2\sqrt{6} - 7 + 2\sqrt{6} = 0$$
$$0 = 0$$

Try checking the root $x = -1 - \sqrt{6}$.

- - - - - - - - - - - - - - - - - -

Does $(-1 - \sqrt{6})^2 + 2(-1 - \sqrt{6}) - 5 = 0$?
$$(1 + 2\sqrt{6} + 6) - 2 - 2\sqrt{6} - 5 = 7 + 2\sqrt{6} - 7 - 2\sqrt{6} = 0$$
$$0 = 0$$

18. Solve the following quadratic equations by completing the square. Simplify all radicals.

(a) $x^2 - 6x + 2 = 0$

(b) $2x^2 + 4x - 8 = 0$

(c) $2x^2 - 3x - 1 = 0$

(a) $x = 3 \pm \sqrt{7}$; (b) $x = -1 \pm \sqrt{5}$; (c) $x = \dfrac{3 \pm \sqrt{17}}{4}$

SOLUTION BY THE QUADRATIC FORMULA

19. The method of completing the square can be used to develop another method for solving quadratic equations. We will apply it to the standard form of the quadratic equation ($ax^2 + bx + c = 0$ where a, b, and c represent real number coefficients and x is the variable) to obtain a general solution.

$$ax^2 + bx + c = 0$$
$$ax^2 + bx = -c$$

Dividing by a:

$$x^2 + \frac{b}{a}x = -\frac{c}{a}$$

Adding $\left(\frac{b}{2a}\right)^2$ to each side to complete the square:

$$x^2 + \frac{b}{a}x + \frac{b^2}{4a^2} = \frac{b^2}{4a^2} - \frac{c}{a}$$

Clearing fractions and forming the binomial squared term:

$$\left(x + \frac{b}{2a}\right)^2 = \frac{b^2 - 4ac}{4a^2}$$

$$x + \frac{b}{2a} = \sqrt{\frac{b^2 - 4ac}{4a^2}} \qquad x + \frac{b}{2a} = -\sqrt{\frac{b^2 - 4ac}{4a^2}}$$

$$x = -\frac{b}{2a} + \frac{\sqrt{b^2 - 4ac}}{2a} \qquad x = -\frac{b}{2a} - \frac{\sqrt{b^2 - 4ac}}{2a}$$

These two solutions can be written

$$x = \frac{-b \pm \sqrt{b^2 - 4ac}}{2a}$$

which is known as the *quadratic formula.* There is no need for you to memorize the proof; it was given merely to increase your ability to follow proofs and to appreciate them. You should, however, memorize the general quadratic formula and be able to use it.

Example: Use the quadratic formula to find the roots of the equation $x^2 - 5x + 6 = 0$.

Quadratic formula: $x = \dfrac{-b \pm \sqrt{b^2 - 4ac}}{2a}$

From our equation, $a = 1$ (the coefficient of x^2), $b = -5$ (the coefficient of x), and $c = 6$ (the constant or third term).

Substituting these coefficients in the quadratic formula:

$$x = \frac{-(\ 5) \pm \sqrt{(-5)^2 - 4\cdot 1\cdot 6}}{2\cdot 1}$$

$$x = \frac{5 \pm \sqrt{25 - 24}}{2}$$

$$x = \frac{5 \pm 1}{2}$$

$$x = 3,\ 2$$

The roots may not always be integers (or even rational) but the procedure is the same.

Use the quadratic formula to solve $x^2 + 2x - 15 = 0$.

$a = 1,\ b = 2,\ c = -15$

$$x = \frac{-2 \pm \sqrt{2^2 - (4)(1)(-15)}}{2\cdot 1}$$

$$x = \frac{-2 \pm \sqrt{4 + 60}}{2} = \frac{-2 \pm 8}{2}$$

$$x = 3,\ -5$$

20. If you plan to continue your study of mathematics you would do well to review the quadratic formula from time to time to make sure you can remember it. You will find frequent use for it and also for the ideas that derive from its use.

To help you gain a little more speed and comfort in using the quadratic formula, here are a few additional problems. Don't be concerned with whether or not they could be solved by factoring. Use the quadratic formula, then verify your results by factoring the equation, if it is possible. Be sure to check your answers by substituting them in the original equation. Write the quadratic formula down before starting your work so that you will have it in front of you for reference at all times.

(a) $x^2 - 10x + 21 = 0$

(b) $x^2 + 2x - 3 = 0$

(c) $a^2 - 25 = 0$ (Hint: $b = 0$)

(d) $2k^2 - 22k + 60 = 0$ (Hint: Divide through by 2 first.)

(e) $y^2 - y - \frac{3}{4} = 0$

(f) $3x^2 - 5x - 2 = 0$ (Hint: $a = 3,\ b = -5,\ c = -2$.)

(a) $x = \frac{-(-10) \pm \sqrt{(10)^2 - 4\cdot 1\cdot 21}}{2} = \frac{10 \pm 4}{2}$, or $x = 7,\ 3$

(b) $x = \frac{-(2) \pm \sqrt{2^2 - (4)(1)(-3)}}{2} = \frac{-2 \pm 4}{2}$, or $x = 1,\ -3$

(c) $a = \frac{0 \pm \sqrt{0 - (4)(1)(-25}}{2} = \frac{\pm 10}{2}$, or $a = 5,\ -5$

(d) $k = \frac{-(-11) \pm \sqrt{121 - (4)(1)(30)}}{2} = \frac{11 \pm 1}{2}$, or $k = 5,\ 6$

(e) $y = \frac{1 \pm \sqrt{1-(4)(1)(-\frac{3}{4})}}{2} = \frac{1 \pm 2}{2}$, or $y = \frac{3}{2}, -\frac{1}{2}$

(f) $y = \frac{5 \pm \sqrt{25-(4)(3)(-2)}}{6} = \frac{5 \pm 7}{6}$, or $x = 2, -\frac{1}{3}$

SOLUTION BY GRAPHING*

21. Now it is time to discuss the last method of solving quadratic equations–graphing. In Chapter 8 we found that to be able to plot a curve on a rectangular coordinate system we needed a pair of coordinates to locate any single point on the curve. These coordinates are the x coordinate (abscissa) and y coordinate (ordinate).

Suppose we have the equation $x^2 - 2x - 3 = 0$ and wish to draw a graph in order to find its roots (that is, the values of x that will make the equation a true statement). The value of the left member, $x^2 - 2x - 3$, depends upon the value of x. Let's use ordered pairs to help us. As different values are assigned to x, the expression $x^2 - 2x - 3$ takes on different values. These may be thought of as values of y (the ordinate) for the corresponding values of x (the abscissa). We prepare a table of x and y values (just as you learned to do in Chapter 8) assigning various values to x and finding the corresponding values of y.

If x =	4	3	2	1	0	−1	−2
Then y (or $x^2 - 2x - 3$) =	5	0	−3	−4	−3	0	5

Plotting these pairs of coordinates on graph paper we get the curve shown below.

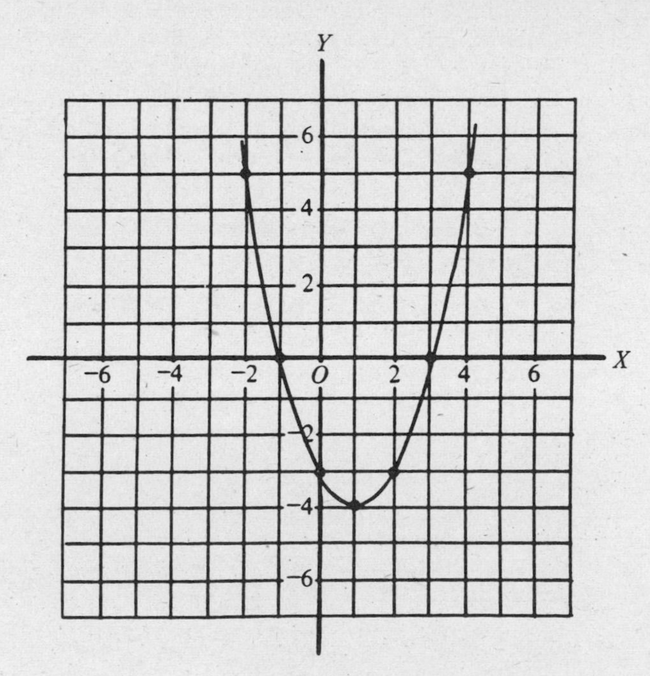

*You will need a few sheets of graph paper on which to graph solutions throughout this chapter.

We analyze the results of the above graph as follows.

1. Our original equation was $x^2 - 2x - 3 = 0$.
2. The solution or roots of the equation are those values of x that will make the equation a true statement (that is, that will make the left member become zero).
3. Since $y = x^2 - 2x - 3$, the value of x that will make the expression $x^2 - 2x - 3$ zero will also make y zero.

Think carefully about the above statements and then see if you can tell from the graph what the roots of the equation are. The roots of the equation are $x =$ ______ and $x =$ ______.

3, −1

22. You could have found the roots of the equation in the last frame by factoring. Hopefully you did not fall back on the factoring method. What you should have observed is that at the two points on the curve where $y = 0$, the curve crosses the X axis. Hence the roots of the equation are simply the values of x at these two points. You could say, then, that the roots represent the values of the two points at which the lines $y = 0$ and $y = x^2 - 2x - 3$ intersect. This is true of all quadratic equations that have real solutions.

Incidentally, the curve plotted in frame 21 is known as a *parabola.* There are many important applications of the parabola in engineering and science. The supporting cables of suspension bridges are in the shape of a parabola and the path of a projectile fired from a gun is also a parabola (or would be if there were no air resistance.)

23. Let's summarize what we know about the solution of a quadratic equation by graphing.

(1) A quadratic equation that has real roots will have either two unequal roots or one root with a multiplicity of two.

(2) Graphically, the real roots are the values of x where the curve crosses the x axis or, expressed differently, the two points of intersection of $y = 0$ (the X axis) and the curve $y = ax^2 + bx + c$.

(3) To solve a quadratic equation by graphing:
(a) Write the equation in the form $ax^2 + bx + c = 0$.
(b) Plot the curve $y = ax^2 + bx + c$.
(c) Find where the curve crosses the X axis. The values of x at these points are the roots of the equation $ax^2 + bx + c = 0$.

(4) (a) The roots are real and unequal if there are two points of intersection.
(b) The roots are equal (that is, they occur with a multiplicity of two) if there is a point of tangency.
(c) There are no real roots if there is no intersection.

Shown below are examples of the situations discussed in 4b and 4c above.

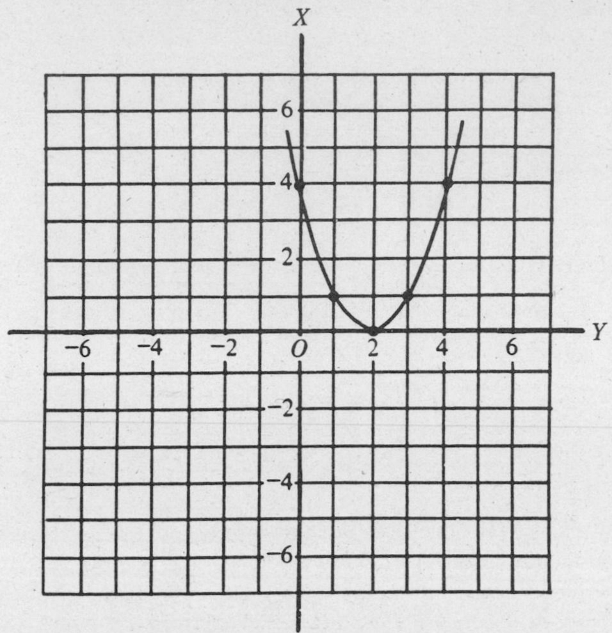

Curve is tangent to (that is, just touching) the X axis. Roots are $x = 2, 2$

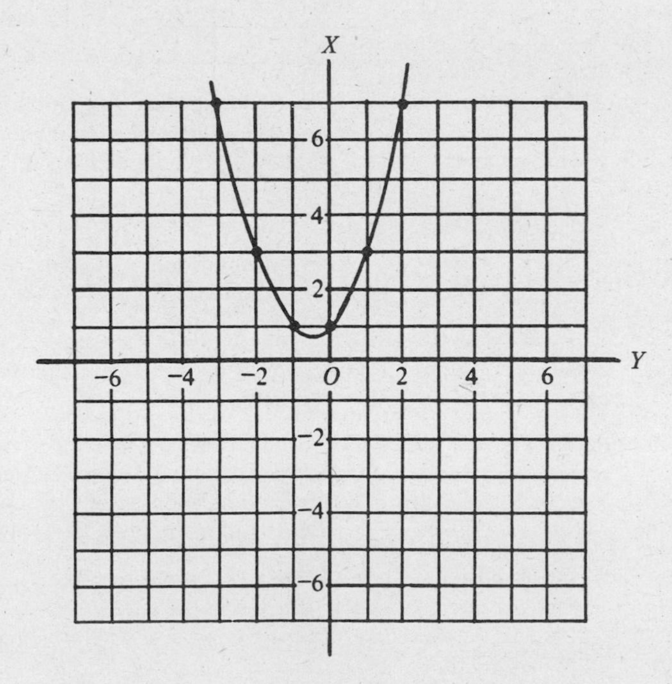

Curve fails to intersect the X axis. Roots (by quadratic formula) are

$$x = \frac{-1 \pm \sqrt{-3}}{2}$$

(No real roots.)

On a separate sheet of graph paper, use what you have learned so far to solve the following quadratic equation graphically.

$x^2 - 5x + 4 = 0$

As you can see from the graph of the curve, $x = 1$ and $x = 4$ are the roots of the equation.

$y = x^2 - 5x + 4 = 0$

x	5	4	3	2	1	0
y	4	0	−2	−2	0	4

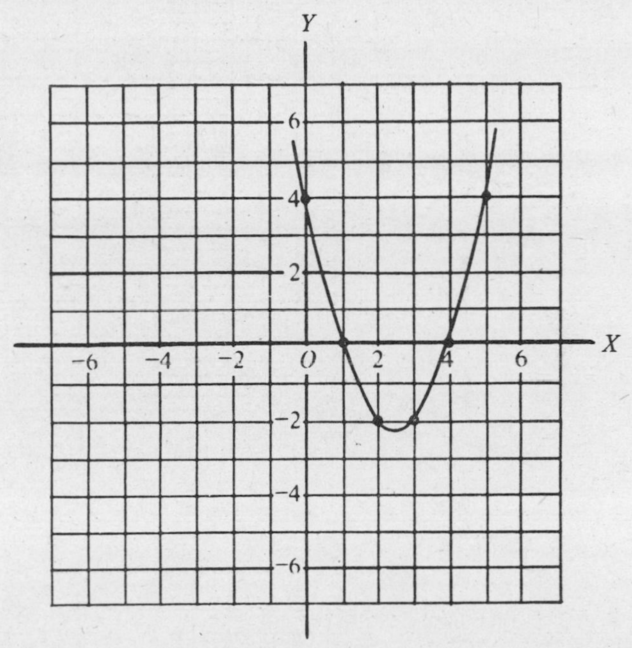

24. In each instance where you are seeking a solution by graphing, when you have plotted enough points to establish the shape of the parabola and to make sure it has crossed the X axis twice (or dipped toward it and turned back), draw a smooth curve through the points plotted. This is important because you may have to estimate the values of the abscissas (roots) from the curve you have drawn. The more accurate your drawing, the more accurate the value of the roots. If you are interested in learning more about the characteristics and uses of such curves as the parabola, the ellipse, and the hyperbola, you should read *Geometry and Trigonometry for Calculus* (another Wiley Self-Teaching Guide) or some good text on plane analytic geometry.

    Now it is time for you to gain more skill in graphing parabolas. Solve the following quadratic equations graphically.

    (a) $x^2 - 4 = 0$

    (b) $x^2 - 3x = 0$

    (c) $x^2 - 3x - 4 = 0$

    (d) $x^2 - 6x + 9 = 0$

    (e) $x^2 + 4 = 0$

(a) $x = 2,\ x = -2$

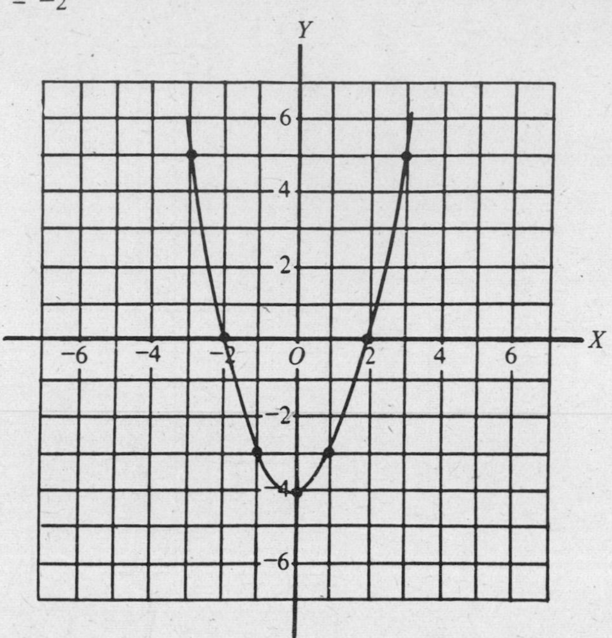

(b) $x = 0,\ x = 3$

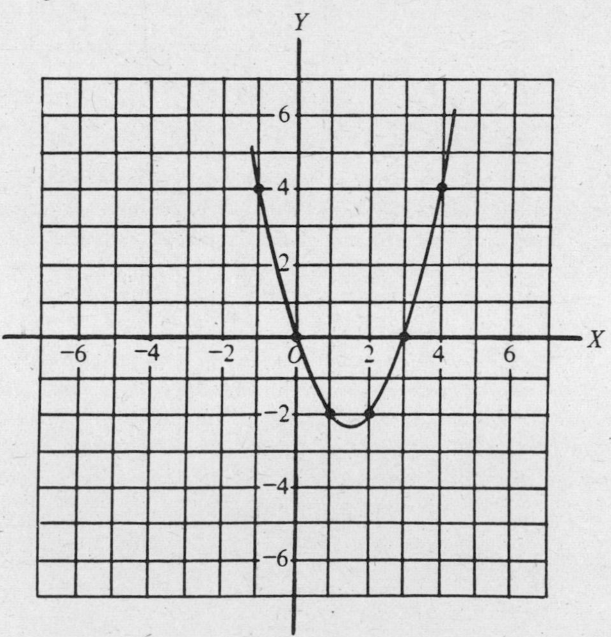

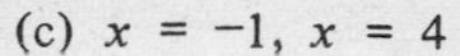
(c) $x = -1$, $x = 4$

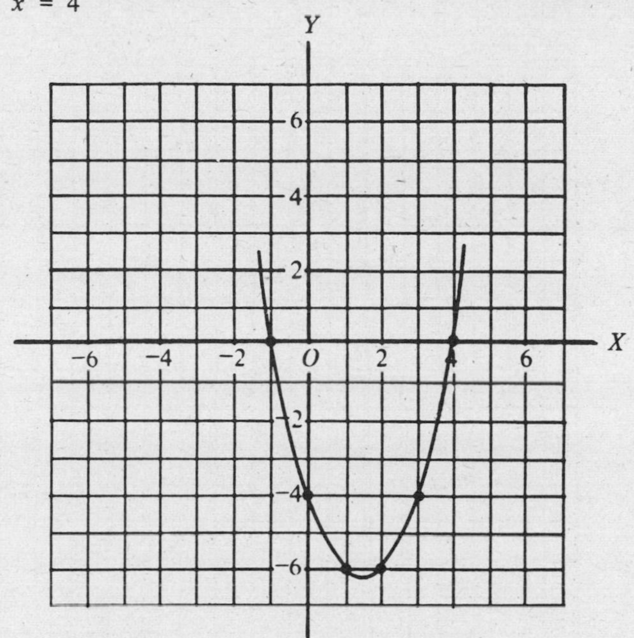

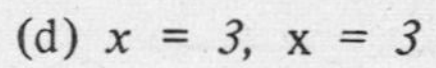
(d) $x = 3$, $x = 3$

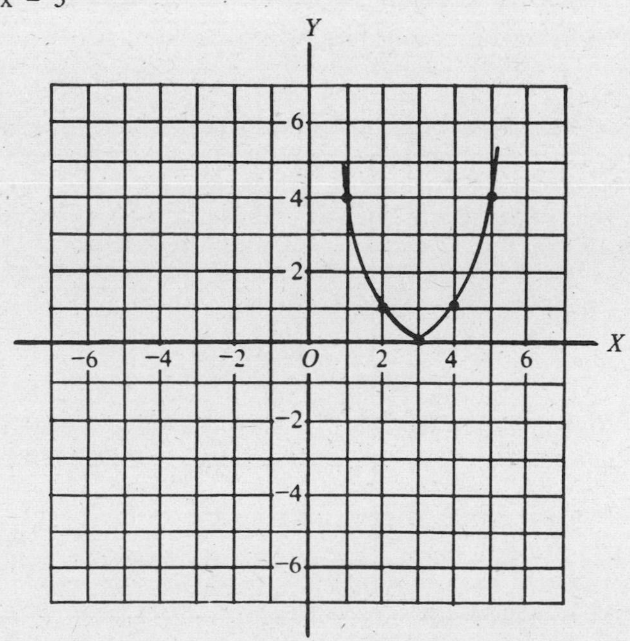

(e) no real roots

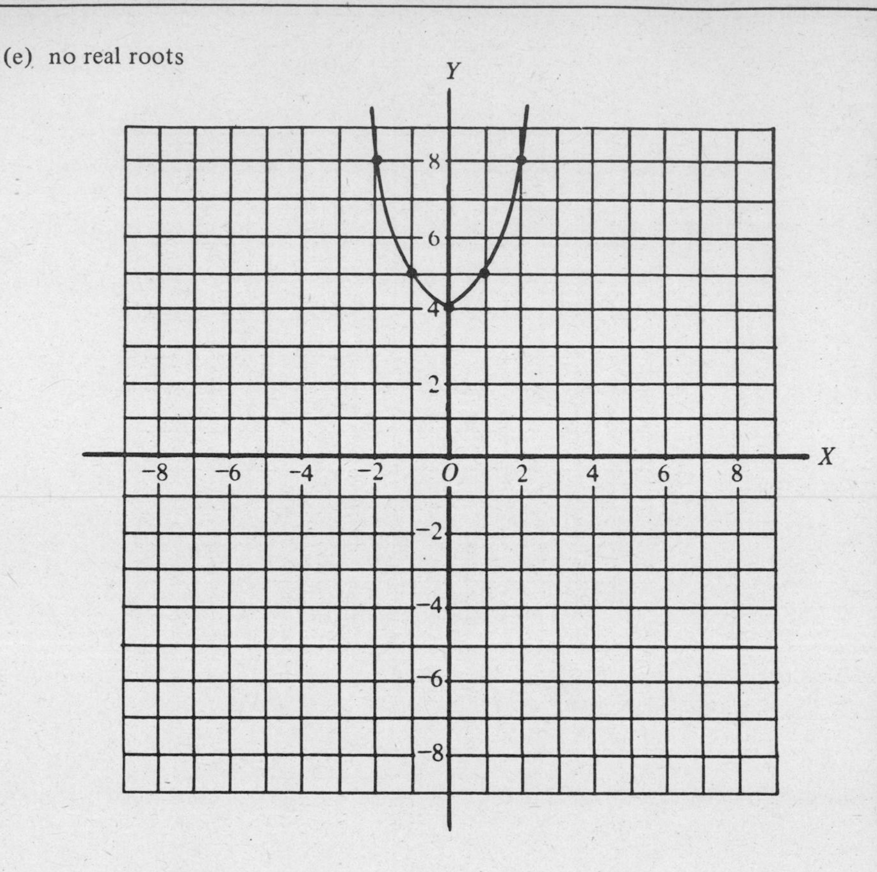

We have now covered four ways of solving quadratic equations: by factoring (including the method of extraction of roots), by completing the square, by using the quadratic formula, and by graphing. Since practice is the secret of assurance with quadratic equations, be sure you can do the Self-Test that follows before taking Review Test 3 or going on to Chapter 10.

SELF-TEST

1. Write the following quadratic equations in standard form and factor the left membe

   (a) $y^2 + 2y = 15$ (d) $x^2 - 2x = 24$

   (b) $x^2 + 6x + 9 = 0$ (e) $x^2 - 6x = 0$

   (c) $a^2 - 2a - 35 = 0$ (f) $b^2 + 2b = 3$ (frame 5

2. Solve for the roots of each of the equations in problem 1.

   (a) $y =$ ________, $y =$ ________ (d) $x =$ ________, $x =$ ________

   (b) $x =$ ________, $x =$ ________ (e) $x =$ ________, $x =$ ________

   (c) $a =$ ________, $a =$ ________ (f) $b =$ ________, $b =$ ________

   (frame 5

3. Check your solutions in problem 2 by substituting the roots back in the original equations in problem 1. (frame 5)

4. Solve and check the following equations.

(a) $x^2 = 9$

(b) $y^2 - 1 = 0$

(c) $\frac{a^2 - 4}{4} = 3$

(d) $\frac{3x^2}{4} - 3 = 6$

(e) $k^2 - 3k = 0$

(frame 7)

5. Write the following quadratic equations in standard form.

(a) $x(x + 3) = 10$

(b) $5a^2 = 125$

(c) $c^2 - 9c = 10$

(d) $2x^2 + 9x = 2x - 3$

(e) $5(y^2 + 2) = 7(y + 3)$

(f) $x - 5 = \frac{7}{x}$

(frame 12)

6. Square the following binomials.

(a) $(x - 4)^2$ = ______________________

(b) $(y + 7)^2$ = ______________________

(c) $(c - 11)^2$ = ______________________

(d) $(d - 5)^2$ = ______________________ (frame 13)

7. Find the principal square root of the following trinomials.

(a) $t^2 - 16t + 64$

(b) $x^2 + 18x + 81$

(c) $y^2 - 3y + \frac{9}{4}$

(d) $36 + x^2 - 12x$

(e) $25 + 10b + b^2$

(f) $c^2 - 26c + 169$

(frame 13)

8. Solve by completing the square.

(a) $a^2 + 12a = 45$

(b) $x^2 - 18x = -65$

(c) $x^2 + 7x = 8$

(d) $z^2 - 11z = -28$

(frame 14)

9. Use the quadratic formula to solve the following equations. Check your answers.

(a) $x^2 - 2x = 24$

(b) $2x^2 + 3x - 2 = 0$

(c) $y^2 + 5y - 36 = 0$

(d) $4x^2 - x - 5 = 0$

(frames 19 and 20)

0. Solve the following quadratic equations graphically and check your results.

(a) $x^2 - x - 2 = 0$

(b) $x^2 - 4 = 0$

(c) $x^2 + 1 = 0$

(d) $x^2 + 2x + 1 = 0$

(frames 21 and 24)

Answers to Self-Test

1. (a) $(y+5)(y-3) = 0$; (b) $(x+3)(x+3) = 0$; (c) $(a-7)(a+5) = 0$; (d) $(x-6)(x+4) = 0$; (e) $x(x-6) = 0$; (f) $(b+3)(b-1) = 0$
2. (a) $y = -5$, $y = 3$; (b) $x = -3$, $x = -3$ (Notice that the root has a multiplici of two.); (c) $a = 7$, $a = -5$; (d) $x = 6$, $x = -4$; (e) $x = 0$, $x = 6$; (f) $b = -3$, $b = 1$
3. (a) $(-5)^2 + 2(-5) = 15$, $25 - 10 = 15$, $15 = 15$
   $3^2 + 2(3) = 15$, $9 + 6 = 15$, $15 = 15$
   (b) $(-3)^2 + 6(-3) + 9 = 9 - 18 + 9 = 0$, $0 = 0$
   (c) $7^2 - 2 \cdot 7 - 35 = 49 - 14 - 35 = 0$, $0 = 0$
   $(-5)^2 - 2(-5) - 35 = 25 + 10 - 35 = 0$, $0 = 0$
   (d) $6^2 - 2(6) = 24$, $36 - 12 = 24$, $24 = 24$
   $(-4)^2 - 2(-4) = 24$, $16 + 8 = 24$, $24 = 24$
   (e) $0 - 0 = 0$, $0 = 0$
   $6^2 - 6 \cdot 6 = 0$, $36 - 36 = 0$, $0 = 0$
   (f) $(-3)^2 + 2(-3) = 3$, $9 - 6 = 3$, $3 = 3$
   $1^2 + 2(1) = 3$, $1 + 2 = 3$, $3 = 3$
4. (a) $z = \pm 3$; check: $(\pm 3)^2 = 9$; $9 = 9$
   (b) $y = \pm 1$; check: $(\pm 1)^2 - 1 = 0$; $0 = 0$
   (c) $a = \pm 4$; check: $\dfrac{(\pm 4)^2 - 4}{4} = 3$; $\dfrac{16-4}{4} = 3$; $3 = 3$
   (d) $x = \pm 2\sqrt{3}$; check: $\dfrac{3(\pm 2\sqrt{3})^2}{4} - 3 = 6$; $\dfrac{3(12)}{4} - 3 = 6$; $6 = 6$
   (e) $k = 0$, $k = 3$; check: $0 - 0 = 0$; $9 - 9 = 0$
5. (a) $x^2 + 3x - 10 = 0$; (b) $5a^2 - 125 = 0$; (c) $c^2 - 9c - 10 = 0$; (d) $2x^2 + 7x + 3 = 0$; (e) $5y^2 - 7y - 11 = 0$; (f) $x^2 - 5x - 7 = 0$ $(x \neq 0)$
6. (a) $x^2 - 8x + 16$; (b) $y^2 + 14y + 49$; (c) $c^2 - 22c + 121$; (d) $d^2 - 10d + 25$
7. (a) $(t-8)$; (b) $(x+9)$; (c) $\left(y - \dfrac{3}{2}\right)$; (d) $(x-6)$; (e) $(b+5)$ or $(5+b)$; (f) $(c-13)$
8. (a) $a^2 + 12a + 36 = 81$; $a = 3$, $a = -15$
   (b) $x^2 - 18x + 81 = -65 + 81$; $x = 5$, $x = 13$
   (c) $x^2 + 7x + \dfrac{49}{4} = 8 + \dfrac{49}{4}$; $x = 1$, $x = -8$
   (d) $z^2 - 11z + \dfrac{121}{4} = -28 + \dfrac{121}{4}$; $z = 7$, $z = 4$
9. (a) $x = \dfrac{-(-2) \pm \sqrt{(-2)^2 - (4)(1)(-24)}}{2 \cdot 1} = \dfrac{2 \pm 10}{2}$; $x = 6, -4$
   (b) $x = \dfrac{-3 \pm \sqrt{3^2 - (4)(2)(-2)}}{2 \cdot 2} = \dfrac{-3 \pm 5}{4}$; $x = \dfrac{1}{2}, -2$
   (c) $y = \dfrac{-5 \pm \sqrt{5^2 - (4)(1)(-36)}}{2 \cdot 1} = \dfrac{-5 \pm 13}{2}$; $y = 4, -9$
   (d) $x = \dfrac{-(-1) \pm \sqrt{(-1)^2 - (4)(4)(-5)}}{2 \cdot 4} = \dfrac{1 \pm 9}{8}$; $x = \dfrac{5}{4}$; -1

10. (a) $y = x^2 - x - 2$

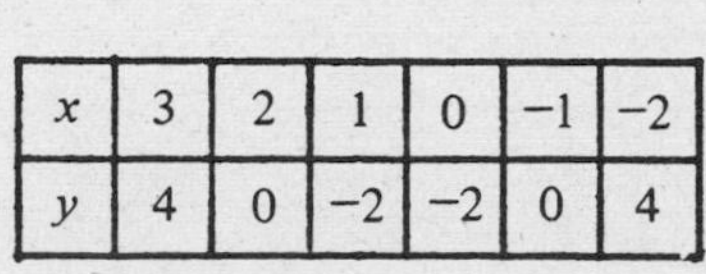

x	3	2	1	0	−1	−2
y	4	0	−2	−2	0	4

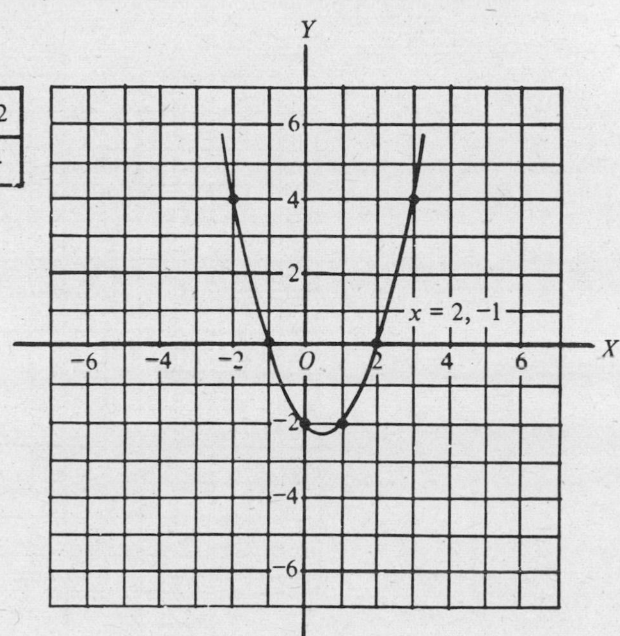

(b) $y = x^2 - 4$

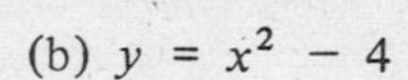

x	3	2	1	0	−1	−2	−3
y	5	0	−3	−4	−3	0	−5

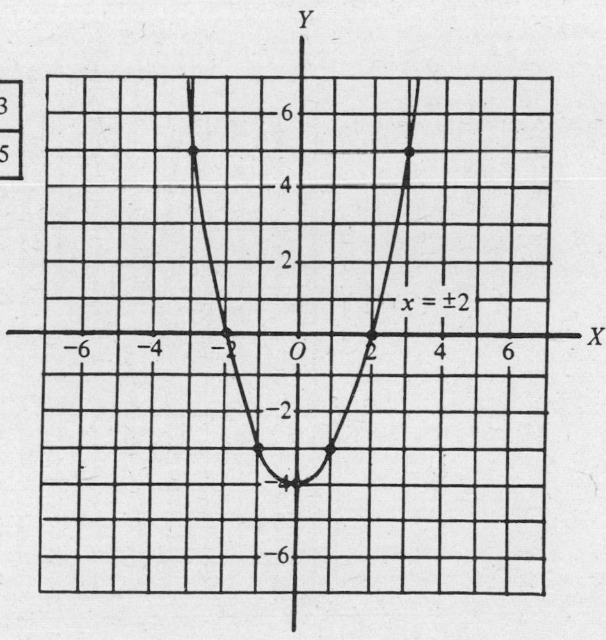

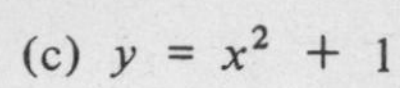

(c) $y = x^2 + 1$

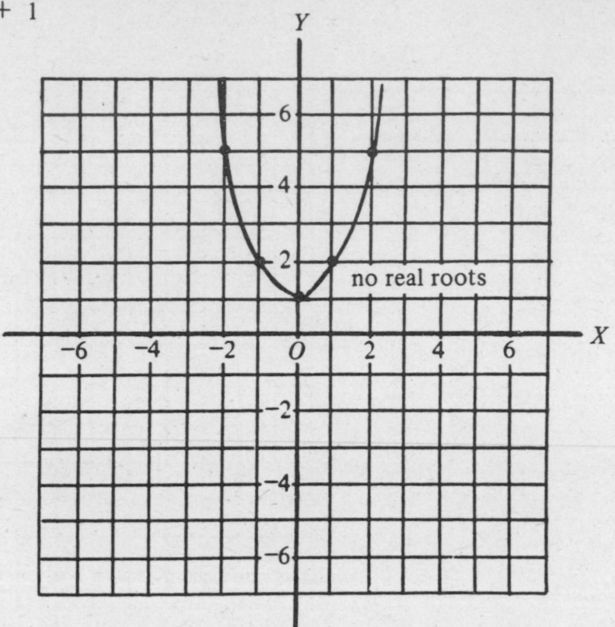

(d) $y = x^2 + 2x + 1$

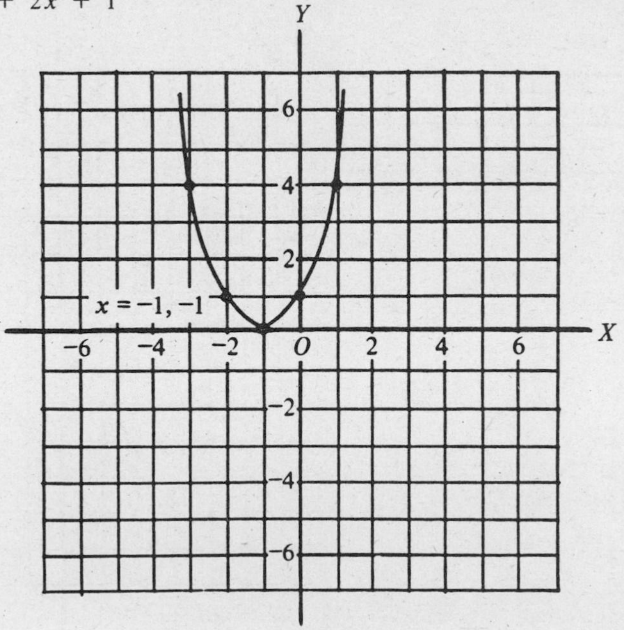

REVIEW TEST 3

Chapter 7

1. An equation is a statement in mathematical short and that two expressions are equal. (True / False)

2. Solving an equation means finding a value of the unknown that satisfies the equation. (True / False)

3. Solve $2b - 3 = 5$ for b. ________

4. Solve $\frac{y}{4} = 3$ for y. ________

5. Solve $2c = \frac{a}{b}$ for c. ________

6. Solve $\frac{xy}{a} = b$ for x. ________

7. Find the root of $4x + 3 = 19$. ________

8. If $k = c + m$, what does c equal? ____________________

9. Solve $\frac{3}{x} = \frac{1}{2}$ for x. ______________

10. A factor may be eliminated from one side of an equation by dividing both members of the equation by that factor. (True / False)

11. Find the value of y in the equation $\frac{3y}{4} = \frac{15}{2}$.

12. $(\frac{2}{3})q$ and $\frac{2q}{3}$ are equivalent terms. (True / False)

13. Solve $4 - 2b = 8$ for b. ______________

14. Solve and check: $7a + 28 = 5a + 6$.

15. Solve and check: $(3y + 2) - (2y - 12) = 18$.

16. Solve and check: $3(b - 5) - 2(3b + 1) = 13$.

17. Solve and check: $(b - 2)(b + 4) = 40 - 7b + b(b - 3)$.

18. Solve $\frac{3}{4x} = \frac{1}{x} - \frac{1}{4}$ for x. ______________

19. Solve $\frac{8}{a-2} - \frac{13}{2} = \frac{3}{2(a-2)}$ for a. ______

20. Find v if $k = \frac{mv^2}{2}$ and $k = 24$, $m = 3$. ______

Chapter 8

21. In a rectangular coordinate system, the x coordinate is called the ______

22. Similarly, the y coordinate is called the ______.

Note: To assist you in answering the next few questions, draw the X and Y axes of a rectangular coordinate system on a piece of graph paper.

23. The sign of the abscissas in the second quadrant is positive. (True / False)

24. In which quadrant are the signs of both abscissas and ordinates negative?

    ______.

25. Plot the two points (−3, 5) and (5, −3) and draw a straight line between them. Through what value on the Y axis does your line pass? ______

26. A set of coordinates used to locate a point (like those in question 25) is called an

    ______.

27. When the values of two quantities can vary depending upon the values assigned either or both of them, they are called ______.

28. In the relationship $y = x^2$ we say that y is a ______ of x.

29. In question 28, since the value of y is determined by x, we say that x is the ______ variable.

30. A linear equation takes its name from the fact that its graph is a ______ line.

31. A characteristic of a linear equation is that it contains no term with degree greater than ______.

32. Plot the equation $y = 3x - 6$. At what point does it cut the X axis? ______

33. The point at which a curve cuts an axis is called an ______.

34. Find the x and y intercepts of the equation $y + 3x - 9 = 0$. $x =$ ______, $y =$ ______

35. What are the x and y intercepts of the equation $2x - y = 6$? $x =$ ________,
$y =$ ________

36. The x and y intercepts of the equation $8 - x = 4y$ are $x =$ ________ and
$y =$ ________.

37. The coordinates of any point on the graph of an equation satisfy the equation (that is, make it a true statement). (True / False)

38. In order to solve linear equations for the values of the unknown, we must have as many equations as there are unknowns. (True / False)

39. Solve the following pair of equations simultaneously: $x + 3y = 4$ and
$2x - 7y = -5$. $x =$ ________, $y =$ ________

40. Solve the following pair of equations by graphing: $x + 3y = 4$ and
$2x - 7y = -5$. $x =$ ________, $y =$ ________

Chapter 9

41. The standard form of a quadratic equation is ________________.

42. What are the values of a, b, and c in the equation $3x^2 + 4x - 7 = 0$?
$a =$ ________, $b =$ ________, $c =$ ________

43. What are the binomial factors of $x^2 + 4x - 21$? ____________ and
____

44. Factor $3x^2 + x - 2$. ________________

45. Factor $9k^2 - 6k + 1$. ________________

46. Is the trinomial $y^2 + 2y + 1$ a quadratic equation? ________________

47. Rewrite the equation $x^2 + 4x = 4$ in standard form. ________________

48. Solve the quadratic equation $x^2 - 5x + 6 = 0$ by factoring.
____

49. What are the roots of the equation in problem 48? ________________

50. Do the roots $x = 3$, $x = -2$ both satisfy the equation $x^2 - 5x + 6 = 0$? (Yes / No)

51. Factor the incomplete quadratic equation $3x^2 = 12x$ to find the roots.
$x =$ ________, $x =$ ________

52. Solve the incomplete quadratic equation $3x^2 - 27 = 0$ by the method of extraction of roots.

53. Express the following quadratic equation in standard form with the value of x positive.

$$\sqrt{1-x} = x\sqrt{6}$$

54. Solve by completing the square: $a^2 - 6a + 2 = 0$.

55. Write the quadratic formula: $x =$ ______________________________

56. Use the quadratic formula to solve the equation $x^2 - x - \frac{3}{4} = 0$ and check your answers.

57. In the expression $y = x^2$, y is said to be a function of x. (True / False)

58. The curve of the quadratic equation $y = ax^2 + bx + c$ is known as a parabola. (True / False)

59. A quadratic equation whose graph neither touches nor intersects the X axis will have no real roots. (True / False)

60. Solve graphically $x^2 - 9 = 0$.

Answers to Review Test 3

Chapter 7

1. True
2. True
3. 4
4. 12
5. $\frac{a}{2b}$
6. $\frac{ab}{y}$
7. 4
8. $c = k - m$
9. $x = 6$
10. True
11. $y = 10$
12. True
13. $b = -2$
14. $a = -11$; check: $-49 = -49$
15. $y = 4$; check: $18 = 18$
16. $b = -10$; check: $13 = 13$
17. $b = 4$; check: $16 = 16$
18. $x = 1$
19. $a = 3$
20. $v = \sqrt{\frac{2k}{m}} = \sqrt{\frac{2 \cdot 24}{3}} = \sqrt{16} = 4$

Chapter 8

21. abscissa
22. ordinate
23. False
24. the third quadrant
25. $y = 2$
26. ordered pair
27. variables
28. function
29. independent
30. straight
31. one
32. (2, 0)
33. intercept
34. $x = 3$, $y = 9$
35. $x = 3$, $y = -6$
36. $x = 8$, $y = 2$
37. True
38. True

39. Multiplying by −2:

$$\begin{aligned} \text{(a)} \quad -2x - 6y &= -8 \\ \text{(b)} \quad 2x - 7y &= -5 \\ \hline -13y &= -13 \\ y &= 1 \end{aligned}$$

Substituting $y = 1$ in (a):

$$\begin{aligned} x + 3 &= 4 \\ x &= 1 \end{aligned}$$

40.

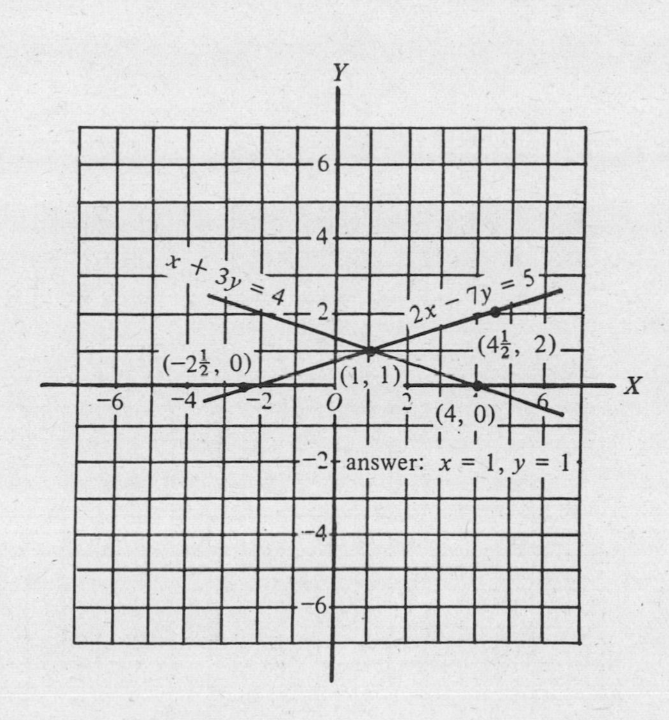

Chapter 9

41. $ax^2 + bx + c = 0$
42. $a = 3,\ b = 4,\ c = -7$
43. $(x + 7)$ and $(x - 3)$
44. $(3x - 2)(x + 1)$
45. $(3k - 1)(3k - 1)$
46. No; because it contains no statement of equality (equal sign). It is simply a quadratic expression.
47. $x^2 + 4x - 4 = 0$
48. $(x - 3)(x - 2) = 0$; the solution is $x = 3$, $x = 2$.
49. 3 and 2
50. No; $x = 3$ does but $x = -2$ does not.
51. $x = 0,\ x = 4$

52. $3x^2 = 27;\ x^2 = 9;\ x = +3, -3$
53. Squaring both sides we get $1 - x = 6x^2$ or $6x^2 + x - 1 = 0$.
54. $a = 3 \pm \sqrt{7}$
55. $x = \dfrac{-b \pm \sqrt{b^2 - 4ac}}{2a}$
56. $x = \dfrac{3}{2}, -\dfrac{1}{2}$
57. True
58. True
59. True
60. $x^2 - 9 = 0$
    $y = x^2 - 9$
    $x = -3,\ x = +3$

x	4	3	2	1	0	−1	−2	−3	−4
y	7	0	−5	−8	−9	−8	−5	0	7

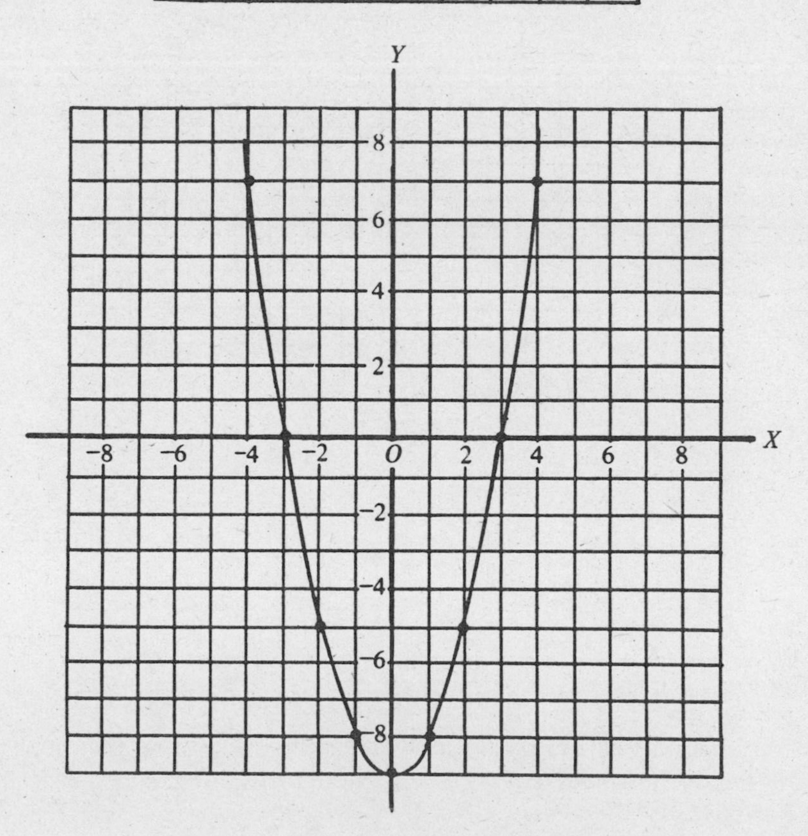

CHAPTER TEN

Inequalities

OBJECTIVES

In the last three chapters we worked extensively with *equalities,* that is, expressions that are equal to one another. We found that there are a variety of ways to find the values of the variables that make equalities true statements. Now we are going to consider ways to work with (solve) expressions that are not equal to each other. We call such expressions *inequalities.* The idea of inequality applies only to real numbers.

You may well wonder why we bother with inequalities. In mathematics we frequently have to deal with situations in which it is necessary to determine values of the variable in which one expression is greater than or less than another. It is possible to put statements of inequality into symbolic form and, by processes that are in many respects similar to those used in solving equations, to find the conditions under which a statement of inequality is true.

In this chapter you are going to learn something about how to put statements of inequality into symbolic form, how to graph such statements, and how to find the values that make statements of inequality true statements. When you have reached the end of this chapter you will be able to:

- tell whether an inequality is true or false;
- perform the fundamental operations of addition, subtraction, multiplication, and division on inequalities;
- simplify inequalities;
- distinguish between an absolute and a conditional statement of inequality;
- represent inequalities graphically on a number line;
- graph inequalities of two unknowns on a rectangular coordinate system.

SOME BASIC CONCEPTS

1. We will need a few definitions relating to inequalities to assist us as we go along. Let's start with this one.

   Definition 1: An inequality is a statement that says that one expression is greater than or less than another expression.

   It is important that you accept the idea that equations and inequalities are either true statements or false statements. They cannot be both at the same time. One of our primary concerns with inequalities (as with equations) is to find the value or values that make the statement concerned a true statement. Finding the set of values that makes the statement true is (as you already know from our work with equations) called *solving* the inequality.

In Chapter 2 we introduced the concept of number scales or number lines to help explain positive and negative numbers. Thus, in the horizontal number scale below, positive numbers are defined as those to the right of zero, and negative numbers as those to the left of zero.

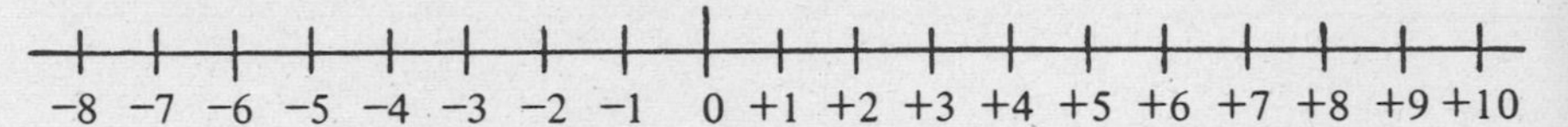

Furthermore, any number to the right of another number on the number scale is the greater of the two numbers. The symbol that means "greater than" is $>$. Looking at our number line above we can say that $+4 > +1$ (meaning that positive four is greater than positive one) or that $+2 > -6$ (meaning that positive 2 is greater than negative 6).

Conversely, the symbol $<$ means "less than." Thus, the two statements above could have been expressed as $+1 < +4$ (which says that positive 1 is less than positive 4) and $-6 < +2$ (meaning that negative 6 is less than positive 2).

See if you can tell whether each of the following statements of inequality is true or false. Use the number scale above to assist you.

(a) $+5 > +4$ __________

(b) $+1 > 0$ __________

(c) $-4 < -6$ __________

(d) $-1 < +1$ __________

(e) $-5 < -1$ __________

- - - - - - - - - - - - - - - - - -

(a) true; (b) true; (c) false; (d) true; (e) true

2. Did you notice that in a true statement the vertex (point) of the inequality symbol always points toward the smaller quantity (and remember that by "smaller" we mean farther to the left on the number scale). With this in mind, review the frame 1 problems. Here they are again. (Use the number scale in frame 1 if necessary.)

$+5 > +4$	a true statement
$+1 > 0$	a true statement
$-4 < -6$	not a true statement (Note that the vertex does not point to the smaller number.)
$-1 < +1$	a true statement
$-5 < -1$	a true statement

If we let the letters a and b represent any two numbers, we can generalize the ideas we have been working with and discuss inequalities without a number scale. Consequently, we define the relationships *greater than* and *less than* as follows.

Definition 2: For any two real numbers a and b, the statement $a > b$ is equivalent to the statement "$a - b$ is positive." Similarly, $a < b$ is equivalent to the statement "$a - b$ is negative."

We can observe that $a - b$ is either positive, zero, or negative. Another way to say the same thing is to note that given two real numbers a and b, only one of the following is true:

$$a > b \qquad a = b \qquad a < b$$

This last statement is known as the *trichotomy principle* (just for the record).

Bearing in mind that the vertex of the inequality symbol always points to the smaller value and without reference to the number scale in frame 1, insert the correct inequality symbol in thc problems below.

(a) -3 ____ -2 (d) $+3$ ____ $+4$

(b) $+1$ ____ -6 (e) 0 ____ $+1$

(c) -1 ____ $+6$ (f) -8 ____ 0

- - - - - - - - - - - - - - - - - -

(a) $-3 < -2$; (b) $+1 > -6$; (c) $-1 < +6$; (d) $+3 < +4$; (e) $0 < +1$; (f) $-8 < 0$

PROPERTIES OF INEQUALITIES

3. Now we need to discuss some of the properties of inequalities so that you can begin to develop some "feel" for the way they can be handled, just as you already have developed such a "feel" for equations.

For example, if $a > b$ and $c > d$, we say that these inequalities have the *same sense* since the signs of inequality point in the same direction. On the other hand, if $a > b$ and $c < d$, we say that the inequalities have an *opposite sense.* This terminology helps us talk about some of the things that happen when the familiar operations of addition, subtraction, multiplication, and division are performed on inequalities.

Property 1: The sense of an inequality is unchanged if the same number is added to or subtracted from both sides of the inequality. For example,

$$4 > 3$$
$$4 + 2 > 3 + 2$$
$$4 - 2 > 3 - 2$$

In general, if $a > b$ then $a + c > b + c$ for any number c.

Property 2: The sense of an inequality is unchanged if both sides of the inequality are multiplied or divided by the same positive number. For example,

$$4 > 3$$
$$4 \cdot 2 > 3 \cdot 2$$
$$4 \div 2 > 3 \div 2$$

In general, if $a > b$ then $a \cdot c > b \cdot c$ if $c > 0$.

Property 3: The sense of an inequality is changed if both sides of the inequality are multiplied or divided by the same negative number. For example,

$$4 > 3$$
$$4(-1) < 3(-1)$$
$$\frac{4}{-1} < \frac{3}{-1}$$

In general, if $a > b$ then $a \cdot c < b \cdot c$ if $c < 0$.

Insert the correct inequality symbols in the following.

(a) $7 + 2$ ____ $5 + 2$

(b) $8 \cdot 3$ ____ $3 \cdot 3$

(c) $12 - 5$ ____ $6 - 5$

(d) $\frac{6}{2}$ ____ $\frac{4}{2}$

(e) $9(-2)$ ____ $4(-2)$

(f) $\frac{11}{-3}$ ____ $\frac{7}{-3}$

- -

(a) $>$; (b) $>$; (c) $>$; (d) $>$; (e) $<$; (f) $<$

4. The second and third properties taken together tell us what happens when both sides of an inequality are multiplied (or divided) by the same numbers. If the multiplier is positive, the inequality sign stays the same; if the multiplier is negative, the direction of the inequality symbol must be reversed.

Because the failure to remember this is such a common source of error in working with inequalities, you should work a few more problems in order to make sure you are aware of this point. In the problems below you are given an inequality and a multiplier. Multiply both sides of the inequality by the multiplier to arrive at a new inequality.

	original inequality	multiplier	new inequality
(a)	$8 > 7$	2	____
(b)	$2 > -1$	-3	____
(c)	$11 > 5$	1	____
(d)	$5 < 7$	5	____
(e)	$12 < 17$	-1	____
(f)	$-8 < 7$	2	____

- -

(a) $16 > 14$; (b) $-6 < 3$; (c) $11 > 5$; (d) $25 < 35$;
(e) $-12 > -17$; (f) $-16 < 14$

WORKING WITH INEQUALITIES

5. The three properties discussed in frame 3 allow us to work with inequalities in the same way we work with equations. We can use these properties to change the form of the inequality until we can see the value (or range of values) of the variable that will make the statement of inequality a true statement. Thus, if we are given the inequality $6x - 4 > 3x + 2$ and wish to find the values that make the statement true, we proceed as follows:

Given:	$6x - 4 > 3x + 2$
Subtracting $3x$ from both members and adding 4 to both members:	$6x - 3x > 2 + 4$
Simplifying both members:	$3x > 6$
Finally, dividing both members by positive 3 (the sense of the inequality remains unchanged):	$x > 2$

The above solution to the inequality tells us that the original statement will be a true statement if the variable x is replaced by any number greater than positive 2. To check this we select any number greater than 2 and replace the variable with that value. If we select the number 3, for example, we would get 14 in the left member and 11 in the right. Since $14 > 11$, the statement is true. If we were to draw a graph of the solution on a number scale it would appear as follows:

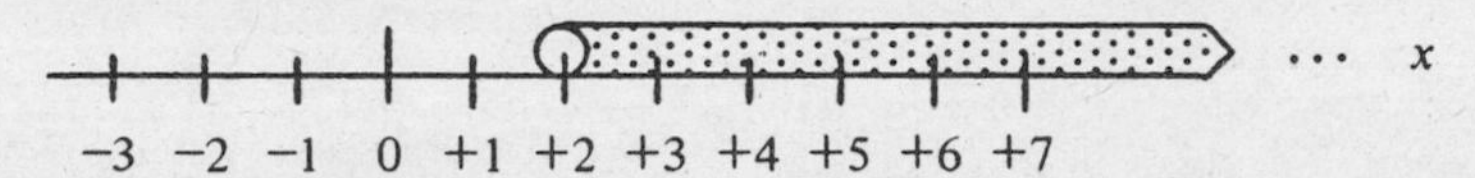

The solution, then, would be the range of values greater than 2. Notice that the hollow circle appears at the number 2 to indicate that the inequality is not true when the value of the variable is 2.

Work the following problems in the same step-by-step manner shown in the example. (You need not draw the graph unless you wish, although it might help to do so.)

(a) $7x - 5 > 3x + 4$

(b) $9x + 2 < 4x - 3$

(c) $5 + 4y > 2y + 1$

(d) $6 + 2y < 5y + 9$

(e) $2y - 7 < y + 3$

(a) $x > \frac{9}{4}$; (b) $x < -1$; (c) $y > -2$; (d) $y > -1$; (e) $y < 10$

6. To emphasize the similar way equalities (equations) and inequalities can be solved, compare the following two solutions:

Equation	Inequality
$x - 5 = 3$	$x - 5 > 3$
$(x-5) + 5 = 3 + 5$	$(x-5) + 5 > 3 + 5$
$x = 8$	$x > 8$

In solving the equation we used the fact that both $x - 5$ and 3 are names for the same number, hence $(x - 5) + 5$ and $3 + 5$ are two ways of writing the same sum. In solving the inequality we could have used our definition of the symbol $>$ (frame 2) to interpret $x - 5 > 3$ to mean $x - 5 - 3$ is a positive number. Then $x - 5 - 3$ is $x - 8$, and since $x - 8$ is positive, $x > 8$.

> *Once again:* For any two real numbers a and b, the statement $a > b$ is equivalent to the statement "$a - b$ is positive." Similarly, $a < b$ is equivalent to the statement "$a - b$ is negative."

TYPES OF INEQUALITIES

7. We are now ready to consider the three basic types of inequalities. They are classed as *absolute, conditional,* and *inconsistent.* Classification of each type derives from the special characteristics of the set (group) of replacement values for each.

> *Definition 3:* An *absolute inequality* is one that is true for all values of the letters involved.

For example, $(a - b)^2 > 0$ (with $a \neq b$) is an example of an absolute inequality. Regardless of whether $a > b$ or $a < b$, the square of the difference $(a - b)$ will be a positive number and therefore greater than zero.

> *Definition 4:* A *conditional inequality* is one that is true only for certain values of the letters involved.

For example, $x - 3 > 0$ is an example of a conditional inequality, for if x is a negative number the relationship is not true. Or, if x is positive but less than 3 the relationship is not true. Even if $x = 3$ it still is not true. But for all values of x greater than 3, $x - 3$ is a positive number and therefore greater than zero.

> *Definition 5:* An *inconsistent inequality* is one which is not a true statement for any replacement value for the variable.

For example, $x^2 < 0$ is not true for any real number because the square of any real number is greater than or equal to zero.

Practice using the above definitions by indicating whether each of the following inequalities is absolute, inconsistent, or conditional.

(a) $a^2 - b^2 > 0$ ____________________

(b) $a^2 > 0$ $(a \neq 0)$ ____________________

(c) $x^2 + 1 > 0$ ____________________

(d) $2x + 3 > 5x - 9$ ____________________

(e) $(x + 3)^2 < 0$ ____________________

- - - - - - - - - - - - - - - - - -

(a) conditional; (b) absolute; (c) absolute; (d) conditional; (e) inconsistent

SOLVING INEQUALITIES OF ONE VARIABLE

8. You will recall that inequalities have solutions. Thus, if $x - 3 > 0$, then (adding 3 to each member) the solution is $x > 3$. This is an algebraic solution of a linear inequality (linear because none of its terms is of a power higher than one). There are, however, *quadratic inequalities* jut as there are quadratic equations. It is important to remember that equations and inequalities can be solved by both algebraic and graphic methods. It is time to discuss the graphic method of solving inequalities. You will recall that we used this method briefly in frame 5.

Think for a moment about the inequality $x > 3$. What does it say? What does it represent? It says that the solution values of x can be selected from the entire set of numbers greater than 3. Can you see how we might indicate this idea on a horizontal number scale? (Refer back to frame 5 if you need help.) Try it on the number scale below.

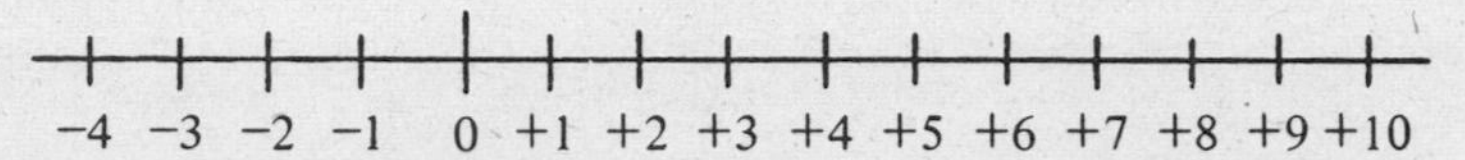

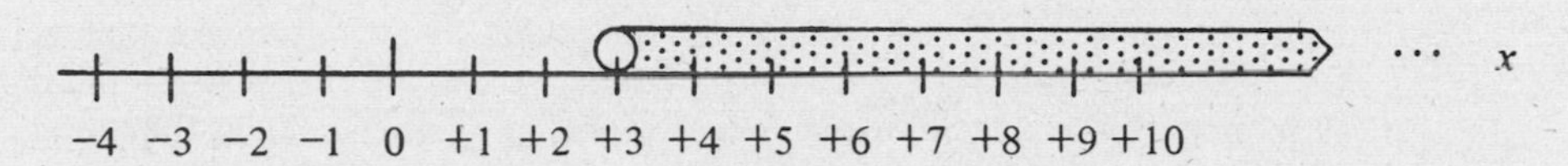

Your graph should look approximately like that shown above. It should indicate clearly that all numbers greater than 3 are included and all numbers less than 3 are excluded. There are several ways of indicating that the number 3 itself is not included. We have used (and will continue to use) the hollow circle placed at the number to show this.

9. In the exercise above if the inequality had been $x \geqslant 3$ (this symbol means "greater than or equal to"), we would have shown that 3 was included by filling in the circle above it. Thus, for $x \geqslant 3$ we would have this graph:

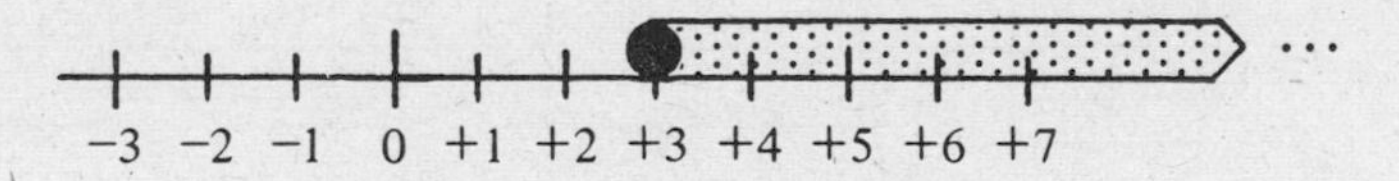

How would you represent the inequality $x \leqslant 2$? (This symbol means "less than or equal to.") Use the number scale given below.

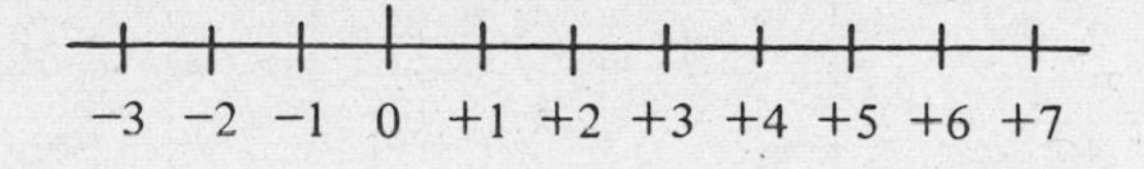

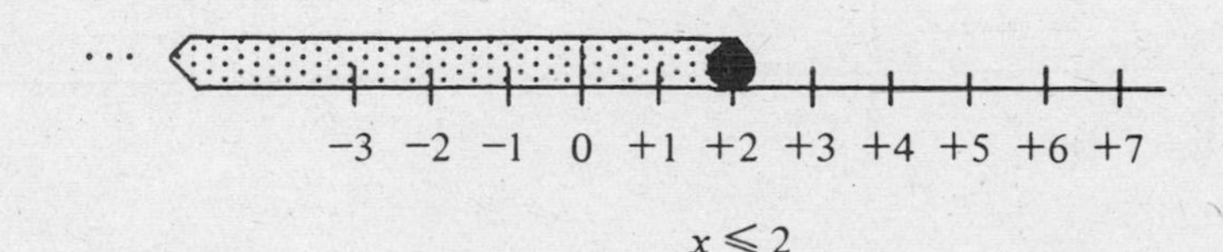

$x \leqslant 2$

Since the group of numbers represented by x is less than 2, the bar graph must start on 2 and run indefinitely to the left. And since x is also equal to 2, we indicate that 2 is included by making the circle above 2 solid.

SOLVING INEQUALITIES OF TWO VARIABLES

10. The concepts we have discussed so far are simple but they form the basis for further work. We will go now to inequalities containing two variables rather than just one. The number line graphs we used to plot inequalities of one variable will not serve for plotting inequalities of two variables. We will use instead the two-dimensional, rectangular coordinate system we worked with in Chapters 8 and 9 when plotting linear and quadratic equations.

Consider, for example, the inequality $2x - y + 4 > 0$. This symbolic statement is telling us that there exists an entire group of points (x, y) such that the value of the expression $2x - y + 4$ is greater than zero. Therefore, if we set this expression equal to zero (that is, $2x - y + 4 = 0$), we will get a linear equation that will graph as a straight line on one side of which lie all the points giving a value greater than zero, and on the other side of which lie all the points giving a value less than zero.

Graph the equation $2x - y + 4 = 0$. (Do it the easy way by just finding the x and y intercepts plus one additional check point, plotting these, and then drawing the straight line.) When you have plotted this line, see if you can invent some way to determine on which side of the line the values are greater than zero.

$2x - y + 4 = 0$
$y = 2x + 4$

x	y
0	4
−2	0
−1	2

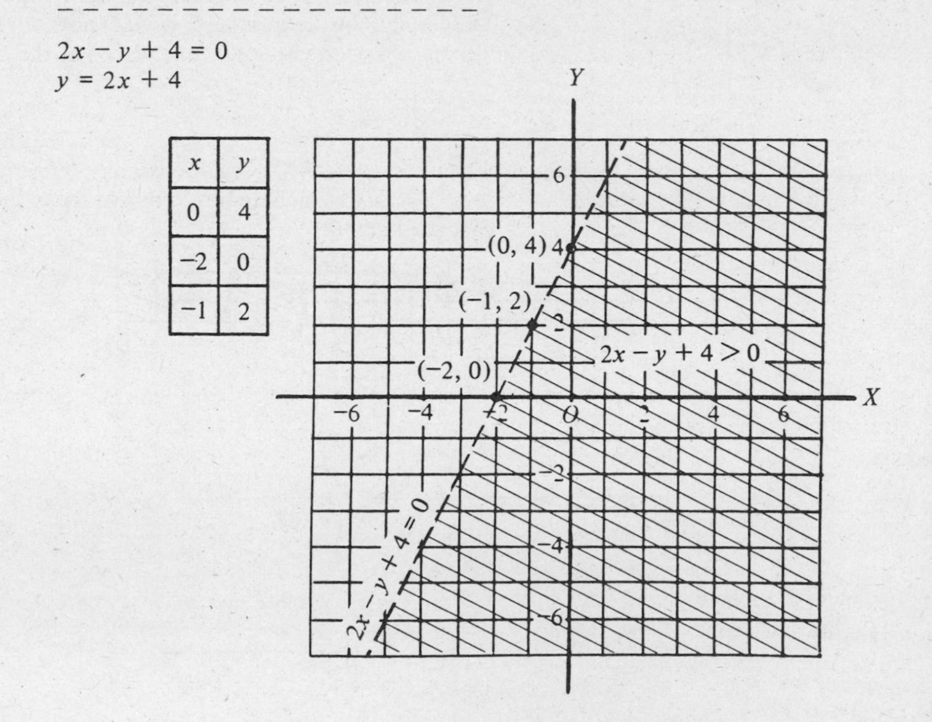

11. Did you find a way to test for values on one side or other of the line graph above? Consider the original inequality again: $2x - y + 4 > 0$. There are many points that could be used to test this inequality, but the easiest is the point (0, 0), the origin (point where the two axes cross). If we substitute the values (0, 0) for x and y in the above inequality we discover that at the origin (which happens to be on the right-hand side of the line) the value of the inequality is 4, which is indeed greater than zero. Thus, for the values $x = 0$, $y = 0$, representing the point (0, 0), the inequality $2x - y + 4 > 0$ becomes $0 - 0 + 4 > 0$ or $4 > 0$.

We can conclude that all points in the coordinate half-plane (that is, half of the total coordinate plane) lying to the right of the line are greater than zero (positive) and, by inference, all the points lying in the left half-plane are less than zero (negative). This is why the right half-plane is shaded, indicating that this is the area containing all the points whose values make the inequality a true statement.

Notice that the graph of $2x - y + 4 = 0$ is a dashed line. This is to show that the points lying along this line are not part of the solution to the inequality.

Use the same general procedure to graph the inequality $3x + y - 3 > 0$ and to determine which half-plane is represented by this inequality.

In order to find the solution set for $3x + y - 3 > 0$, we first draw the graph of $3x + y - 3 = 0$.

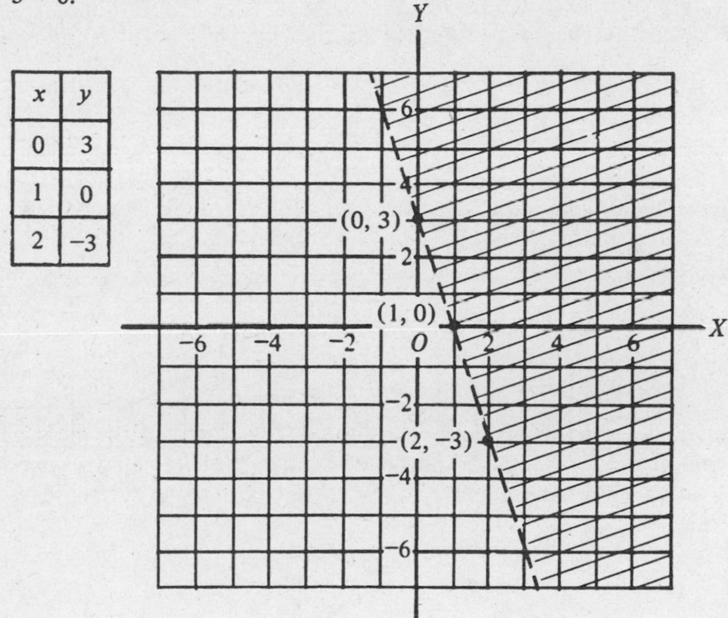

x	y
0	3
1	0
2	−3

From the graph it is apparent that $3x + y - 3$ is positive on the side of the line that does not contain (0, 0), that is, on the right-hand side, as shown by the shading. Therefore, this is the half-plane that represents the inequality $3x + y - 3 > 0$.

12. Here is a summary of the steps involved in graphing a linear inequality in two unknowns.

    (1) Set the left member of the inequality equal to the right member (zero). This gives a corresponding linear equation.

    (2) Find the x and y intercepts (by alternately letting x and then y equal zero) and one other point on the line to obtain three ordered pairs.

    (3) Plot the graph of the linear equation.

    (4) Select a point near the straight line (usually the origin is most convenient) and substitute its coordinates for the x and y terms in the original inequality.

    (5) If the resultant value of the left term makes the inequality a true statement, the solution points of the inequality all lie on the same side of the line as the point tested. If the test point does not make the inequality true, the solution points lie on the opposite side of the line.

Before leaving this chapter be sure to work the Self-Test for further practice.

SELF-TEST

1. Indicate in each case whether the inequality is true or false. (Draw a horizontal number scale to assist you if you feel you need it.)

   (a) $0 < -2$ (d) $9 < -10$

   (b) $-6 < -8$ (e) $-1 > -8$

   (c) $4 > 0$ (frame 1)

2. In each of the following problems, the original terms of the inequality have had the same amount added to or subtracted from each side or they have been multiplied or divided by the same amount. Your job is to insert the correct inequality symbol.

   (a) If $4 < 7$, then $4 + 3$ ____ $7 + 3$.

   (b) If $-2 > -7$, then $-2 + 5$ ____ $-7 + 5$.

   (c) If $4 < 5$, then $4(-3)$ ____ $5(-3)$.

   (d) If $8 > 4$, then $8 - 2$ ____ $4 - 2$.

   (e) If $-4 < -2$, then $-4(3)$ ____ $-2(3)$.

   (f) If $9 > 7$, then $\frac{9}{3}$ ____ $\frac{7}{3}$. (frame 3)

3. Complete the following.

   (a) Multiplying both sides of $7 > 2$ by -2 results in ____________________.

   (b) Subtracting 7 from both sides of $2 < 5$ results in ____________________.

   (c) Adding 6 to both sides of $-4 > -8$ results in ____________________.

   (d) Dividing both sides of $-6 > -8$ by -2 produces the result ____________________.

   (e) Solving an inequality means finding the values that make it a ____________________

   ______________________________________. (frame 4)

4. Simplify the following inequalities.

   (a) $3x + 4 > 2x - 1$ (b) $2x + 6 < 7x - 4$

   (b) $4k - 2 < k + 7$ (d) $-3 + m > 7 - 2m$

   (frame 5)

5. Indicate which of the following inequalities are absolute and which are conditional.

   (a) $a^2 > 0$ (if $a \neq 0$) ______________________________

   (b) $2x + 3 > 5$ ______________________________

   (c) $x^2 - 9 > 4$ ______________________________

   (d) $y^2 + 1 > 0$ ______________________________ (frame 7)

6. Draw a graph (on a horizontal number scale) to represent each of these inequalities.

   (a) $n > 3$ (where n represents some number)

   (b) $n < -1$

   (c) $n \geqslant -2$
   (d) $n \leqslant 4$ (frames 8 and 9)

7. Plot the graphs of the following inequalities.

   (a) $2x + y - 3 > 0$ (c) $2x - 5y < 0$

   (b) $2x - y - 2 > 0$ (frames 10 and 11)

Answers to Self-Test

1. (a) false; (b) false; (c) true; (d) false; (e) true
2. (a) $<$; (b) $>$; (c) $>$ (changes $<$ to $>$ because multiplier is negative); (d) $>$; (e) $<$; (f) $>$
3. (a) $-14 < -4$ (changes $>$ to $<$ because multiplier is negative); (b) $-5 < -2$; (c) $2 > -2$; (d) $3 < 4$ (direction of inequality symbol changes because divisor is negative); (e) true statement
4. (a) $x > -5$; (b) $k < 3$; (c) $x > 2$ (direction of inequality sign changes because both sides are divided by a negative number); (d) $m > 3\frac{1}{3}$
5. (a) absolute; (b) conditional; (c) conditional; (d) absolute

6. (a) 

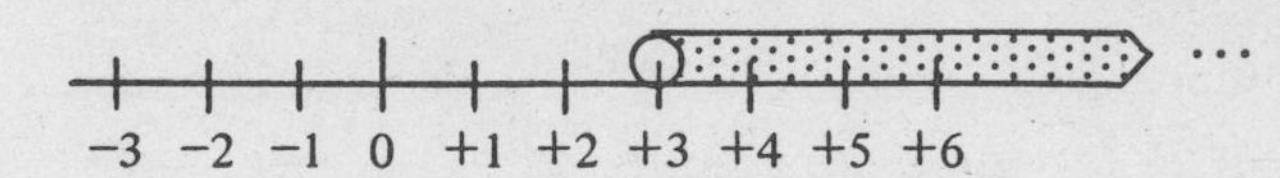

(b)

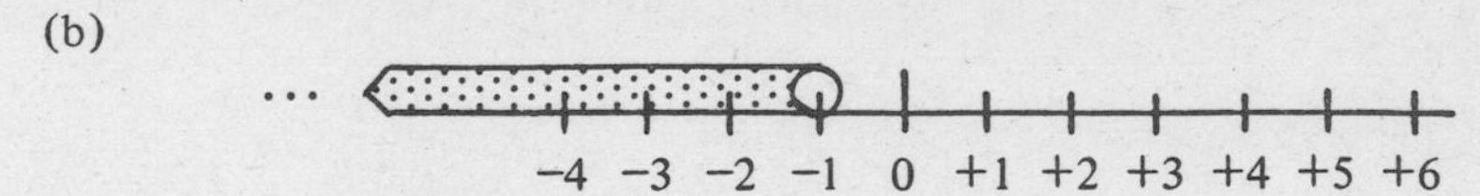

(c)

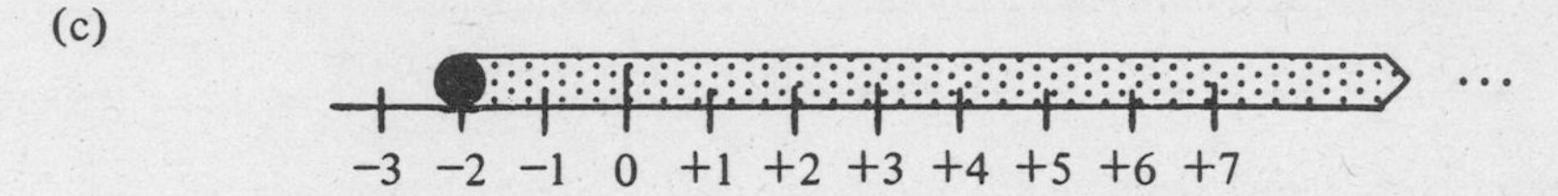

(d)

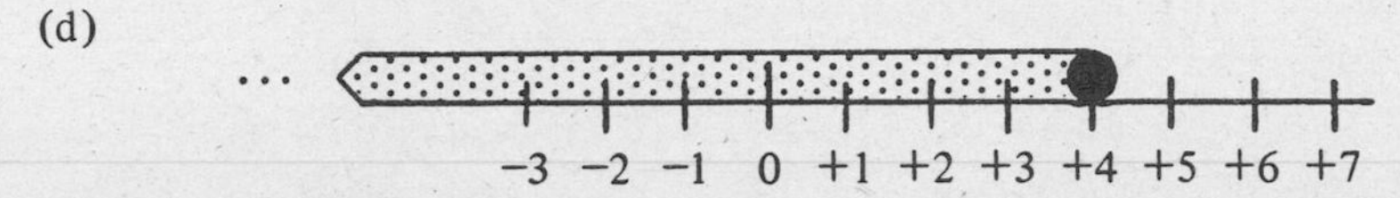

7. (a) $2x + y - 3 > 0$
$2x + y - 3 = 0$
$y = 3 - 2x$
At origin (0, 0) we get:
$2(0) + 0 - 3 \overset{?}{>} 0$
$-3 \overset{?}{>} 0$
$-3 \not> 0$
Therefore, $2x + y - 3 > 0$ is true only in the right half-plane.

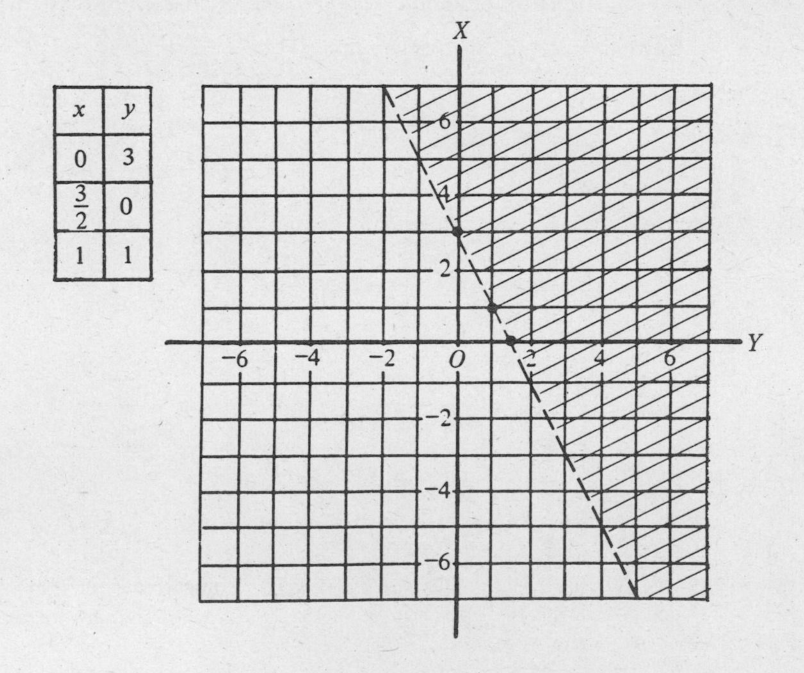

x	y
0	3
$\frac{3}{2}$	0
1	1

(b) $2x - y - 2 > 0$

$2x - y - 2 = 0$

$y = 2x - 2$

At origin (0, 0) the inequality becomes:

$$2(0) - 0 - 2 \overset{?}{>} 0$$

$$-2 \not> 0$$

Therefore, $2x - y - 2 > 0$ is true only in the right half-plane.

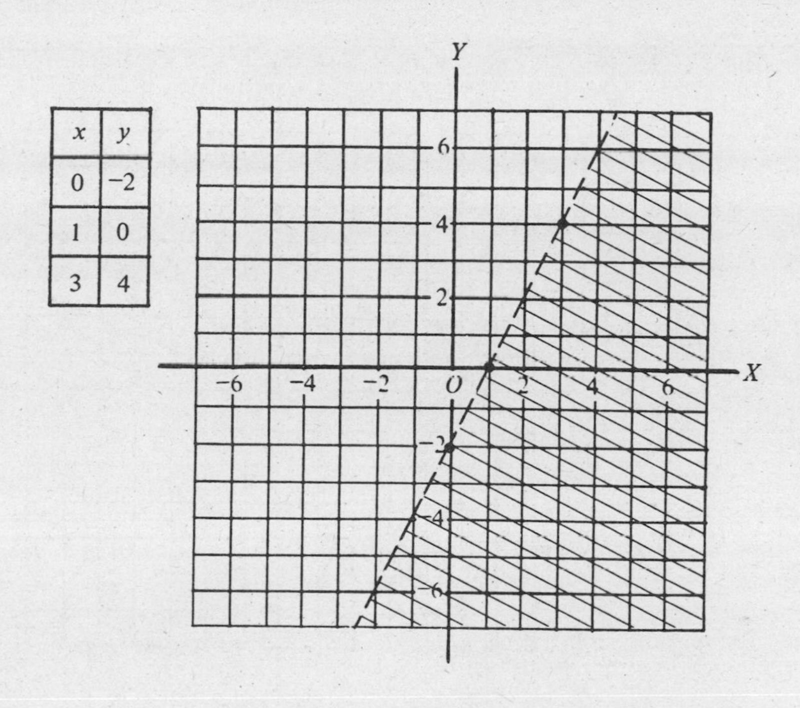

x	y
0	−2
1	0
3	4

7. (c) $2x - 5y < 0$

$2x - 5y = 0$

$x = \frac{5y}{2}$

Since the line passes through the point or origin, you could try substituting the coordinates of any point near (0, 0).

Using the point (1, 1), for example, gives us $2(1) - 5(1) \overset{?}{<} 0$, or $-3 \overset{\checkmark}{<} 0$. Hence $2x - 5y < 0$ is true only in the half-plane to the left of the graph-line.

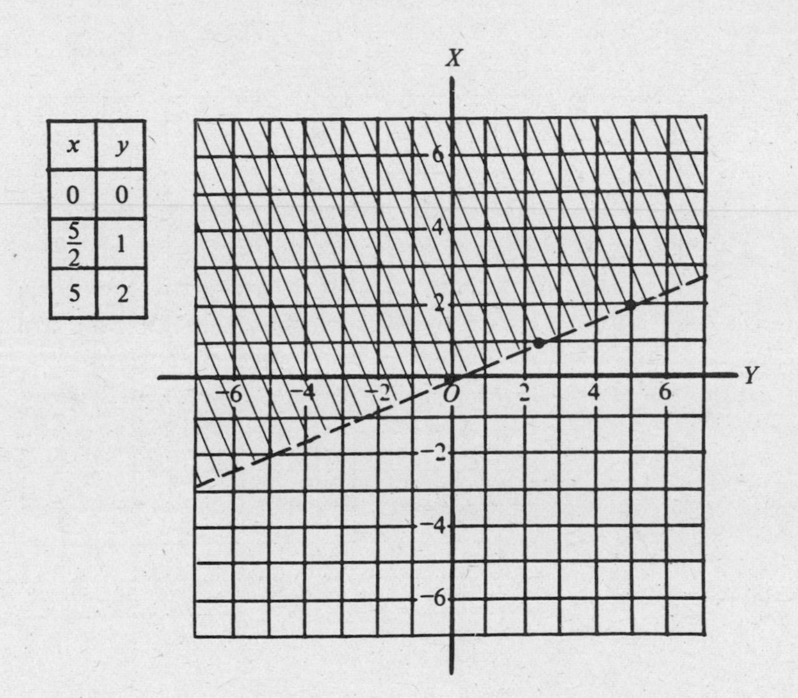

x	y
0	0
$\frac{5}{2}$	1
5	2

CHAPTER ELEVEN

Ratio, Proportion, and Variation

OBJECTIVES

This chapter deals with a variety of useful concepts and applications relating to ratio, proportion, and variation. When you have completed the work here you will be able to:

- write ratios and reduce them to their lowest terms;
- recognize the best expression of a ratio;
- recognize the means and the extremes of a proportion;
- recognize examples of direct proportion and solve problems involving the use of direct proportion;
- establish a proportion from the information given in a statement problem and solve for the missing value;
- recognize and solve problems involving inverse proportion;
- recognize and solve problems involving direct, inverse, joint, and combined variation.

RATIO

1. The need to count led men to invent the concept of a number. If you observe people you may feel that another early use for numbers might have been that of comparing various quantities. As you watch two young children you will see that they are vitally interested in whose glass of milk is larger, whose all-day sucker is the longer, whose dad is bigger, whose mom is smarter, and so on.

   One method of comparing quantities is to find the difference in the two quantities. For example, if one person is 22 and the other is 33, the difference in their ages is 11 years. We could say, therefore, that the first person is 11 years younger than the second or that the second is 11 years older than the first.

   A second method of comparing these same two ages would be to say that one person is $\frac{2}{3}$ the age of the other. When we compare quantities by this method we are using the concept of *ratio*. We are saying, in effect, that for each two years of life for the younger person, the older person has had three years of life. Or, we could say that the ratio of the first person's age in years to the second person's age in years is 2 to 3. Using symbols we could write the ratio of their ages as $\frac{2}{3}$, 2:3, 2/3, or $2 \div 3$. The ratio of two like quantities is defined as the quotient of the first quantity divided by the second. Notice that this quotient is usually expressed in lowest terms. Thus, the ratio of 20 to 50 is written 2:5, not 20:50.

Express the ratio of 15 days to 35 days. ______________________

- - - - - - - - - - - - - - - - - - -

The ratio of 15 days to 35 days can be expressed as $\frac{3}{7}$, 3:7, or $3 \div 7$.

2. Here we will generalize a bit and state the following definition.

Definition 1: If a is a quantity expressed in some unit of measure and b is some quantity expressed in that same unit of measure, then the ratio of a to b is the quotient $\frac{a}{b}$.

As we have already discussed, there are three ways to write the ratio of a to b: $\frac{a}{b}$ (or a/b), $a{:}b$, and $a \div b$. Notice that the ratio of two quantities is defined only if the units of measure are the same, but you do not have to include these units in the final form of the ratio. In the case of our earlier reference to the ratio of two ages, for example, both ages were given in years, but the ratio of the ages was simply 2:3.

Select one of the three answers given to complete the following statement. A ratio is:

___ (a) the quotient of two quantities

___ (b) the difference between two like quantities

___ (c) the quotient of two like quantities.

- - - - - - - - - - - - - - - - - - -

(c)

3. Choice (c) is correct because it includes the two ideas of *quotient* and *like quantities.* This is important. It tells us, for example, that the ratio of 3 days to one week would be found by changing 1 week to 7 days and writing the ratio as 3:7, which we can now do because the units are the same.

We have defined ratio as the quotient of two like quantities. However, you will often see a ratio expressed between two quantities that are entirely different in nature. For example, in Chapter 7 (frame 27) we expressed speed (S) as the ratio of distance (D) to time (T):

$$S = \frac{D}{T}$$

If a and b do not represent quantities of the same kind, the ratio $a{:}b$ simply represents a portion of a that corresponds to one unit of b, as one mile per hour.

Write the ratio of the following (in simplest form).

(a) 7 dozen to 6 dozen

(b) 3 yards to 7 feet

(c) 85 pounds per square inch to 150 pounds per square inch

(d) 36 square feet to 3 square yards

(e) 25 gallons to 50 gallons

(f) 15 days to 36 quarts

(g) 12 pounds to 36 yards

(h) 45 dimes to 12 quarters

(i) 18° F to 30° F (F indicates Fahrenheit scale)

(j) 3 cubic feet to 1 cubic yard

- - - - - - - - - - - - - - - - - -

(a) 7:6, (b) 9:7; (c) 85:150 or 17:30; (d) 36:27 or 4:3; (e) 25:50 or 1:2; (f) cannot be done, units not compatible; (g) same as (f); (h) 45:30 or 3:2; (i) 18:30 or 3:5; (j) 3:27 or 1:9

PROPORTION

4. You have worked with ratios enough to have some understanding of them. You will become more proficient in forming and using them (and converting units as required) as you begin to work with proportions. This is true because *a proportion is merely a statement that one ratio is equal to another ratio.* Thus, $\frac{1}{2} = \frac{3}{6}$ or 1:2 = 3:6 is a proportion.

Definition 2: If $\frac{a}{b}$ and $\frac{c}{d}$ are equal ratios, then $\frac{a}{b} = \frac{c}{d}$ is a proportion. (The form $a:b = c:d$ is also commonly used.)

Which of the following pairs of ratios could be used to form a true proportion?

___ (a) $\frac{2}{3}$ and $\frac{3}{4}$

___ (b) $\frac{3}{4}$ and $\frac{6}{8}$

___ (c) $\frac{2}{7}$ and $\frac{6}{21}$

___ (d) $\frac{3}{7}$ and $\frac{5}{9}$

- - - - - - - - - - - - - - - - - -

(b) and (c)

5. You must learn to recognize a proportion when you see one *and* be able to solve for some value that would make a statement become a true proportion. If one term of a proportion is missing, it can frequently be found by inspection. However, since not all proportions can be solved so easily, we need to learn something about the methods of solution.

An important point to keep in mind is that a proportion is an equation. Consequently, all the rules of equations can be used to solve for missing terms in proportions. Also, since ratios are fractions, all the rules relating to fractions apply. All you need to learn are a few special names and some important properties of proportions.

In the proportion $a:b = c:d$ (which we usually read "a is to b as c is to d"), the letters a, b, c, and d are called the first term, second term, third term, and fourth term, respectively. The terms a and d are called the *extremes* (since they are farthest apart), and b and c are called the *means* (since they are the middle terms).

In the proportion $5:7 = 10:14$, the extremes are _____ and _____ , and the means are _____ and _____.

- - - - - - - - - - - - - - - - - - -

5, 14, 7, 10

6. One of the basic properties of a proportion is that the product of the extremes is equal to the product of the means. We can express this idea algebraically as follows:

Theorem 1: If b and $d \neq 0$ and if $\frac{a}{b} = \frac{c}{d}$, then $a \cdot d = b \cdot c$.

Notice that this theorem is the same as the first part of our definition of equal fractions (Chapter 5, frame 5). Although we have not stressed proofs in this book it might be useful for you to see the proof for this theorem.

Prove: If $\frac{a}{b} = \frac{c}{d}$, then $a \cdot d = b \cdot c$ $(b \neq 0,\ d \neq 0)$.

Statements	Reasons
(1) $\frac{a}{b} = \frac{c}{d}$, $b \neq 0$, $d \neq 0$	(1) Given
(2) $\frac{a \cdot d}{b} = c$	(2) Both members multiplied by d.
(3) $a \cdot d = b \cdot c$	(3) Both members multiplied by b.

Answer the following questions.

(a) In the proportion $2:7 = 3:x$, the product of the extremes is _____ and the product of the means is _____.

(b) If $3:x = 4:7$ is a proportion, then $4x =$ ________.

(c) If $n:a = c:r$ is a proportion, then ________ $= nr$?

- - - - - - - - - - - - - - - - - - -

(a) $2x$ and 21; (b) 21; (c) ac

DIRECT PROPORTION

7. Now that we have defined proportions and discussed some of their properties, we will illustrate some of the various types of proportions. One kind of proportion is illustrated by the relation shown in the table on page 267 which was developed from information given in the following problem:

Suppose you are traveling at a constant rate (r) of 60 miles per hour and want to make a table of distances (d) covered according to time (t) traveled. Using the formula $d = rt$ you could develop a table with the following entries:

t	= elapsed time in hours	1	2	3	6	10	12
d	= distance in miles	60	120	180	360	600	720

Notice that the ratio of any time to its distance is always 1:60 (read "one to sixty"). We can write this as a continued proportion.

$$\frac{t}{d} = \frac{1}{60} = \frac{2}{120} = \frac{3}{180} = \ldots = \frac{12}{720}$$

If any pair of ratios are considered together we find the definition of a proportion established. When the quotient of any two quantities (such as the ordered pairs in the table above) stays constant, we have what is known as a *direct proportion* between two quantities.

Three terms of a proportion are usually known; it is required to find the missing term. Notice how Theorem 1 is used to solve the proportion in the following example.

Example: Solve the proportion $\frac{3}{x} = \frac{9}{18}$.

$$9x = 54$$
$$x = 6$$

Use the fact that if $a{:}b = c{:}d$ then $a \cdot d = b \cdot c$ to find the missing term in each of the following proportions.

(a) $3{:}4 = x{:}8$

(b) $a{:}7 = 7{:}3$

(c) $(a - 3){:}6 = 3{:}12$

(d) $5{:}(x - 2) = 4{:}7$

(e) $3{:}x = x{:}12$

(f) $x{:}5 = 5{:}x$

(g) $3a{:}5 = 6{:}1$

(h) $3{:}(x - 2) = x{:}1$

(i) $3{:}(x - 1) = x{:}2$

(j) $4{:}3 = x{:}(x + 1)$

(a) $x = 6$; (b) $a = 16\frac{1}{3}$; (c) $a = 4\frac{1}{2}$; (d) $x = 10\frac{3}{4}$; (e) $x = \pm 6$; (f) $x = \pm 5$; (g) $a = 10$; (h) $x = 3, -1$; (i) $x = 3, -2$; (j) $x = -4$

8. The problems given above were designed to range from an elementary level to one that would require the application of several of the principles covered in preceding chapters. If you had some difficulty with problem (e), (f), or (i), you probably have forgotten some of the work covered in Chapter 9. If so, you should return to that chapter and review the handling of quadratic equations. Remember that you must always check your answer to see if the solution value you have found makes the proportion a true statement, just as you would with any equation you are solving.

You may have noticed as you worked the problems in frame 7 that the position of the extremes and/or means may be interchanged. The following definition covers this situation.

Definition 3: If $a{:}b = c{:}d$, then $a{:}c = b{:}d$, $d{:}b = c{:}a$ and $d{:}c = b{:}a$.

In the light of this definition which of the following statements would you say is true?

(a) If $2{:}3 = x{:}y$, then $2{:}x = 3{:}y$. ____________

(b) If $2{:}5 = x{:}y$, then $5{:}2 = y{:}x$. ____________

(c) If $3{:}4 = 6{:}8$, then $8{:}4 = 6{:}3$. ____________

- - - - - - - - - - - - - - - - - - -

All these statements are true.
Watch for applications of Definition 3 as you study the rest of this chapter.

9. Proportions are used to solve many practical problems in a wide variety of fields. Any situation that involves the comparison of one ratio to another ratio is a possible application of proportion. Here is an example of an application of direct proportion that occurs frequently in plane geometry.

Similar triangles are triangles having the same shape but not necessarily the same size. Similar triangles have congruent corresponding angles, but not necessarily congruent corresponding sides. Use the figure given below as we discuss the relation between the corresponding sides of two similar triangles.

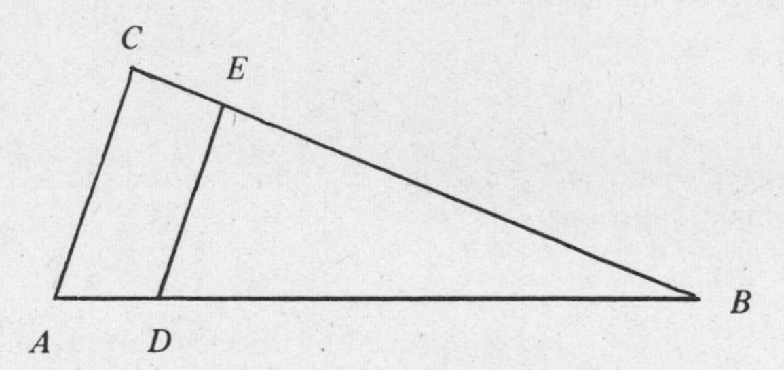

If triangle ABC is given as shown above and triangle DBE is formed by drawing DE parallel to AC, then the corresponding angles of the two triangles are congruent and triangle ABC is similar to triangle DBE. In geometry it is shown that the corresponding sides of two similar triangles are proportional in size. Thus, we have the continued proportion

$$\frac{BC}{BE} = \frac{AC}{DE} = \frac{AB}{DB}$$

If $AC = 12$ inches, $AB = 22$ inches, and $DB = 14$ inches, how long is DE?

- - - - - - - - - - - - - - - - - - -

$$\frac{12}{DE} = \frac{22}{14}$$

$$22(DE) = 168$$

$$DE = 7\tfrac{7}{11}$$

10. The idea of proportional corresponding sides is used in drawing architectural and engineering plans, making scale models, and solving many other kinds of geometric problems. Your own practical experience will enable you to add to the uses named

above. The type of proportion illustrated in the previous frame is an instance of *direct proportion.* This means that if one side of triangle DBE is smaller than its corresponding side in triangle ABC, then each side of triangle DBE will be smaller than its corresponding side in triangle ABC. Remember that *in a direct proportion the two ratios are equal fractions.*

The following problem is another example of the use of direct proportion to solve a practical problem.

Example: If a certain parking lot contains Fords and Plymouths in the ratio 5:3 and there are 900 Plymouths in the lot, how many Fords are there?

Let x equal the number of Fords in the lot

$$5:3 = x:900$$
$$3x = 4500$$
$$x = 1500\text{, the number of Fords}$$

Solve the following problems using the idea of direct proportion.

(a) An aircraft has a wing span of 130 feet. The height from the ground to the top of the tail is 25 feet. If on a drawing the wing span is 42 inches, how many inches will be needed to show the distance from the ground to the top of the tail?

(b) If an automatic billing machine can process 21,000 cards in 14 hours, how many cards could be processed in one 8-hour shift?

(c) The vibrating frequency of a string is directly proportional to the square root of the tension on the string. If a bass violin string vibrates 254 times per second when the tension is 5 pounds, what would the vibrating frequency be when the tension is 20 pounds?

(d) The distance that a falling body drops is directly proportional to the square of the time it has fallen. If a body falls 64 feet in 2 seconds, how far will it fall in 10 seconds?

(a) $130:42 = 25:x$; $x = 8\frac{1}{13}$ inches
(b) $21{,}000:14 = x:8$; $x = 12{,}000$ cards
(c) $254:\sqrt{5} = x:\sqrt{20}$; $x = 1016$ times per second
(d) $64:(2)^2 = x:(10)^2$; $x = 1600$ feet

Note: Although the two numbers in a ratio must be in the same units, the two ratios comprising a proportion can be, and usually are, in different units, as in the previous problems.

INVERSE PROPORTION

11. The counterpart of direct proportion is *inverse proportion.* In an inverse proportion the product of one ordered pair is the same as the product of the two numbers in the other ordered pair. We might represent this in symbols as $a \cdot b = c \cdot d$. A more formal and somewhat more general definition follows:

If a set of ordered pairs (a, b) has the property that a is inversely proportional to b, then $a \cdot b$ is the same value for all of the ordered pairs (a, b). Or, using symbols, if $a \cdot b = k$ for any of the ordered pairs (a, b), then a is inversely proportional to b.

The value of k is called the *constant of proportionality.*

The table below, showing the relationship between the pressure and volume of a gas, illustrates how inverse proportion differs from direct proportion. (For the purpose of this example we assume that the container can expand and that the temperature of the gas remains constant.)

p	= pressure in pounds per square inch	50	40	20	15	10
v	= volume in cubic inches	200	250	500	$66\frac{2}{3}$	1000

Notice that we can no longer write a continued proportion such as $\frac{p}{v} = \frac{50}{200} = \frac{40}{250} = \ldots = \frac{10}{1000}$ because the definition does not hold. For example, if we write $50{:}200 = 20{:}500$, we note that $50 \cdot 500 \neq 200 \cdot 20$. As a consequence we do not have a true proportion.

Look at the table again and see if you can find a relationship there that might permit us to use our definition of a proportion. For example, what do you notice about the value of $p \cdot v$ for any pair of entries in the table? Check the answer below that best represents your conclusions.

___ (a) I don't see anything special about the value of $p \cdot v$.

___ (b) I notice that $p \cdot v$ is always 10,000 but this fact doesn't help me set up a proportion.

___ (c) I know that $p \cdot v = 50 \cdot 200 = 40 \cdot 250 = \ldots = 10 \cdot 1000$, but I can't set up a proportion.

___ (d) I notice that the statement of equality at the beginning of the frame is true and I can establish a proportion.

- - - - - - - - - - - - - - - - - -

Hopefully you selected either answer (c) or (d). You need only remember that if $a \cdot b = c \cdot d$, then $\frac{a}{c} = \frac{d}{b}$ is a proportion. Thus, we could write a proportion $\frac{p}{50} = \frac{200}{v}$, or a proportion $\frac{50}{40} = \frac{250}{200}$, or a proportion $\frac{50}{250} = \frac{40}{200}$.

12. Whenever we have two quantities that behave in the manner exhibited in the table shown in frame 11 (that is, one in which the product of the two entries in any given column is the same as the product of the two entries in any other column) we have an *inverse proportion* relationship. The table below is another example of an inverse proportion between two quantities.

a	6	4	12	24	
b	8	12	4		−12

Use the information in this table to solve the following problems.

(a) Set up a proportion and find b when a has a value of 24.

(b) Set up a proportion and find the value of a when b is -12.

(c) Write the definition of inverse proportion.

- - - - - - - - - - - - - - - - - -

(a) Some possible proportions are:

$$\frac{6}{24} = \frac{b}{8} \quad \frac{6}{b} = \frac{24}{8} \quad \frac{4}{24} = \frac{b}{12} \quad \frac{12}{24} = \frac{b}{4}$$

There are many other proportions that could be established by selecting various ordered pairs (a, b) and/or interchanging the extremes or interchanging the means. In any of the proportions, however, the value of b will be 2 when the value of a is 24.

(b) The proportion could be $\frac{6}{a} = \frac{-12}{8}$ or some other proportion obtained from the table of ordered pairs and using the definition of inverse proportions. The value of a is -4 when b is -12.

(c) If a set of ordered pairs (a, b) has the property that a is inversely proportional to b, then $a \cdot b$ is the same value for all of the ordered pairs (a, b). In symbols we might write the definition as follows:

If $a \cdot b = k$ for any of the ordered pairs (a, b), then a is inversely proportional to b.

The value of k for the relation of a to b in the present case (that is, as shown in the table) is 48.

13. Now that you have had a little practice recognizing inverse proportion and setting up such relationships you are going to get a chance to work a few problems involving inverse proportion. Set up a proportion and solve the following problems.

(a) If x is inversely proportional to y and $x = 6$ when $y = 12$, find the value of x when $y = 4$. Show the proportion and the solution. Hint: Remember that if x is inversely proportional to y, then the product of every (x, y) pair is a constant (that is, the same). Thus, in this problem we know that $x \cdot y = 6 \cdot 12 = x \cdot 4$ and we can use any pair of these equal ratios to set up a proportion.

(b) If r is inversely proportional to d^2 and $r = 3$ when $d = 5$, find r if $d = 8$.

(c) If y is inversely proportional to $4x + 5$ and $y = 4$ when $x = 8\frac{1}{2}$, find x when $y = 12$.

(d) The intensity of light (i) on a surface is inversely proportional to the square of the distance (d) between the surface and the light source. A light is 3 feet from a surface which receives a certain amount of light. What distance will cause the surface to receive $\frac{1}{4}$ the present amount of light? Hint: What is true about the pairs of values (i, d^2)? Also, remember that $\frac{i}{\frac{1}{4}i} = \frac{1}{\frac{1}{4}}$.

- - - - - - - - - - - - - - - - - -

(a) The proportion should be some form of $\frac{6}{4} = \frac{x}{12}$; $x = 18$ is the answer.

(b) The proportion should be some form of $\frac{3}{8^2} = \frac{r}{5^2}$; the answer is $r = \frac{75}{64}$.

(c) The proportion should be some form of $\frac{4}{12} = \frac{4x+5}{39}$; the solution is $x = 2$.

(d) The proportion should be some form of $\frac{i}{\frac{1}{4}i} = \frac{d^2}{3^2}$, which is equivalent to $\frac{1}{\frac{1}{4}} = \frac{d^2}{9}$. The solution is $d = 6$ feet. (Note: The variable i is an example of a *dummy variable*; it has no bearing on the solution.)

DIRECT VARIATION

14. It is important for you to notice that while you may not have had the identical proportion as those given in the answers to frame 13, you should have been able to change your proportion to the one given by interchanging the means and/or interchanging the extremes. The problems you have worked with should have given you some understanding of the concept of proportion. You will find more problems of this kind in the Self-Test at the end of the chapter. Work as many of these as you find necessary to bring your understanding up to the level you desire.

Now let us turn our attention to the subject of *variation*. The idea of proportion was introduced to illustrate the relationships that can exist between two variables. Another method of showing this mutual dependence between two variables is known as variation. While there is no essential difference between proportion and variation, the language and the equations involved are somewhat different, and some problems are easier to work by using variation. Let's define first what we mean by *direct variation*.

Definition 4: If two variables, a and b, are related in such a way that the quotient $\frac{a}{b}$ has the same value for any ordered pair (a, b), a is said to vary directly as b.

The relation described in Definition 4 can be written symbolically as $\frac{a}{b} = k$, where k represents the *constant of variation*.

The following examples illustrate how Definition 4 is used.

Example 1: Using symbols write the statement "x varies directly as y and the constant of variation is 26."

By direct application of the information in Definition 4 we can write $\frac{x}{y} = 26$. Naturally, by applying the axiom of multiplication we could change this to $x = 26y$. Both forms are acceptable and each has its own advantages for certain situations.

Example 2: If x varies directly as y^2 and $x = 6$ when $y = 4$, find k (the constant of variation).

By Definition 4 and the information in the problem we can write $\frac{x}{y^2} = k$. Since we know that $x = 6$ when $y = 4$, we can write $\frac{6}{4^2} = k$. The answer is $k = \frac{6}{16}$ or $\frac{3}{8}$.

Example 3: If a varies directly as b and $a = 12$ when $b = 3$, find a when $b = 16$.

By using Definition 4 we can write $\frac{a}{b} = k$. Now we know that $a = 12$ when $b = 3$ so we can write $\frac{12}{3} = k$ or $k = 4$. The next step is to return to the equation $\frac{a}{b} = k$ and write $\frac{a}{16} = 4$, from which $a = 64$ when $b = 16$.

All of these examples may make the use of variation appear more difficult than it actually is. This is often the case with "explanations." In actual practice you will find that you will use proportion for some problems and variation for others. At this time, however, you should solve the problems given by using variation so that you will become increasingly familiar with the method.

Solve the following problems by using the idea of direct variation.

(a) Find the constant of variation, k, if a varies directly as b and a equals 6 when $b = 2$.

(b) If $r = 56$ when $s = 4$ and if r varies directly as s, find r when $s = 3$.

(c) If x varies directly as y^3 and if $x = 64$ when $y = 2$, find x when $y = 1$.

- - - - - - - - - - - - - - - - - - -

(a) $k = 3$; (b) $r = 42$; (c) $x = 8$

INVERSE VARIATION

15. You might suspect that since we have both direct and inverse proportion we also have both direct and inverse variation. Therefore, let us define *inverse variation.*

Definition 5: If two variables, a and b, are related in such a way that the product $a \cdot b$ has the same value for all ordered pairs (a, b), a is said to vary inversely as b.

In symbols we can write this definition as follows:

If $a \cdot b = k$, then a varies inversely as b and k is the *constant of variation.*

The following examples should help you understand the use of inverse variation in solving problems.

Example 1: Find k if a varies inversely as b and if $a = 5$ when $b = 8$.

Since a varies inversely as b, we can use Definition 5 to write $a \cdot b = k$. Also, we know that $a = 5$ when $b = 8$, so we can write $5 \cdot 8 = k$ or $k = 40$.

Example 2: If x varies inversely as y and if $x = 6$ when $y = 2$, find the value of x when $y = 24$.

Since x varies inversely as y, we can use Definition 5 to write $x \cdot y = k$. But since $x = 6$ when $y = 2$, we can write $6 \cdot 2 = k$ or $k = 12$. Using this value of k we can now write $x \cdot y = 12$, so $x \cdot 24 = 12$ is solved to find that $x = \frac{1}{2}$ when $y = 24$.

Solve the following.

(a) x varies inversely as y. If $x = 9$ when $y = 2$, find x when $y = 3$.

(b) p varies inversely as v. If $p = 30$ when $v = 10$, find p when $v = 15$.

(a) $x \cdot y = 9 \cdot 2 = 18 = k$; hence $x \cdot 3 = 18$ or $x = 6$ when $y = 3$.
(b) $p \cdot v = 30 \cdot 10 = 300 = k$; hence $p \cdot 15 = 300$ or $p = 20$ when $v = 15$.

JOINT AND COMBINED VARIATION

16. An important aspect of the language of variation is that it allows us to discuss quite easily the relation between more than two variables. We will discuss briefly two kinds of variation involving more than two variables. The first, known as *joint variation*, is defined below.

Definition 6: If x varies directly as the product of two or more variables, x is said to vary jointly as these variables.

As an example of Definition 6 we could write $\frac{x}{abc} = k$, where k is the *constant of variation*. We would read this "x varies jointly as a, b, and c." Notice that x varies directly as any variable (a, b, or c) when the other two variables are held constant.

A second kind of variation involving more than two variables is *combined varition*. An example of combined variation is $\frac{xy}{z} = k$ where k is again the constant of variation. Notice that in this case x varies directly as z and inversely as y. Similarly y varies directly as z and inversely as x. The wording of the problem will tell you when to use combined variation.

Each of the following questions involves a type of variation we have discussed. Supply the missing words in each case. (Note: x is the variable related to the other variables and k is the constant of variation.)

(a) $\frac{x}{y} = k$, x varies ________________ as y

(b) $xy = k$, x varies ________________ as y

(c) $\frac{x}{ab} = k$, x varies ________________ as the ________________ of a and b

(d) $\frac{xy}{z} = k$, x varies directly as ______ and inversely as ______; this is an example of ________________ variation.

(e) $\frac{xy^2}{ab} = k$, x varies jointly as ________________ and inversely as ________

(a) directly; (b) inversely; (c) jointly, product; (d) z, y, combined;
(e) a and b, y^2

17. Express the constant of variation as an integer or a fraction in simplest form in each of the following problems. (Do not perform any additional steps.)

(a) $x = 7$ and $y = 14$ when x varies directly as y. $k =$ ____________

(b) $x = 7$ and $y = 14$ when x varies inversely as y. $k =$ ____________

(c) $x = 3$, $y = 7$, and $z = 6$ when x varies jointly as y and z. $k =$ ____________

(d) $x = 6$, $y = 7$, and $z = 3$ when x varies directly as y and inversely as the cube of z.

$k =$ ____________

(e) $x = 12$, $y = 3$, and $z = 2$ when x varies jointly as y and the square of z.

$k =$ ____________

- - - - - - - - - - - - - - - - - -

(a) $k = \frac{1}{2}$; (b) $k = 98$; (c) $k = \frac{1}{14}$; (d) $k = \frac{162}{7}$; (e) $k = 1$

18. After the constant of variation is found, the next step is usually to express the equation of variation in the most convenient form to complete the problem. For example, let's consider problem (a) from frame 17. We will write the equation of variation in a way convenient to finding x.

Step 1: Since we know that $k = \frac{1}{2}$ we can write $\frac{x}{y} = \frac{1}{2}$.

Step 2: Solving for x we get $x = \frac{1}{2}y$.

It now would be an easy task to find the value of x for any given value of y. Although the steps we took above were directed at finding x as a function of y, we could as easily have written y as a function of x had we so desired.

Using the equation of variation found in Step 2 above ($x = \frac{1}{2}y$) find the value of x for the given value of y.

(a) $y = 2$, $x =$ ____________

(b) $y = 12$, $x =$ ____________

(c) $y = 20$, $x =$ ____________

(d) $y = \frac{1}{2}$, $x =$ ____________

(e) $y = \frac{2}{3}$, $x =$ ____________

(f) $y = 25$, $x =$ ____________

- - - - - - - - - - - - - - - - - -

(a) $x = 1$; (b) $x = 6$; (c) $x = 10$; (d) $x = \frac{1}{4}$; (e) $x = \frac{1}{3}$; (f) $x = \frac{25}{2}$

19. The previous problems should serve to illustrate that once the value of k has been found and the equation of variation established, it is a simple matter to find the value of the desired variable when the values of the other variables are given.

Write the equation of variation you would use to find x in the set of problems (b) through (e) in frame 17. These equations should be such that each one expresses x as a function of the other variables in the problem.

(b) $x =$ ____________ (d) $x =$ ____________

(c) $x =$ ____________ (e) $x =$ ____________

- - - - - - - - - - - - - - - - - -

(b) Since $xy = k$ (inverse variation) and $k = 98$, we can write $xy = 98$, $x = \frac{98}{y}$.

(c) Since $\frac{x}{yz} = k$ (joint variation) and $k = \frac{1}{14}$, we can write $\frac{x}{yz} = \frac{1}{14}$, $x = \frac{yz}{14}$.

(d) Since $\frac{xz^3}{y} = k$ (combined variation) and $k = \frac{162}{7}$, we can write $\frac{xz^3}{y} = \frac{162}{7}$, $x = \frac{162y}{7z^3}$.

(e) Since $\frac{x}{yz^2} = k$ (joint, inverse variation) and $k = 1$, we can write $\frac{x}{yz^2} = 1$, $x = yz^2$.

EVALUATING THE VARIABLE

20. The next step in solving problems by the use of variation is to find the value of the variable in question when the values of the other variables are known. For our first example of a complete solution by the method of variation we will use the facts given in problem (a) of frame 17 but restate them in a slightly different way.

Example: If $x = 7$ when $y = 14$ and it is known that x varies as y, find x when $y = 6$. (Note: When no different kind of variation is indicated, assume that the problem is one of direct variation.)

Step 1: Set up the equation involving x, y, and k. ________ = ________

Step 2: Find k. $k =$ ________

Step 3: Write the equation of variation that will express x as a function of y.

$x =$ ________

Step 4: Compute x by replacing y with the value 6 in the equation established in Step 3. $x =$ ________

- - - - - - - - - - - - - - - - - -

Step 1: $\frac{x}{y} = k$

Step 2: $\frac{7}{14} = k$, $k = \frac{1}{2}$

Step 3: $\frac{x}{y} = \frac{1}{2}$, $x = \frac{1}{2}y$

Step 4: $x = \frac{1}{2}(6)$, so $x = 3$ when $y = 6$

21. Solve the following problems by the use of variation. Show the four steps used in the previous example. Your primary interest here should be in learning to use the language of variation to establish an equation.

(a) If x varies inversely as y and $x = 12$ when $y = 3$, find x when $y = 12$.

(b) The relation between x and y is shown in the table below. Find x when $y = 48$.

x	3	2	4	...
y	8	12	6	...

(c) If the temperature remains constant, the resistance of a wire varies directly as the length of the wire and inversely as the square of its diameter. If a wire whose diameter is 10 mils has a resistance of 0.3 ohms when its length is 400 feet, what is the resistance of a wire whose diameter is 15 mils and whose length is 2000 feet if the second wire is made from the same material as the first?

(d) If y varies directly as x and if the graph of $y = f(x)$ passes through the point (6, 18), find the abscissa if the ordinate is 24.

(a) Since $xy = k$ and we know that one pair of (x, y) values is (12, 3), we can find $k = 36$. Therefore, $x = 3$ when $y = 12$.

(b) You should observe that $x \cdot y$ is always 24, therefore we have an instance of inverse variation. If $xy = 24$, then $x = \frac{24}{48}$ or $x = \frac{1}{2}$ when $y = 48$.

(c) The equation would be $\frac{rd^2}{l} = k$ (where r = resistance, l = length, and d = diameter). We find $k = \frac{0.3}{4}$ or 0.075. Therefore, the resistance of the 2000-foot wire would be 0.666 . . . ohms.

(d) We have $\frac{y}{x} = k$, so $k = 3$ and x (the abscissa) is 8 if y (the ordinate) is 24.

You will find a fairly representative group of problems in the Self-Test that follows. If you are faithful in working your way through them your knowledge of the concepts of ratio, proportion, and variation should be greatly enhanced, along with your ability to apply successfully the procedures associated with these concepts.

SELF-TEST

1. Write each ratio in lowest terms.

(a) 4:8

(b) $\frac{45}{15}$

(c) $45r{:}15r$

(d) $\frac{45x^2y}{15x^2y}$

(e) 3 days to 45 hours

(f) 35 revolutions per second to 45 rps

(g) The area of an 8-inch by 40-inch rectangle to the area of a rectangle that is 5 inches by 24 inches

(h) The number of men employees to women employees if there are 4,000 men in a total of 7,000 employees

(i) If Tom, Dick, and Harry receive a total of $675 for a week's work, how much should each receive if they are paid in the ratio of 3:4:3? (Hint: Let $3x$, $4x$, and $3x$ represent their shares, and then set up an equation.)

(frames 1 and 3)

2. (a) Is the following statement mathematically correct? If you compare two quantities of the same kind by expressing their measures in the same unit and then compute their quotient (as a decimal fraction), you have formed the ratio of the two quantities. (True / False)

   (b) Which of the following is the *best* statement?

   ___ (1) The ratio of 5 to 4 is $\frac{5}{4}$.

   ___ (2) The ratio of 5 to 4 is 5:4.

   ___ (3) The ratio of 5:4 is 1.25:1.

   ___ (4) The ratio of 5 to 4 can be expressed as $\frac{5}{4}$, 5:4, or 1.25:1. (frame 2)

3. An equality of two ratios is called a ______________________ (frame 4)

4. (a) If we have the proportion $x:y = a:b$, then x and b are called the __________ and y and a are called the ______________.

   (b) In the preceding problem the third term is ______.

   (c) The proportion $a:b = c:d$ is read ____.

   (d) The statement "In a proportion, the product of the extremes equals the product of the means" is

   ___ (1) always true

   ___ (2) sometimes true

   ___ (3) never true (frames 5 and 6)

5. Does the table below show that t is directly proportional to r? (Yes / No)

t	1	2	3	4
r	8	16	24	32

(frame 7)

6. Solve the following proportions for the missing term.

(a) $2:3 = a:18$

(b) $b:21 = 3:63$

(c) $(3 + c):4 = 8:16$

(d) $(3 - d):5 = 4:15$

(e) $\frac{1}{2}:e = 3:\frac{3}{4}$

(f) $(f - 3):4 = (f + 3):3$

(g) $g:(g - 3) = 7:4$

(h) $h:4 = 3:6$ (frame 7)

7. Establish a proportion from the information given in each problem; then solve the proportion.

   (a) If it requires 3000 calories to melt a mass of 15 grams, how many grams can be melted by 2500 calories if the heat required to melt the substance is directly proportional to the mass of the substance?

   (b) If the power required to move a certain boat is directly proportional to the cube of the speed of the boat, how much power is necessary to propel a boat at 40 knots if 60 horsepower is necessary to propel it at 20 knots?

   (c) If the distance required to stop a car is directly proportional to the square of the speed of the car, how much distance is required to stop a car traveling 80 miles per hour if it requires 240 feet to bring the car to a stop at 60 miles per hour?

   (d) The price of a certain type of precious gem is directly proportional to the cube of its weight. If a 4-carat gem costs $600, how much should you pay for a gem that weights $1\frac{1}{2}$ carats?

   (e) If one inch equals 24 miles on a certain map, and a certain county appears as a $2\frac{3}{8}$-inch by $1\frac{3}{4}$-inch rectangle, what is the area of the county in square miles?

   (f) What length represents $14\frac{3}{5}$ feet on a scale drawing if the drawing is scaled at $\frac{5}{8}$-inch equals 1 foot?

(frames 7 and 10)

8. What type of proportion is shown in the following table?

m	3	2	6	$\frac{1}{2}$	24
n	8	12	4	48	1

(frame 11)

9. (a) Does the force required to remove a bolt by using a wrench vary inversely or directly as a function of the length of the handle of the wrench?

   (b) AB represents a lever with the fulcrum (pivot point) at C. A force of 15 pounds at A will lift a weight of 45 pounds at B. How many pounds of force at A will lift 90 pounds at B if the distance from A to C is halved and the distance from B to C remains the same?

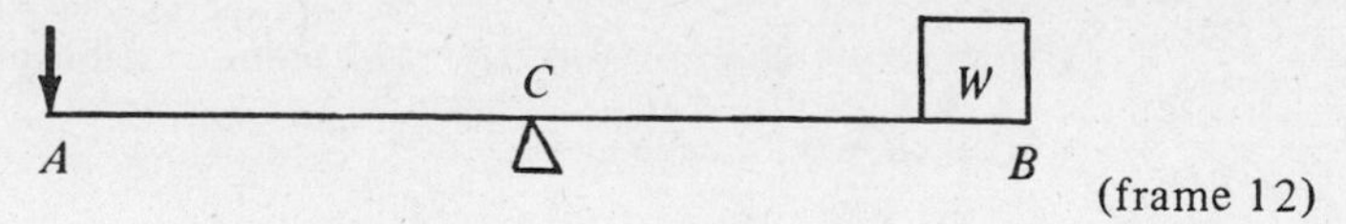

(frame 12)

10. (a) If the components of (x, y) are related in such a way that $\frac{x}{y} = k$, the x is

    ______________ (inversely / directly) proportional to y, and k is

    called the ______________.

    (b) Which of the following equations would represent the fact that the components of the ordered pair (x, y) vary inversely?

    ___ (1) $xy = k$ ___ (3) $y = kx$

    ___ (2) $x = ky$ ___ (4) all the equations

(c) Within certain limits it could be assumed that the number of weeks required to build a large building would be inversely proportional to the number of men assigned to the project. If we stayed within these limits what would happen to the number of weeks required to complete a project if the number of men assigned to the project was as follows:
(1) doubled
(2) halved
(3) tripled
(4) decreased by 6
(5) increased by 3
(6) increased by 5 when it was to be started with 10

(d) Which of the following equations indicate that m varies inversely with n? (Hint: More than one is correct.)

___ (1) $\frac{m}{k} = n$ ___ (3) $m \cdot n = k$

___ (2) $\frac{m}{n} = k$ ___ (4) $m = \frac{k}{n}$

(frames 13 and 14)

11. (a) A wire 0.3 inches in diameter has 14 ohms resistance. How much resistance would there be in a wire of the same material and length if the resistance varies inversely as the square of the diameter and the new wire has a diameter of 0.25 inches?

(b) A gear with 24 teeth turns at 600 revolutions per minute and drives a gear havin 36 teeth. Find the revolutions per minute of the second gear if the rpm of the second gear varies inversely as the number of teeth on the gear.

(frame 15)

12. (a) Is it mathematically correct to say "joint variation is always direct variation"? (Yes / No)

(b) If x varies jointly as r and t, and $x = 12$ when $r = 42$ and $t = \frac{1}{2}$, find x when $r = 21$ and $t = \frac{1}{4}$.

(c) If the volume of a rectangular solid varies jointly as its length, width, and depth, what would happen to the volume if the length, width, and depth were each doubled?

(d) If one variable varies directly as a second variable and inversely as a third variable, we would have an instance of ______________________ variation.

(frame 16)

13. (a) List the four kinds of variation discussed in this chapter.

(b) If x varies directly as y, what happens to x when y is doubled?

(c) If x varies directly as y^3, what happens to x when y is tripled?

(d) If x varies inversely as y, what happens to x when y is halved?

(e) If g varies inversely as t^2 and $g = 800$ when $t = 100$, find the value of g when $t = 50$.

(f) If $\frac{ar}{t} = k$, then a varies directly as _____ and inversely as _____.

(g) If x varies directly as m and inversely as n, and $x = 30$ when $m = 45$ and $n = 5$, find x if $m = 90$ and $n = 10$.

(h) Find x in the following variation problems.

(1) $\frac{x}{3(6)} = 4$ (3) $\frac{x}{12} = 3$

(2) $\frac{3x}{7} = 12$ (4) $\frac{\frac{1}{2}x}{3(\frac{1}{4})} = \frac{1}{3}$

(i) If the volume of a right circular cylinder is $V = \pi r^2 h$ (where V is the volume, r is the radius, h is the height, and π is a constant), would it be true to say that the volume of the cylinder varies jointly as its height and the square of the radius? (Yes / No)

(frames 16 and 21)

Answers to Self-Test

1. (a) 1:2; (b) 3:1; (c) 3:1; (d) 3:1; (e) 8:5; (f) 7:9; (g) 8:3; (h) 4:3; (i) Since $3x + 4x + 3x = 675$, Tom receives \$202.50, Dick receives \$270.00, and Harry receives \$202.50
2. (a) True; (b) 4
3. proportion
4. (a) extremes, means; (b) a; (c) a divided by b equals c divided by d, or a is to b as c is to d; (d) 1
5. Yes. This is true because the ratio of t to r is 1 to 8 for all of the entries in the table.
6. (a) 12; (b) 1; (c) −1; (d) $\frac{5}{3}$; (e) $\frac{1}{8}$; (f) −21; (g) 7; (h) 2
7. (a) $3000{:}2500 = 15{:}x$, $x = 12\frac{1}{2}$ grams
   (b) $60{:}x = 20^3{:}40^3$, $x = 480$ horsepower
   (c) $60^2{:}80^2 = 240{:}x$, $x = 426\frac{2}{3}$ feet
   (d) $600{:}x = 4^3{:}(1\frac{1}{2})^3$, $x =$ approximately \$31.65
   (e) $1{:}24 = (2\frac{3}{8}){:}x$, $1{:}24 = (1\frac{3}{4}){:}x$. The length is 57 miles; the width is 42 miles; therefore, the area is 2394 square miles.
   (f) $1{:}(\frac{5}{8}) = (14\frac{3}{5}){:}x$, $x = 9\frac{1}{8}$ inches
8. This is an inverse proportion because the product of the pairs of entries remains constant (is 24 in each case).
9. (a) inversely; (b) 60 pounds of force
10. (a) directly; constant of proportionality, or the constant of variation
    (b) 1
    (c) (1) the weeks would be halved
        (2) the weeks would be doubled
        (3) one-third the number of weeks would be required
        (4) This question can't be answered because we don't know the ratio of decrease in the number of men.
        (5) Again, this can't be answered because we don't know the ratio of the increase in the number of men.
        (6) It would take two-thirds the number of weeks.
    (d) 3 and 4

11. (a) approximately 20 ohms; (b) 400 rpm
12. (a) Yes; (b) $x = 3$; (c) The volume would be multiplied by 8; (d) combined
13. (a) direct, inverse, joint, and combined; (b) x is doubled; (c) x becomes 27 times as large; (d) x is doubled; (e) $g = 3200$; (f) t, r; (g) $x = 30$; (h) (1) $x = 72$, (2) $x = 28$, (3) $x = 36$, (4) $x = \frac{1}{2}$, (i) Yes

CHAPTER TWELVE

Solving Everyday Problems

OBJECTIVES

Way back in Chapter 1 we touched on the subject of translating word statements into numerical relationships or mathematical shorthand. With the knowledge and skills you have acquired through the intervening chapters you are ready to explore this area further. You should find this study challenging, interesting, and rewarding since it will sharpen your powers of analysis and give you the opportunity to develop your own approach to solving everyday problems.

When you have completed the work in this chapter you will be able to:

- translate verbal expressions of mathematical relationships into symbols;
- translate word statements into equations;
- solve verbal number problems of one unknown;
- solve verbal number problems of two unknowns;
- represent the age of a person by a mathematical expression derived from related facts and solve for the age value;
- formulate and solve percentage problems;
- formulate and solve time-speed-distance problems involving a separation situation, a closure situation, a round trip situation, an overtake situation, and motion problems of more than one variable;
- set up and solve mixture problems.

PRACTICE WITH ALGEBRAIC REPRESENTATION

1. The first (and sometimes the only) problem many individuals encounter in working with word problems is that of turning words into mathematical symbols. Since most of the words we use to express the relations between various quantities have either an explicit or implied mathematical operation associated with them, we are going to start by restating some of these in the form of a chart we first used in Chapter 1.

+	−	×	÷	=	x or another letter
sum and plus added to increased by	minus subtract less less than difference remainder decreased by	times product multiplied by	quotient divided by ratio	equals is is equal was equal is as much as	an unknown number

Refer to this chart as frequently as you need to in working with the concepts and problems throughout the rest of this chapter.

Start by translating the following statement into a mathematical expression.

If the sum of two numbers is 7 and the letter n represents one of them, represent the other number in terms of n.

$7 - n$

2. The previous problem might have been obvious to you. Since it is the most basic kind of algebraic representation, it is important as a starting point.

Many problems can be solved either by arithmetic *or* algebra. You may feel at times that we are doing things the hard way by solving a problem algebraically when the solution could be found more easily by arithmetic. We need practice with easy problems before we can hope to solve the harder problems, the ones that *cannot* be solved by arithmetic. So be patient and work the practice problems carefully.

Here is another one for you to try. If n stands for a certain number, represent the numbers described below.

(a) Five times as large as n ________

(b) Seven larger than n ________

(c) Three less than n ________

(d) One-half as large as n ________

(a) $5n$; (b) $n + 7$; (c) $n - 3$; (d) $\frac{1}{2}n$ or $\frac{n}{2}$

3. When translating words into symbols you must be sure you know what the words mean. The way writers of textbooks use words to express relationships between quantities often appears confusing to the reader. The best way to avoid confusion is to learn the meaning of the expressions used most commonly. Here are two such expressions which you will recognize from the chart in frame 1.

Increased by means *added to.* Thus, "8 increased by 3 equals 11" is written $8 + 3 = 11$.

Is often means *equals.* Thus, "half of 8 is 4" is written, $\frac{1}{2}(8) = 4$.

Translate the following statements into symbols.

(a) Two increased by 5 is 7. ______________

(b) n increased by 3 equals 8. ______________

(c) If a number (n) is increased by 9, the result is 16. ______________

(d) x increased by y is z. ______________

- - - - - - - - - - - - - - - - - -

(a) $2 + 5 = 7$; (b) $n + 3 = 8$; (c) $n + 9 = 16$; (d) $x + y = z$

4. Two more expressions commonly found in word problems are *decreased by* and *diminished by*; both mean subtract. Thus, "9 decreased by 4 is 5" is written $9 - 4 = 5$.

Translate the following statements into symbols.

(a) Seven decreased by 6 equals 1. ______________

(b) If a number (n) is decreased by 6, the result is 5. ______________

(c) x diminished by 12 is 7. ______________

(d) Find the number (n) that results when 5 is diminished by 2. (Hint: The equation will begin with $n = \ldots.$) ______________

- - - - - - - - - - - - - - - - - -

(a) $7 - 6 = 1$; (b) $n - 6 = 5$; (c) $x - 12 = 7$; (d) $n = 5 - 2$

5. Another way to indicate subtraction is by use of the word *exceeds.* Thus, "5 exceeds 3 by 2" means 5 is 3 more than 2, which we write $5 - 3 = 2$ or $5 - 2 = 3$.

The following exercises include both addition and subtraction operations expressed in various ways. Read them carefully and think about what they mean before you attempt to translate the word statements into algebraic equations. Where possible solve the equations and check your solutions by substitution back in the original word statements.

(a) Eight exceeds 1 by 7. ______________

(b) n exceeds 6 by 7. ______________ $n =$ ________

(c) Fifteen exceeds a certain number by 5. ______________

$n =$ ________

(d) Six increased by 7 equals 17 decreased by 4. ______________

(e) A number added to half itself equals 30. ______________

$n =$ ________

(f) A number diminished by 4 equals the sum of 3 and 6. ____________

$n =$ ________

(g) The sum of 5 and 8 exceeds a certain number by 2. ____________

$n =$ ________

(h) n increased by 9 exceeds 7 by 10. ____________ $n =$ ________

(a) $8 - 1 = 7$; (b) $n - 6 = 7$, $n = 13$; (c) $15 - n = 5$, $n = 10$;
(d) $6 + 7 = 17 - 4$; (e) $n + \frac{n}{2} = 30$, $n = 20$;
(f) $n - 4 = 3 + 6$, $n = 13$; (g) $(5 + 8) - n = 2$, $n = 11$;
(h) $(n + 9) - 7 = 10$, $n = 8$

MATHEMATICAL MODELS

6. We often refer to the algebraic representation of a verbal statement as a *mathematical model.* Below are some further examples to help familiarize you with the meaning of certain operational words and the way they are used.

Word Statement	Mathematical Model
Five *increased by* three means three *added to* five.	$5 + 3 = 8$
Half of ten *is* five means half of ten *equals* five.	$\frac{10}{2} = 5$
Twelve *decreased by* five means five *subtracted from* twelve.	$12 - 5 = 7$
Seven *exceeds* five by two.	$7 - 5 = 2$
Four *times* five is twenty.	$4 \cdot 5 = 20$
The positive *difference* of eight and three is five.	$8 - 3 = 5$

Although these expressions (and the rest of those shown in frame 1) occur commonly in word problems, they are by no means all-inclusive. They are intended only to represent the kinds of phrases or words you are apt to encounter. When you come across such words or phrases, make a real effort to think out what they mean. Remember that meaning is all-important. You must preserve the meaning when you write the mathematical model.

Here is some more practice. Translate each of the following statements into an equation and then solve the equation.

(a) If a number (n) is decreased by 7, the result is 8.

(b) Four times a number, increased by 3, equals 19.

(c) Three-eighths of a number equals 12.

(d) A number added to one-third of itself equals 24.

(e) When four times a number is diminished by 10, the remainder is 26.

(f) Seven exceeds one-half a number by 4.

(a) $n - 7 = 8$, $n = 15$; (b) $4n + 3 = 19$, $n = 4$; (c) $\frac{3n}{8} = 12$, $n = 32$;

(d) $n + \frac{n}{3} = 24$, $n = 18$; (e) $4n - 10 = 26$, $n = 9$;

(f) $7 - \frac{n}{2} = 4$, $n = 6$

SOLVING WORD PROBLEMS

7. Now let's talk for a moment about the steps involved in problem solving. Assuming you have read the problem carefully and determined the specific question you are being asked to answer, the four essential steps of problem solving are as follows.

Step 1: A *representation* of the unknown(s).

Step 2: A *translation* of the relationships about the unknowns into an equation or system of equations.

Step 3: A *solution* of the equation or system of equations to find values of the unknowns.

Step 4: *Checking* the values found to make sure that they satisfy the original problem. (Do not check your values in your equation as the equation itself may be wrong!)

The above steps are not new to you. They summarize some ideas we have worked with before. It is important now that you see these steps as necessary and sequential parts of the overall process of problem solving. Let us apply them, therefore, in solving a number problem having one unknown.

Example: Twice a certain number increased by 12 equals 26. Find the number.

Step 1, representation: Let n = number

Step 2, translation:

Twice a certain number	$2n$
increased by 12	$+ 12$
equals 26	$= 26$
or	$2n + 12 = 26$

Step 3, solution: $2n = 26 - 12 = 14$, $n = 7$

Step 4, check:

Twice a certain number	$2 \cdot 7 = 14$
increased by 12	$+ 12$
equals 26	$= 26$
or	$14 + 12 \stackrel{?}{=} 26$
	$26 \stackrel{\checkmark}{=} 26$

The sum of a certain number and 8 is 17. Find the number. (Follow the procedure given above.)

Step 1: Let n = the number
Step 2: $n + 8 = 17$
Step 3: $n = 17 - 8 = 9, n = 9$
Step 4: The sum of a certain number (9) and 8 is 17; $17 \overset{\checkmark}{=} 17$.

NUMBER PROBLEMS OF ONE UNKNOWN

8. It is always important to check your solution by substituting the values you have found for the unknown back in the original statement of the problem. Any root of an equation will, when substituted back in that equation, always reduce it to an identity. However, if the equation itself is not a correct mathematical model of the original problem, the solution values of the equation will not satisfy the conditions of the problem. Therefore, your answer may be wrong even though it satisfies the equation that you thought represented the problem correctly.

   Solve the following problems having one unknown.

   (a) A certain number is equal to 35 decreased by 21. What is the number?

   (b) 13 is equal to a number increased by 7. What is the number?

   (c) The sum of a number and 11 is 23. What is the number?

   (d) Seven added to three times a number is equal to 22. What is the number?

   (e) Four times a certain number is equal to 35 decreased by the number. What is the number?

   (f) If 25 is subtracted from a certain number, the difference is one-half the number What is the number?

   (g) Four times a certain number, decreased by 5, equals 25 diminished by 6 times the number. What is the number?

(a) $n = 35 - 21, n = 14$
check: $14 \overset{?}{=} 35 - 21, 14 \overset{\checkmark}{=} 14$
(b) $13 = n + 7, n = 6$
(c) $n + 11 = 23, n = 12$
check: $12 + 11 \overset{?}{=} 23, 23 \overset{\checkmark}{=} 23$
(d) $7 + 3n = 22, 3n = 15, n = 5$
(e) $4n = 35 - n, 5n = 35, n = 7$
check: $28 \overset{?}{=} 35 - 7, 28 \overset{\checkmark}{=} 28$
(f) $n - 25 = \frac{n}{2}, n = 50$
(g) $4n - 5 = 25 - 6n, n = 3$
check: $12 - 5 \overset{?}{=} 25 - 18, 7 \overset{\checkmark}{=} 7$

Note: It may appear that we are checking by substituting the values found back in the equation. It *appears* this way because our equations are correct models.

9. Did your checks prove out your solutions in each of the previous problems? If you are having difficulty with checking, study the examples below before going on. Otherwise proceed at once to frame 10.

Example 1: Does the number value 20 check in the statement "one-fifth of a number increased by 3 is 7"?

Check: $\frac{20}{5} + 3 \stackrel{?}{=} 7,\ 4 + 3 \stackrel{?}{=} 7,\ 7 \stackrel{\checkmark}{=} 7$ Answer: Yes

Example 2: Does the number value 3 check in the statement "seven times a number less 12 is 9"?

Check: $7(3) - 12 \stackrel{?}{=} 9,\ 21 - 12 \stackrel{?}{=} 9,\ 9 \stackrel{\checkmark}{=} 9$ Answer: Yes

Example 3: Is 24 the correct number if "three-fourths of a number decreased by 10 equals 8"?

Check: $\frac{3}{4}(24) - 10 \stackrel{?}{=} 8,\ 18 - 10 \stackrel{?}{=} 8,\ 8 \stackrel{\checkmark}{=} 8$ Answer: Yes

Example 4: Is 21 the correct number if "five times the sum of a number and 8 is three times the number less 2"?

Check: $5(21 + 8) \stackrel{?}{=} 3(21) - 2,\ 5(29) \stackrel{?}{=} 63 - 2,\ 145 \neq 61$ Answer: No

NUMBER PROBLEMS OF TWO UNKNOWNS

10. Having worked a bit with number problems involving one unknown, let's extend the procedures used to number problems having two unknowns. In number problems of two unknowns, two relationships concerning the unknowns are needed to achieve a solution (as you learned in Chapter 7).

Example: One number is 3 more than twice the other. What are the numbers if their sum is 18?

Method 1: Use one relationship to represent the two unknowns in terms of one letter. Use the other relationship to obtain a single equation. Then solve for both unknowns.

(1) Let s = the smaller number
Then $2s + 3$ = the larger number (since the larger number is 3 more than twice the smaller)
(2) Since the sum of the numbers is 18, then $s + (2s + 3) = 18$
(3) Solving: $3s = 15$, $s = 5$ (smaller number)
$2 \cdot 5 + 3 = 13$ (larger number)
(4) Checking: Sum of the numbers is 18, therefore $5 + 13 = 18$, or $18 = 18$; also, 13 is 3 more than twice 5.

Method 2: Represent each unknown by a different letter. Use the two relationships to obtain two separate equations which can then be solved by the algebraic methods for systems of equations (which you learned about in Chapter 8).

(1) Let s = smaller number
Let l = larger number
(2) Then $l = 2s + 3$, and $s + l = 18$.
(3) By substitution, $s + (2s + 3) = 18$, from which $s = 5$ and $l = 13$.
(4) Check: same as Method 1.

The second method should look familiar to you since it is one of the methods we used for solving a pair of linear equations in Chapter 8. Do you recall that such a pair of equations can be solved simultaneously by either substitution or by graphing? If you are vague about these methods, review Chapter 8 (beginning with frame 16) before continuing with the material that follows.

Below are some representative number problems having two unknowns. Solve them by either of the methods illustrated in the previous example and follow the four procedural steps described in frame 7.

(a) The sum of two numbers is 13 and their difference is 5. What are the two numbers?

(b) The larger of two numbers is equal to twice the smaller number increased by five. The smaller number equals the larger number decreased by 16. What are the numbers?

(c) Five times the smaller of two numbers decreased by seven is equal to twice the larger number increased by two. If the sum of the two numbers is 13, what are the numbers?

(d) The smaller of two numbers is equal to two less than half the larger number. If the larger number, increased by four, is equal to six times the smaller, what are the two numbers?

(e) The difference between two numbers is ten. Four times the larger decreased by 30 equals three times the smaller increased by twenty. What are the numbers?

(f) Three times the smaller of two numbers is equal to twice the larger. If the larger number plus two equals twice the smaller number less four, what are the numbers?

- - - - - - - - - - - - - - - - - -

(a) $s = 4$, $l = 9$; (b) $s = 11$, $l = 27$; (c) $s = 5$, $l = 8$;
(d) $s = 2$, $l = 8$; (e) $s = 10$, $l = 20$; (f) $s = 12$, $l = 18$

11. Do not check your answers in any equation because the equation may be incorrect! Check them in the original problem. We have repeated this warning again on the theory that you can't be reminded of it too often. Not following the advice contained in it is a common source of errors.

The process of turning word problems into algebraic equations in order to solve them is largely a matter of interpreting the meaning of the words correctly and then assembling the proper arrangement of symbols to represent them. This is not alway as easy as it sounds. We have discussed the meaning of such terms as decreased by and increased by. Now let's consider the word *exceeds* which is often used in place of the words "is greater than."

For example, we say that 9 *exceeds* 5 by 4 instead of saying that 9 *is greater than* 5 by 4. Also, the amount by which one quantity exceeds another is called the *excess.* Therefore, the excess is the difference between two numbers (that is, the amount by which one number exceeds another). Nearly everyone has trouble with expressions of this kind at some time or other. If you find you have difficulty forming equations for problems that contain the words exceeds or excess, mentally substitute *is greater than* for *exceeds,* and *difference of* for *excess over.*

Check to see if you have caught the right idea by completing the statements below.

(a) Another way of saying "is greater than" is ________________.

(b) Another way of saying "excess over" is ________________.

- - - - - - - - - - - - - - - - - -

(a) exceeds; (b) difference of

12. Before going on let's explore the three common ways of expressing number relationships: a larger number, a smaller number, or the difference between the numbers.

$$\text{larger} - \text{smaller} = \text{difference}$$
$$\text{larger} = \text{smaller} + \text{difference}$$
$$\text{larger} - \text{difference} = \text{smaller}$$

Apply these relationships in the following exercises.

(a) Write the expression "7 exceeds 4 by 3" in equation form in three different ways.

(b) Write the expression "b exceeds c by 9" in equation form in three different ways.

(c) Bill's age exceeds Tom's age by 5 years. Let x = the number of years for Tom's age. Is Bill's age greater than or less than Tom's age? Which of the following expressions will represent Bill's age: $x - 5$, $5 - x$, or $x + 5$?

(d) A's age exceeds B's age by 6 years. If x stands for B's age, how will we represent A's age?

(e) Separate 60 into two parts such that one part will exceed the other part by 36. (Use x for the larger and y for the smaller.)

(f) The rate of one train exceeds by 12 twice the rate of another. If x is the rate of the slower train in miles per hour, what is the rate of the faster train? (All that is required here is the correct algebraic expression for the rate of the faster train, not the solution.)

- - - - - - - - - - - - - - - - - -

(a) $7 - 4 = 3$, $7 = 4 + 3$, $7 - 3 = 4$; (b) $b - c = 9$, $b = c + 9$, $b - 9 = c$; (c) greater, $x + 5$; (d) $x + 6$; (e) $x + y = 60$, $x - y = 36$, $x = 48$, $y = 12$; (f) Let y = rate of the faster train in miles per hour; then $y = 2x + 12$.

AGE PROBLEMS

13. Here are two useful rules to remember when working age problems.

Rule 1: To find a person's future age, add the given number of years to his present age. Thus, in 5 years a person 20 years old will be 25 years old $(20 + 5)$.

Rule 2: To find a person's past age, subtract the given number of years from his present age. Thus, 5 years ago a person 20 years old was 15 years old $(20 - 5)$.

Represent the age of the person in each of the following problems.

(a) Ten years hence, if his present age is 25. Answer: $25 + 10$

(b) Ten years ago, if his present age is x years. ______________________

(c) Fifteen years hence, if his present age is y years. ______________________

(d) In x years, if his present age is 35 years. ______________________

(e) In y years, if his present age is x years. ______________________

(f) x years ago, if his present age is 50 years. ______________________

(g) y years ago, if his present age is k years. ______________________

- - - - - - - - - - - - - - - - - - - -

(b) $(x - 10)$ years; (c) $(y + 15)$ years; (d) $(35 + x)$ years; (e) $(x + y)$ years
(f) $(50 - x)$ years; (g) $(k - y)$ years

14. In the problems of frame 13 you were required to represent a person's age on the basis of his present age. Age problems often require that you represent a person's age on the basis of some past or future age.

For example, you may be asked to represent the age of a person (in years) 8 years hence, if he was 15 years old 5 years ago. The idea here is always to start by arriving at an expression for his present age, then adding (or subtracting) the additional years given. In this case we would represent his present age as $(15 + 5)$, to which we would then add 8 years to arrive at his age 8 years from now. This gives us $(15 + 5) + 8 = 28$.

Represent the age of the person in each of the following situations.

(a) Five years hence, if he was 20 years old 10 years ago. ______________________

____

(b) x years hence, if he was 10 years old 15 years ago. ______________________

(c) Nine years ago, if he will be 30 in 3 years. ______________________

(d) Ten years ago, if he will be 40 in x years. ______________________

(e) y years ago, if he will be 20 in x years. ______________________

(f) Ten years ago, if he was 20 two years ago. ______________________

- - - - - - - - - - - - - - - - - - -

(a) $(20 + 10) + 5 = 35$; (b) $(10 + 15) + x$; (c) $(30 - 3) - 9$;
(d) $(40 - x) - 10$; (e) $(20 - x) - y$; (f) $(20 + 2) - 10$

15. Now let's consider some typical age problems and see how we would go about solving them.

Example 1: A man is 9 times as old as his son. In 3 years the father will be only 5 times as old as his son. What is the age of each?

It is advisable to make four preliminary statements in the solution of a problem of this kind:

Let x = the son's age now (in years)
$9x$ = the father's age now (also in years, of course)
$x + 3$ = the son's age 3 years hence
$9x + 3$ = the father's age 3 years hence

Then $9x + 3 = 5(x + 3)$ (from the statement of the problem)
$$9x + 3 = 5x + 15$$
$$9x - 5x = 15 - 3$$
$$4x = 12$$
$$x = 3 \quad \text{(the son's age now)}$$
$$9x = 27 \quad \text{(the father's age now)}$$

Check: $9(3) + 3 \stackrel{?}{=} 5(3 + 3)$, $30 \stackrel{\checkmark}{=} 30$

Example 2: Bill is 10 years older than Hank. In 8 years twice Bill's age will equal 3 times Hank's age. What are their present ages?

Let x = Hank's age now
$x + 10$ = Bill's age now
$x + 8$ = Hank's age 8 years hence
$x + 18$ = Bill's age 8 years hence

Then $2(x + 18) = 3(x + 8)$
$$2x + 36 = 3x + 24$$
$$x = 12 \quad \text{(Hank's age now)}$$
$$x + 10 = 22 \quad \text{(Bill's age now)}$$

In the above problems we made use of the fact that the difference between two people's ages remains constant during their lives.

Using these examples as a guide, solve the following age problems.

(a) Walter is 8 years older than Ray. In 6 years, 5 times Walter's age will equal 9 times Ray's age. How old is each at present?

(b) The sum of the ages of Mary and her mother is 60 years. In 20 years, twice Mary's age increased by her mother's age (then) will equal 138 years. How old is each now?

(c) Robert is 14 years old and his father is 38 years old. How many years ago was the father exactly seven times as old as his son? (Hint: Let x = the number of years ago that the father's age was 7 times his son's age. Start by representing both son's and father's previous ages in terms of their present ages minus x.)

(d) Two years ago a woman was four times as old as her son. Three years from now the mother will be only 3 times as old as the son. How old is each at present? (Hint: Let x and $4x$ represent their ages 2 years ago.)

(e) A man was 30 years of age when his daughter was born. The father's age now exceeds 3 times the daughter's age by 6 years. How old is each at present?

(f) Fred is four times as old as Cliff. In 10 years he will be only twice as old as Cliff is then. How old is each?

\- - - - - - - - - - - - - - - - - -

(a)
$$\begin{aligned} \text{Let } x &= \text{Ray's age now} \\ x + 8 &= \text{Walter's age now} \\ x + 6 &= \text{Ray's age 6 years hence} \\ x + 14 &= \text{Walter's age 6 years hence} \\ \text{Then } 5(x + 14) &= 9(x + 6) \\ 5x + 70 &= 9x + 54 \\ 4x &= 16 \\ x &= 4 \quad \text{(Ray's age now)} \\ x + 8 &= 12 \ \text{(Walter's age now)} \end{aligned}$$

(b) x and $60 - x$ represent their ages now. $x + 20$ and $80 - x$ represent their ages 20 years hence. Equation is $2(x + 20) + (80 - x) = 138$ from which $x = 18$ and $60 - x = 42$, their present ages.

(c)
$$\begin{aligned} \text{Let } x &= \text{number of years ago father's age was 7 times the son's age} \\ 14 - x &= \text{son's age then} \\ 38 - x &= \text{father's age then} \\ \text{Equation: } (38 - x) &= 7(14 - x) \\ x &= 10 \text{ years} \end{aligned}$$

(d)
$$\begin{aligned} \text{Equation: } 4x + 5 &= 3(x + 5) \\ x &= 10 \ \text{(son's age two years ago)} \\ x + 2 &= 12 \ \text{(son's age now)} \\ 4x + 2 &= 42 \ \text{(mother's age now)} \end{aligned}$$

(e)
$$\begin{aligned} \text{Let } x &= \text{daughter's age now} \\ 30 + x &= \text{father's age now} \\ \text{Equation: } (30 + x) - 6 &= 3x \\ x &= 12 \\ 30 + x &= 42 \end{aligned}$$

(f)
$$\begin{aligned} \text{Let } x &= \text{Cliff's age now} \\ 4x &= \text{Fred's age now} \\ x + 10 &= \text{Cliff's age 10 years hence} \\ 4x + 10 &= \text{Fred's age 10 years hence} \\ \text{Equation: } 4x + 10 &= 2(x + 10) \\ x &= 5 \quad \text{(Cliff's age now)} \\ 4x &= 20 \ \text{(Fred's age now)} \end{aligned}$$

PERCENTAGE PROBLEMS

6. Now let's turn our attention to percentage problems. From your study of arithmetic you should be generally familiar with the idea of percentage. But just in case you may have forgotten, let's review briefly some of the fundamentals.

A *percent* of a certain number is called *percentage* (P). The number of which the percent is taken is called the *base* (B). The percent itself is called the *rate* (R). Percent means hundredths and is abbreviated %. Thus, 5% means 0.05 or $\frac{5}{100}$. When solving percentage problems always keep in mind that a common way of expressing a fraction whose denominator is 100 is as a certain number of percent. The familiar relationship (from arithmetic) between these three quantities is $P = RB$.

There are three kinds of percentage problems. Most people who try to do them with arithmetic become confused because they do not know what operations are to be performed with the numbers. Algebra simplifies the solution of such problems. The three kinds of percentage problems are as follows.

(1) *Finding a percent of a number.* For example, what is 5% of 240?

(2) *Finding what percent one number is of another.* For example, 34 is what percent of 85?

(3) *Finding a number when a percent of it is known.* For example, 117 is 65% of what number?

For a review, apply the procedures you learned in arithmetic to solve the problems given in the examples above.

(1) ________________ (3) ________________

(2) ________________

- - - - - - - - - - - - - - - - - -

(1) 12; (2) 40%; (3) 180

7. If you got the correct answer to the previous problems you remember the proper procedures from arithmetic. This will help you in comparing those procedures with the ones used in algebra. Examples for solving each of the foregoing problems by the algebraic method are given below.

Example (1): What is 5% of 240?

The word "what" really means "what *number*," and the word "of" means "times." Therefore, the problem is asking "5% of 240 (or 0.05×240) equals what number?"

$$x = \text{the required number}$$
$$x = 0.05 \times 240$$
$$x = 12$$

Example (2): 34 is what percent of 85?

"What percent" means "what number percent."

$$x = \text{the required number of percent}$$
$$34 = \frac{x}{100} \times 85$$
$$85x = 3400$$
$$x = 40 \text{ (the number of percent)}$$

Check: 40% of 85 = $0.40 \times 85 = 34$

Example 3: 117 is 65% of what number?

$$x = \text{the required number}$$
$$0.65x = 117 \text{ (or } 117 = 0.65x)$$
$$65x = 11700 \text{ (multiplying both sides by 100)}$$
$$x = 180 \text{ the number}$$

Check: $0.65 \times 180 = 117$

Use the correct algebraic procedure to solve each of the following problems.

(a) 8% of 450 = ______________________________

(b) 17 is what percent of 68? ______________________________

(c) 24 is 20% of what number? ______________________________

- - - - - - - - - - - - - - - - - -

(a) $x = .08 \times 450 = 36$; (b) $17 = \frac{x}{100} \times 68$, $68x = 1700$, $x = 25$;

(c) $.20x = 24$, $20x = 2400$, $x = 120$

18. Below you will find an assortment of percentage problems. It is up to you to determine and apply the correct procedures for solving each of them. Refer to frame 17 as often as necessary for help. At this point we are mainly concerned with the method of solution rather than the results.

(a) 3% of $500

(b) 35% of $200

(c) 7% of $800

(d) 7 is what percent of 14?

(e) 15 is what percent of 25?

(f) 60 is what percent of 40?

(g) 20 is 40% of what number?

(h) 12 is what percent of 16?

(i) 12 is 12% of what number?

(j) 68 is 200% of what number?

(k) What is $12\frac{1}{2}$% of 96?

(l) A baseball team won 9 games and lost 3. What percent of its games did it win?

(m) A man received $24 on an investment of $400. What rate did he receive?

(n) A merchant bought a chair for \$100 and marked the price up 40% for resale. What price did he put on it?

(o) What is 0.1% of \$5000?

- - - - - - - - - - - - - - - - - -

(a) \$15; (b) \$70; (c) \$56; (d) 50%; (e) 60%; (f) 150%; (g) 50; (h) 75%; (i) 100; (j) 34; (k) 12; (l) 75%; (m) 6%; (n) \$140; (o) \$5

19. A slightly different variety of percentage problems appears below.

Example: 10% of a number increased by 28% of the same number equals 57. What is the number?

$$
\begin{aligned}
x &= \text{the number} \\
0.10x + 0.28x &= 57 \\
0.38x &= 57 \\
38x &= 5700 \\
x &= 150
\end{aligned}
$$

Check: $0.10(150) + 0.28(150) \stackrel{?}{=} 57;\ 15 + 42 \stackrel{?}{=} 57;\ 57 \stackrel{\checkmark}{=} 57$

Use this approach to solve the following problems.

(a) 8% of a number plus 12% of the number is 62. What is the number?

(b) A number decreased by 10% of itself equals 405. What is the number?

(c) 10% of a number plus 8% of the number, decreased by 6% of the number, equals 42. What is the number?

(d) A radio was sold for \$68.00 after discounts of 10% and 5% off the list price were allowed. What was the list price of the radio?

(e) A baseball team with a standing of 0.839 has won 94 games. How many games has it lost? (Hint: Let x = total number of games played.)

(f) How cheaply can a grocer afford to sell berries that cost 12¢ a quart if he must make a profit of 20% based on the selling price? (Hint: Let x = selling price in cents.)

- - - - - - - - - - - - - - - - - -

(a) $0.08x + 0.12x = 62$; $x = 310$, the number
(b) $x - 0.10x = 405$; $0.9x = 405$; $x = 450$, the number
(c) $0.10x + 0.08x - 0.06x = 42$; $0.12x = 42$; $x = 350$
(d) $x - 0.10x - 0.05x =$ \$68; $0.85x =$ \$68; $x =$ \$80, the list price of the radio
(e) $0.839x = 94$; $x = 112$, total games played; $112 - 94 = 18$, number of games lost
(f) $x - 0.20x =$ 12¢; $0.8x =$ 12¢; $x =$ 15¢, selling price

GENERAL APPROACH TO STATEMENT PROBLEMS

20. No doubt you have observed by now that the process of solving a stated problem by means of an equation is not always simple, nor is the procedure always obvious. Now that you have come this far and gained some experience in solving word problems, perhaps it would be helpful to summarize what your general approach to problems of this kind should be. The following should reinforce what you have been doing and help guide you in your subsequent problem solving.

    (1) Read the problem carefully. Study it until the situation is clear to you.
    (2) Identify the quantities (both known and unknown) that are involved in the problem.
    (3) Select one of the unknowns and represent it by a letter (usually x) and then express the other unknown in terms of this letter if you wish to use one equation. If you decide to use more than one letter to represent the two unknowns, you will need to develop at least as many equations as you have unknowns.
    (4) Search the problem for the information that tells you which quantities (or what combinations of them) are equal.
    (5) When the desired combinations have been found, set them equal to each other, thus obtaining an equation.
    (6) Solve the equation thus obtained and check the resulting number value in the original statement of the problem.

MOTION PROBLEMS

21. Now it is time for us to look at a different type of problem, the kind usually known as a motion problem. There are three elements in the motion problem: time (t), rate (r), and distance (d). You will recall that we worked with these quantities in Chapter 7 and that we found they are related as follows:

$$d = rt \qquad r = \frac{d}{t} \qquad t = \frac{d}{r}$$

Thus, we can say that a car moving at a uniform speed of 50 miles per hour will travel 200 miles in 4 hours because $4 \cdot 50 = 200$.

When working with motion problems the following two points must be kept firmly in mind:

    (1) Speed (or rate) means either uniform speed or average speed. (We assume constant speed.)
    (2) The units you select to express time, speed, and distance must be consistent. If the rate is in miles per hour (mph), as it usually is, then you must use miles for distance and hours for time. If you come across a problem in which time is given in minutes, or in hours *and* minutes, you must convert the minutes to a fractional part of an hour by dividing by 60 (since there are 60 minutes in an hour).

For practice in using the time-speed-distance formula, solve the following easy problems.

(a) A car travels at a speed of 60 mph for 3 hours. How far does it go?

(b) An airplane flies 900 miles in 2 hours. What is its average rate of speed?

(c) A train travels 250 miles at an average speed of 50 mph. How long does it take to make the trip?

(a) (60 mph)(3 hours) = 180 miles; (b) $\frac{900 \text{ miles}}{2 \text{ hours}} = 450$ mph;

(c) $\frac{250 \text{ miles}}{50 \text{ mph}} = 5$ hours

SEPARATION SITUATION

22. The previous problems were simply exercises in using the time-speed-distance formula. Now let's try solving some uniform rate problems that require the application of this formula.

Example 1: Suppose you get in your car and drive east at 40 mph for 4 hours. At the same time you leave, a friend gets in his car and drives west (leaving from the same point) at 50 mph during the same 4 hours. How far apart will the two of you be at the end of the 4 hours?

Since motion problems are easier to visualize—and, consequently, to solve—with the aid of a diagram, we have illustrated the situation below. From this illustration you can see that the problem virtually solves itself.

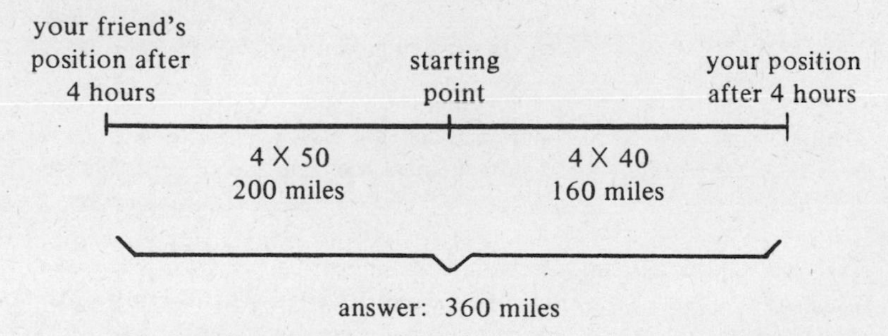

Example 2: Two planes leave the same airport at the same time and fly in opposite directions. The speed of the faster plane is 100 mph faster than that of the slower plane. At the end of 5 hours they are 2000 miles apart. Find the rate of each plane.

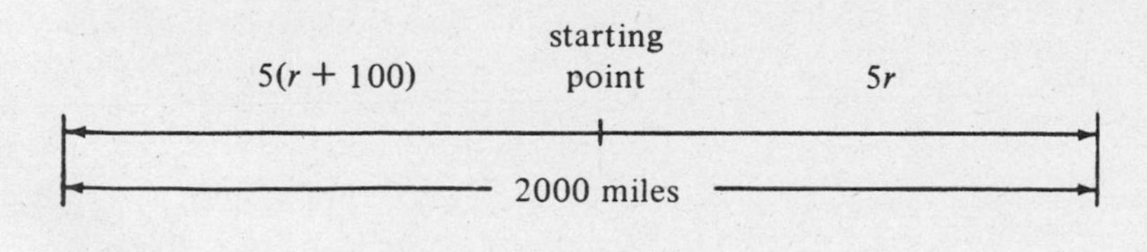

Let r = rate of slower plane
$r + 100$ = rate of faster plane

$$5r + 5(r + 100) = 2000 \text{ (since the sum of the distances each flew in 5 hours equals 2000 miles)}$$
$$r = 150 \text{ mph (rate of slower plane)}$$
$$r + 100 = 250 \text{ mph (rate of faster plane)}$$

The kind of problem discussed in these two examples is known as a *separation situation* or separation problem. Below are three problems of this type. Be sure you draw a simple diagram of the situation in each case to assist you in visualizing it. In each of these problems two people start from the same place at the same time and travel in opposite directions. Use r for rate, t for time, and d for distance, and then develop a mathematical model (relationship between the known and unknown elements) for each problem.

(a) After 3 hours two drivers are 330 miles apart; one is traveling 10 mph faster than the other. Find the rate of speed of the slower driver.

(b) At speeds of 40 mph and 20 mph two cars travel the same amount of time until they are 420 miles apart. How many hours do they travel?

(c) At speeds in the ratio of 7:3, two drivers are 360 miles apart at the end of 3 hours. Find the rate of each. (Hint: Use $7r$ and $3r$ to represent their respective rates.)

- - - - - - - - - - - - - - - - - -

(a) $3r + 3(r + 10) = 330$; $r = 50$ mph
(b) $40t + 20t = 420$; $t = 7$ hours
(c) $3(7r) + 3(3r) = 360$; $r = 12$ mph (Faster rate is 84 mph; slower rate is 36 mph.)

CLOSURE SITUATION

23. The next type of motion problem is called a *closure situation.* In this case two travelers start from distant points at the same time and travel toward each other until they meet.

Example: Tom and Jerry, who are 568 miles apart, start driving toward each other in their cars. Tom drives 40 mph and Jerry drives 36 mph. How soon will they meet if Tom has an accident and is delayed one hour before continuing his trip?

What we are seeking here is time (t), the interval between the instant at which they start and the moment they finally meet. What effect does Tom's delay have on this? Very little. We represent his time by $(t - 1)$. Therefore, we can state the following:

distance Jerry travels = $36t$
distance Tom travels = $40(t - 1)$

$$36t + 40(t-1) = 568$$
$$76t - 40 = 568$$
$$t = 8 \text{ hours}$$

Here is a similar problem for you to solve. Two trains start at the same time from towns 385 miles apart and meet in 5 hours. If the rate of one train is 7 mph less than the rate of the other train, what is the rate of each?

- - - - - - - - - - - - - - - - - -

Let x = rate of faster train
$x - 7$ = rate of slower train
$5x + 5(x-7) = 385$
x = 42 mph (faster train)
$x - 7$ = 35 mph (slower train)

ROUND TRIP SITUATION

24. A third type of motion problem is known as the *round trip situation.* In this situation a traveler leaves and returns to the starting point along the same road. Below is a typical problem of this kind.

Example: Paul O'Shea drove from his home to Boston and back again (along the same road) in 10 hours. His average speed going was 20 mph and his average speed returning was 30 mph. How long did he take in each direction and what distance did he cover each way?

The key fact to use here is that the distance going must equal the distance returning since Paul O'Shea returned by the same route.

Let t = time going
$10 - t$ = time returning
$20t$ = distance one way (going)
$30(10-t)$ = distance one way (returning)

$20t = 30(10-t)$
t = 6 hours (time going)
$10 - t$ = 4 hours (time returning)

Both $20t$ and $30(10-t)$ = 120 miles (distance each way)

Using this same general approach, solve the following problems.

(a) A traveler took 2 hours more returning from a trip than going. He averaged 50 mph out and 45 mph back. Find the time going. (Use t to represent this.)

(b) Mr. Goldsby drove his car from his home to Ventura at a rate of 35 mph and returned at a rate of 40 mph. Find his time going and returning if the time returning was one hour less than the time going.

- - - - - - - - - - - - - - - - - -

(a) $50t = 45(t+2)$, t = 18 hours
(b) t = time going; $35t = 40(t-1)$; t = 8 hours (time going); $t - 1$ = 7 hours (time returning)

OVERTAKE SITUATION

25. A fourth type of motion problem often encountered is known as the gain or *overtake situation.* The essence of this situation is that after a traveler has begun his trip, a second traveler starts from the same place and, going in the same direction, overtakes the first traveler.

Example: An hour after Mike left on a week-end bicycle trip his family noticed he had forgotten to take his sleeping bag. His brother Ed jumped into his car and started after Mike. If Mike was traveling at the rate of 8 mph and Ed drove at the rate of 40 mph, how long did it take Ed to catch up with Mike?

The key to this kind of problem is (as in the case of the round trip situation) that the distance traveled by each individual is the same. The times traveled will differ but the distances will not.

Let x = Ed's time
$x + 1$ = Mike's time (since he started an hour before his brother)

$40x = 8(x + 1)$ (They traveled equal distances.)

$x = \frac{1}{4}$ hour (the time it took Ed to catch up with Mike)

Use this approach to solve this problem. A traveler begins a trip traveling at the rate of 20 mph. Three hours later a second traveler, proceeding at 40 mph, sets out to overtake him. How long will it take him to do so?

– – – – – – – – – – – – – – – – – – –

Let t = time of second traveler
$t + 3$ = time of first traveler
$40t = 20(t + 3)$
$t = 3$ hours

MOTION PROBLEMS OF MORE THAN ONE UNKNOWN

26. Sometimes you will encounter problems that require the solution for two unknowns and that contain enough information to enable you to write two equations. Such a situation permits a solution by simultaneous equations. Learn to recognize such problems. Their solution will be easier if you know how to make the best use of the data available to you. Here is one such problem, another type of motion problem.

Example: Dr. Byrkit covered 310 miles by traveling for 4 hours at one speed and then for 5 hours at another speed. Had he gone 5 hours at the first speed and 4 hours at the second speed he would have covered 320 miles. Find the two speeds.

Obviously, the two speeds are our two unknowns. Therefore, if we let x = the first speed and y = the second speed, we get:

$$\begin{aligned} (1) \quad 4x + 5y &= 310 \\ (2) \quad 5x + 4y &= 320 \end{aligned}$$

Multiplying (1) by 5: $20x + 25y = 1550$

Multiplying (2) by 4: $20x + 16y = 1280$

Subtracting (2) from (1): $9y = 270$

$y = 30$ mph

Substituting $y = 30$ in (1): $4x + 5 \cdot 30 = 310$

$x = 40$ mph

Solve the following using this procedure. By going 20 mph for one period of time and 30 mph for another, Mr. Smith traveled 280 miles. Had he gone 10 mph faster in each case he would have covered 390 miles. How long did he travel at each speed?

This problem differs from the example only in the fact that the two unknown values are time (hours) rather than speed.

Let x = first time (at 20 mph)

y = second time (at 30 mph)

$$\begin{aligned} (1) \quad 20x + 30y &= 280 \\ (2) \quad 30x + 40y &= 390 \text{ (10 mph faster in each case)} \end{aligned}$$

Multiplying (1) by 3: $60x + 90y = 840$

Multiplying (2) by 2: $60x + 80y = 780$

Subtracting (2) from (1): $10y = 60$

$y = 6$ hours

$x = 5$ hours

27. Below is one example of each of the five kinds of motion problems we have just discussed. Solve each referring to the appropriate frame for help if you need it.

(a) An airplane traveled from its base to a distant point and back again along the same route in a total of 8 hours. Its average rate going was 180 mph and its average rate returning was 300 mph. How long did it take in each direction and what was the distance covered each way?

(b) Two trains leave the same terminal at the same time and travel in opposite directions. After 8 hours they are 360 miles apart. The speed of the faster train is 3 mph less than twice that of the slower train. Find the rate of speed of each train.

(c) A boat travels at 24 mph. A patrol boat starts 3 hours later from the same place and travels at 32 mph in the same direction. How long will it take to overtake the first boat?

(d) Two planes leave from points 1925 miles apart at the same time and fly towards each other. Their average speeds are 225 mph and 325 mph, respectively. How soon will the planes meet?

(e) By traveling for 5 hours at one speed and then for 3 hours at another, Mr. Stewart covered 250 miles. Had he traveled for 2 hours longer at each speed he would have covered 370 miles. Find the two rates at which he traveled.

(a) Time going was 5 hours; time returning was 3 hours; distance covered each way was 900 miles.
(b) Rates were 16 mph and 29 mph.
(c) Time needed to overtake is 9 hours.
(d) Planes will meet in $3\frac{1}{2}$ hours.
(e) Rates were 35 mph and 25 mph.

MIXTURE PROBLEMS

28. Now we will leave time-speed-distance problems and investigate the *mixture problem.* This type of problem can appear rather confusing unless you know how to handle it. Here is one such problem.

Example: A seedsman has clover seed worth 30¢ a pound and timothy seed worth 12¢ a pound. How many pounds of each should he use to make a mixture of 300 pounds worth 18¢ a pound?

The important point to bear in mind when working mixture problems (and they come in a great variety) is that:

$$\left.\begin{matrix}\text{the total value of}\\ \text{a number of units}\\ \text{of the same kind}\end{matrix}\right\} = (\text{number of units}) \times (\text{value of one unit})$$

We know the value of the clover (30¢ per pound) and of the timothy seed (12¢ per pound). But since we do not know the weight of either we will have to let x equal the weight of the clover seed and $(300-x)$ equal the weight of the timothy seed (since together they form a mixture that will weight 300 pounds).

$$30x = \text{value of clover in cents}$$
$$12(300-x) = \text{value of timothy in cents}$$
$$18(300) = \text{total value of the mixture}$$

$$30x + 12(300-x) = 18(300)$$
$$18x = 1800$$
$$x = 100 \text{ (pounds of clover)}$$
$$300 - x = 200 \text{ (pounds of timothy)}$$

Notice that we could have solved this using two unknowns (letting y represent the weight of the timothy seed) since we have as a second equation $x+y=300$. This would allow a solution by simultaneous equations. The result would have been the same. Try it for yourself if you are interested in verifying this.

Here are a few mixture problems for you to solve. Some are quite similar to the example above and others are slightly different. They can all be solved by following the basic principles and procedures used in the example. Don't be alarmed or discouraged if each new problem you read sounds different from the preceding one. Remember that you are learning to apply basic principles and procedures. In any case, if word problems were all alike they would be pretty dull!

(a) A dealer in tea has one brand worth 50¢ a pound and another worth 80¢ a pound. How many pounds of each kind must he use to make a 60-pound mixture worth 72¢ per pound?

(b) A coffee merchant blended coffee worth 93¢ a pound with coffee worth $1.20 a pound. The mixture of 30 pounds was valued by him at $1.02 a pound. How many pounds of each grade did he use?

(c) Maple syrup worth $6.00 a gallon and corn syrup worth 80¢ a gallon are used to make a mixture worth $2.36 a gallon. How many gallons of each kind of syrup are needed to make 50 gallons of the mixture?

(d) An order of candy cost $14. It contained one kind of candy worth 60¢ a pound and another worth 50¢ a pound. If there were 5 more pounds of the 60¢ candy, how many pounds of each kind were in the order? (Hint: Let x = pounds of kind worth 50¢, $x + 5$ = pounds of kind worth 60¢.)

(a) x = pounds of 50¢ tea, $60 - x$ = pounds of 80¢ tea; then $50x + 80(60 - x) = 72(60)$; $x = 16$ pounds, $60 - x = 44$ pounds
(b) x = pounds of cheaper coffee, $30 - x$ = pounds of better coffee; then $93x + 120(30 - x) = 102(30)$; $x = 20$ pounds, $30 - x = 10$ pounds
(c) x = gallons of maple syrup, $50 - x$ = gallons of corn syrup; then $6x + .80(50 - x) = 2.36(50)$; $x = 15$ gallons, $50 - x = 35$ gallons
(d) x = pounds of 50¢ candy, $x + 5$ = pounds of 60¢ candy; then $.50x + 60(x + 5) = 14$; $x = 10$ pounds, $x + 5 = 15$ pounds

You will find additional problems of this kind in the Self-Test that follows. Be sure to work some of them if you feel you need more practice in recognizing the various approaches to solving them.

SELF-TEST

1. Translate the following statements into symbols.

(a) Three increased by y is 9. ______

(b) n increased by 11 is 19. ______

(c) The result of increasing y by 7 is 12. ______

(d) z plus 2 increased by 4 is 8. ______

(frames 2 and 3)

2. Translate the following into equations.

(a) Eight decreased by k is 3. ______

(b) A number decreased by half itself equals 8. ______

(c) y decreased by c equals k. ______

(d) The value of p is equal to 15 diminished by 7. ______

(frame 4)

3. Translate these word statements into algebraic equations.

(a) 9 exceeds y by 3. ______

(b) A number diminished by 5 equals half itself. ______

(c) x increased by 6 equals y decreased by 3. ______________

(d) Four increased by c exceeds 11 by 5. ______________

(frame 5)

4. Write equations for the following statements.

(a) Three times n decreased by 7 equals 13. ______________

(b) Twenty-two exceeds m by 18. ______________

(c) y increased by 30 is half of m times 12. ______________

(d) Twice the difference of b and 4 equals three times the sum of m and n.

____

(e) Seventeen exceeds 11 by twice the difference of x and 2.

____

(f) Three times $(z + 2)$ is half the sum of y and 8.

____

(frame 6)

5. Solve the following number problems having one unknown.

(a) The sum of a certain number and 12 is 21. What is the number?

(b) Twenty increased by twice k is 32. What is the value of k?

(c) Sixteen is the difference obtained when n is subtracted from 28. What is the value of n?

(d) The product of 3 and $n + 6$ equals 20. Find the value of n.

(e) The difference between 15 and half of x is 6. What is the value of x?

(frames 7 and 8)

6. Solve the following number problems having two unknowns.

(a) Find the two integers whose sum is 15 and whose difference is 5.

(b) Find the two integers whose sum is 40 and the larger is 10 more than the smaller.

(c) The sum of two numbers is 12. Twice the larger plus the smaller equals 21. Find the numbers.

(d) Separate 50 into two parts such that the larger is nine times the smaller.

(e) Separate 42 into two parts such that the smaller is 3 less than one-half the larger.

(frame 10)

7. Represent the age of the person in each of the following.

(a) 6 years hence, if he was 30 years old 7 years ago.

(b) 10 years ago, if he will be 49 in 6 years.

(c) 9 years hence, if he was 22 years old 9 years ago.

(d) 17 years ago, if he will be 62 in 11 years.

(frame 13)

8. Solve the following age problems.

(a) A man is 28 years older than his son. In 3 years the father will be 5 times as old as the son. Find their present ages.

(b) The sum of the ages of Bill and John is 39 years. John is 3 years more than twice Bill's age. What are their ages?

(c) Mike is twice as old as Ed. In 11 years Ed's age will be $\frac{2}{3}$ Mike's age. What are their present ages?

(d) At present Lyle is 6 times as old as Jerry. Two years hence Lyle will be 10 times as old as Jerry was 3 years ago. What are their present ages?

(frame 15)

9. Solve the following percentage problems.

(a) 5% of $800 = ________

(b) 23% of $1150 = ____________

(c) 23 is what percent of 92? ________

(d) 8 is what percent of 64? ________

(e) 18 is 20% of what number? ________

(f) 40.5 is 15% of what number? ________

(g) Eighteen students in a class of 45 made a grade of B or better. What percent of the students made a grade of less than B? ________

(h) What would be your rate of return if you received $18 interest on an investment of $400? ________

(frame 17)

10. Solve these percentage problems.

(a) A number increased by 20% of itself equals 72. What is the number?

(b) 10% of a number increased by 24% of the same number equals 142.8. What is the number?

(frame 19)

11. (a) A boat travels 18 mph for $4\frac{1}{2}$ hours. How far does it travel?

(b) A man walks 8.8 miles in four hours. What was his average rate of travel?

(c) A racing car travels a distance of 420 miles at an average speed of 105 miles per hour. What was its travel time?

(frame 21)

12. (a) Two people start from the same place at the same time and travel in opposite directions. If one of them travels 20 mph faster than the other and they are 600 miles apart at the end of 5 hours, at what rate was each of them traveling?

    (b) Two cars travel in opposite directions for the same amount of time. The rate of one is 35 mph and the rate of the other is 50 mph. If they are 255 miles apart when they stop, how long did they travel?

(frame 22)

13. (a) Two cars start at the same time from towns 448 miles apart and meet in 4 hours. If the rate of one is 8 mph more than the other, what is the rate of each?

    (b) Starting 297 miles apart, two drivers travel toward each other at rates of 38 mph and 28 mph until they meet. What is the travel time of each?

(frame 23)

14. (a) A traveler takes 3 hours less to travel back to his starting place than he did to go out. If he averaged 45 mph out and 54 mph back, find his time to return.

    (b) After taking 3 hours to go out from his starting place, a traveler returns in 5 hours, averaging 28 mph slower on the way back. What was his average rate going?

(frame 24)

15. (a) Two and one-half hours after a traveler has begun his trip a second traveler starts from the same place and overtakes him after traveling for 6 hours at a rate of 34 mph. Find the rate of the first traveler in mph.

    (b) The first traveler travels 18 mph slower than twice the speed of the second traveler. If the second traveler starts out 2 hours later than the first traveler and overtakes him in 5 hours, find the rate of the second traveler in mph.

(frame 25)

16. (a) By traveling for 2 hours at one speed and then 7 hours at another, Mr. Gerber completed a trip of 258 miles. Had his first rate been 10 mph faster and his second rate twice as fast, he would have gone 488 miles. What were his two rates of speed?

(frame 26)

17. Find the number of pounds of each grade of coffee in each of the following problems.

    (a) A mixture of coffee worth $1.20 a pound and of coffee worth $1.80 a pound is valued at $18. If the price per pound of each grade is increased 30¢, the new value of the mixture would be $21.30.

    (b) A mixture of coffee worth $1 a pound and of coffee worth $1.40 a pound is valued at $17. Had there been twice as many pounds of the second coffee, the value of the mixture would have been $24.

(frame 28)

Answers to Self-Test

Note: Many of these answers can be expressed in different but equivalent ways. Check to see that yours has the same value.

1. (a) $3 + y = 9$; (b) $n + 11 = 19$; (c) $y + 7 = 12$; (d) $(z + 2) + 4 = 8$
2. (a) $8 - k = 3$; (b) $n - \frac{n}{2} = 8$; (c) $y - c = k$; (d) $p = 15 - 7$
3. (a) $9 - y = 3$; (b) $n - 5 = \frac{n}{2}$; (c) $x + 6 = y - 3$;
   (d) $(4 + c) - 11 = 5$
4. (a) $3n - 7 = 13$; (b) $22 - m = 18$; (c) $y + 30 = \left(\frac{m}{2}\right)12$ or $\frac{12m}{2}$;
   (d) $2(b - 4) = 3(m + n)$ (c) $17 - 11 = 2(x - 2)$; (f) $3(z + 2) = \frac{1}{2}(y + 8)$
5. (a) 9; (b) 6; (c) 12; (d) $\frac{2}{3}$; (e) 18
6. (a) 5 and 10; (b) 15 and 25; (c) 3 and 9; (d) 5 and 45; (e) 12 and 30
7. (a) $(30 + 7) + 6$; (b) $(49 - 6) - 10$; (c) $(22 + 9) + 9$;
   (d) $(62 - 11) - 17$
8. (a) 4 and 32; (b) 12 and 27; (c) 11 and 22; (d) 8 and 48, equation: $10(x - 3) = 6x + 2$
9. (a) \$40; (b) \$264.50; (c) 25%; (d) $12\frac{1}{2}$%; (e) 90; (f) 270; (g) 60%;
   (h) $4\frac{1}{2}$%
10. (a) 60; (b) 420
11. (a) 81 miles; (b) 2.2 mph; (c) 4 hours
12. (a) 50 and 70 mph; (b) 3 hours
13. (a) 52 and 60 mph; (b) $4\frac{1}{2}$ hours
14. (a) 15 hours; (b) 70 mph
15. (a) 24 mph; (b) 14 mph
16. (a) 24 mph for 2 hours and 30 mph for 7 hours
17. (a) 3 pounds at \$1.20 and 8 pounds at \$1.80; (b) 10 pounds at \$1.00 and 5 pounds at \$1.40

REVIEW TEST 4

Chapter 10

1. The symbol for "greater than" is ________.

2. Use an inequality symbol to express the fact that +3 is less than +7. ________

3. $-3 < -6$ (True / False)

4. For any two real numbers, the statement $a > b$ is equivalent to the statement $a - b$ is positive. (True / False)

5. Insert the proper inequality sign as needed so that these two inequalities will have the same sense.

$$a < b \qquad c \_\_\_ d$$

6. The sense of an inequality is unchanged if the same number is added to or subtracted from both sides. (True / False)

7. In the inequality $5 > 2$, the sense of the inequality will be unchanged if both sides are multiplied or divided by 3. (True / False)

8. In the inequality above, the sense would remain unchanged if both sides were multiplied or divided by -3. (True / False)

9. Multiply $2 > 1$ by -3. ________

10. Squaring both sides of the inequality $3 > 2$ gives us the new inequality ________

11. Simplify the inequality $3x - 2 > 2x + 1$. ________

12. $a < b$ is equivalent to the statement $a - b$ is negative. (True / False)

13. If x is a real number not equal to zero, then $x^2 > 0$ is an absolute inequality. (True / False)

14. $x - 7 > 0$ is an example of a conditional inequality. (True / False)

15. $x - 3 > 0$ is considered a linear inequality because it contains no term of degree higher than one. (True / False)

16. The algebraic solution of $x - 3 > 0$ is ________.

17. Use the number scale below to indicate the graphic solution of the inequality $x + 3 > 0$.

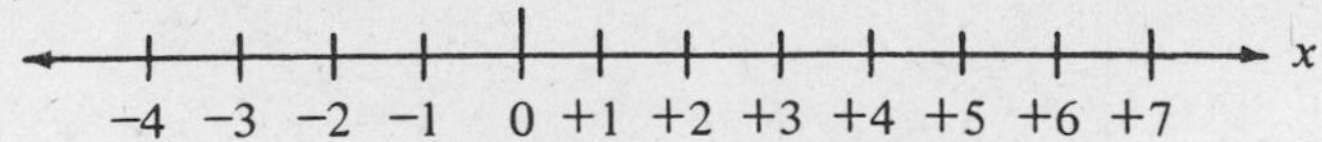

18. What is the symbol that means "greater than or equal to"? ________

19. Show how you would represent graphically the inequality "x greater than or equal to minus one."

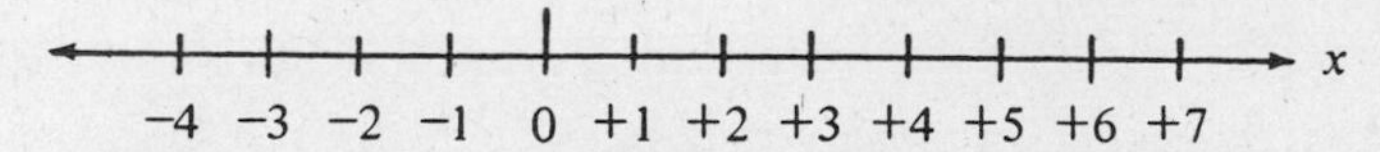

20. Graph the inequality $2x - y + 4 > 0$. Where does the coordinate half-plane lie that contains all the points whose coordinates make the value of the left member greater than zero?

Chapter 11

21. Express the ratio of 10 horses to 25 horses. ____________

22. Write the ratio of 3 feet to 7 feet. ____________

23. $\frac{2}{5} = \frac{4}{10}$ is a proportion. (True / False)

24. Write a matching ratio that will complete the proportion $\frac{3}{4} =$ ____________

25. 2:3 = 8:12 is a correct expression of a proportion. (True / False)

26. In the proportion $2:5 = 4:x$, what is the product of the extremes? ________

 Of the means? ________

27. Find the missing term in the proportion $2:5 = 4:x$. ________

28. Find the value of a in the expression $a:4 = 4:a$. ____________

29. Find the value of y in the proportion $3:(y-2) = y:1$. ____________

30. Use the idea of direct proportion to solve the following problem. If, at a certain time of day, a tree 10 feet tall casts a shadow 20 feet long, how tall will a building be that casts a shadow 100 feet long?

31. The following table is a correct example of inverse proportion. (True / False)

Area = (length)(width)

l	$2\frac{1}{2}$	5	10	20	50	100
w	200	100	50	25	10	5

32. What is the product of $l \cdot w$ for any pair of entries in the table above? ________

33. Referring to your answer in problem 32, what does this tell us about the area in problem 31? ____________

34. When $a \cdot b = k$ for any of the ordered pairs (a, b), then a is inversely proportional to b. (True / False)

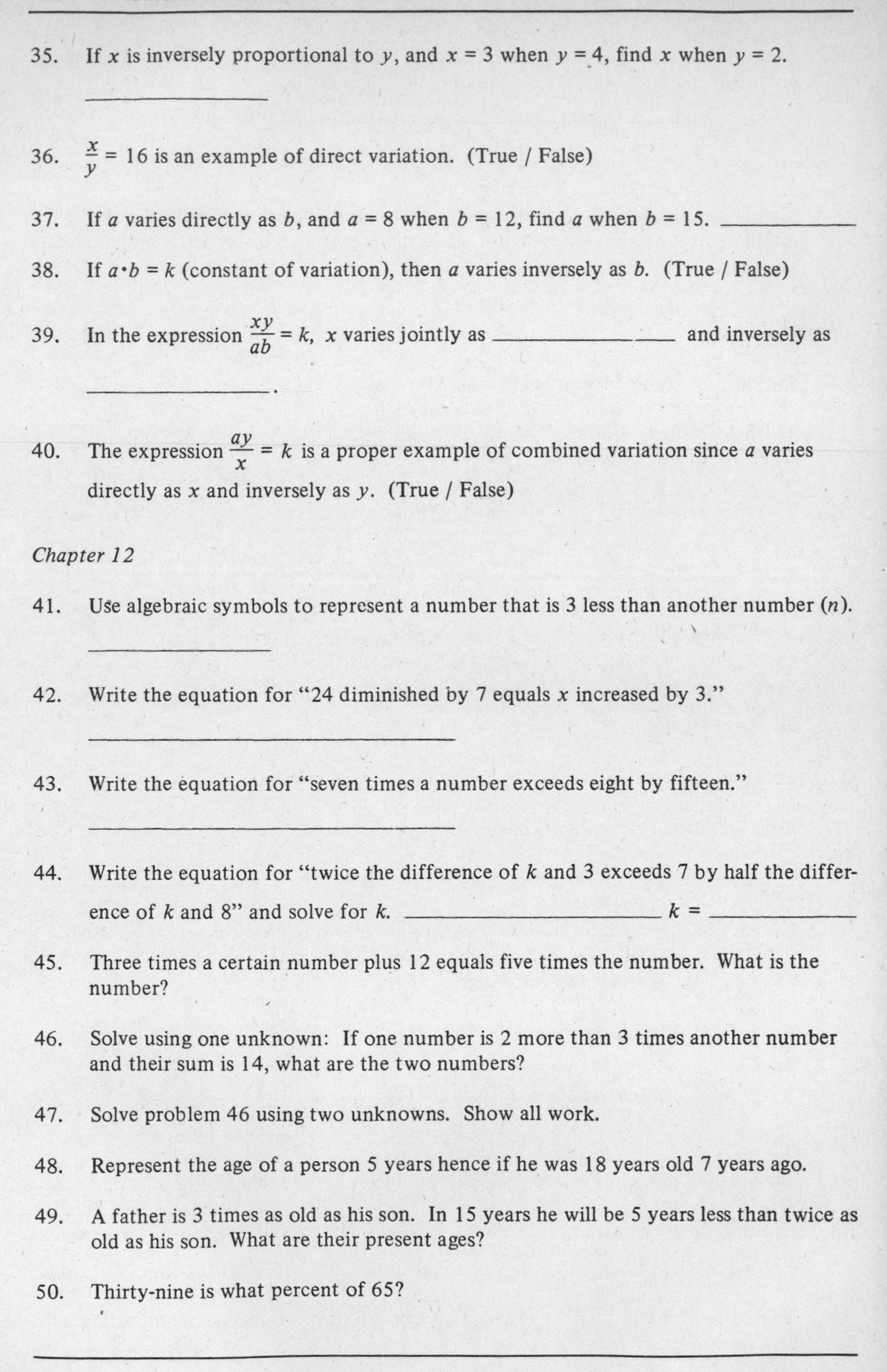

35. If x is inversely proportional to y, and $x = 3$ when $y = 4$, find x when $y = 2$.

____

36. $\frac{x}{y} = 16$ is an example of direct variation. (True / False)

37. If a varies directly as b, and $a = 8$ when $b = 12$, find a when $b = 15$. __________

38. If $a \cdot b = k$ (constant of variation), then a varies inversely as b. (True / False)

39. In the expression $\frac{xy}{ab} = k$, x varies jointly as __________________ and inversely as __________________.

40. The expression $\frac{ay}{x} = k$ is a proper example of combined variation since a varies directly as x and inversely as y. (True / False)

Chapter 12

41. Use algebraic symbols to represent a number that is 3 less than another number (n).

____

42. Write the equation for "24 diminished by 7 equals x increased by 3."

____

43. Write the equation for "seven times a number exceeds eight by fifteen."

____

44. Write the equation for "twice the difference of k and 3 exceeds 7 by half the difference of k and 8" and solve for k. __________________________ k = __________

45. Three times a certain number plus 12 equals five times the number. What is the number?

46. Solve using one unknown: If one number is 2 more than 3 times another number and their sum is 14, what are the two numbers?

47. Solve problem 46 using two unknowns. Show all work.

48. Represent the age of a person 5 years hence if he was 18 years old 7 years ago.

49. A father is 3 times as old as his son. In 15 years he will be 5 years less than twice as old as his son. What are their present ages?

50. Thirty-nine is what percent of 65?

51. A basketball team won 26 games and lost 14. What percentage of their games did they win?

52. If a plane travels 900 miles at an average speed of 450 miles per hour, how long will it take to reach its destination?

53. A car travels for $4\frac{1}{2}$ hours at an average speed of 50 mph. How far does it go?

54. A boy travels 10 miles on his bicycle in $2\frac{1}{2}$ hours. What is his average rate of speed?

55. Two cars travel in opposite directions for the same amount of time. The rate of one is 45 mph and the rate of the other 55 mph. If they are 500 miles apart when they stop, how long did they travel?

56. Two cars start at the same time from towns 450 miles apart and meet in 5 hours. If the rate of one is 10 mph more than the other, what is the rate of each?

57. Starting 280 miles apart, two drivers travel toward each other at rates of 35 mph and 45 mph respectively until they meet. What is the travel time of each?

58. A traveler takes 3 hours longer to travel back to his starting point than he did to go out from it. If he averaged 55 mph out and 22 mph back, find his time to return.

59. Dr. Lynn covered 310 miles by traveling for 4 hours at one speed and then for 5 hours at another speed. Had he gone 5 hours at the first speed and 4 hours at the second speed, he would have covered 320 miles. Find the two speeds.

60. A tea merchant blended tea worth 80¢ a pound with tea worth $1.10 a pound. The mixture of 20 pounds was valued by him at 92¢ a pound. How many pounds of each grade did he use?

Answers to Review Test 4

Chapter 10

1. $>$
2. $+3 < +7$
3. False (-3 is farther to the right on the number line than -6 hence it is greater than -6)
4. True
5. $<$
6. True
7. True
8. False
9. $-6 < -3$
10. $9 > 4$
11. $3x - 2x > 1 + 2$, $x > 3$
12. True
13. True (It will be true no matter what real value is assigned to x.)
14. True (It will be true only if x is greater than 7.)

15. True
16. $x > 3$
17. 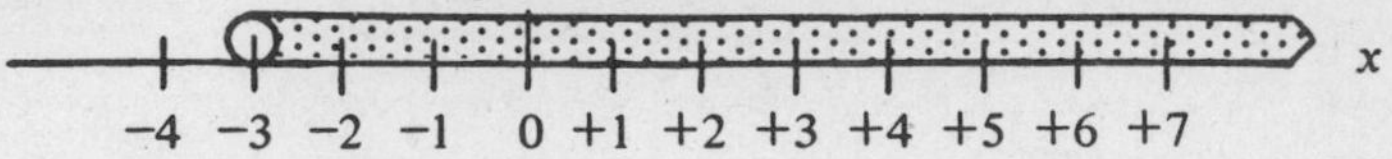
x
−4 −3 −2 −1 0 +1 +2 +3 +4 +5 +6 +7
18. $\geqslant$
19. 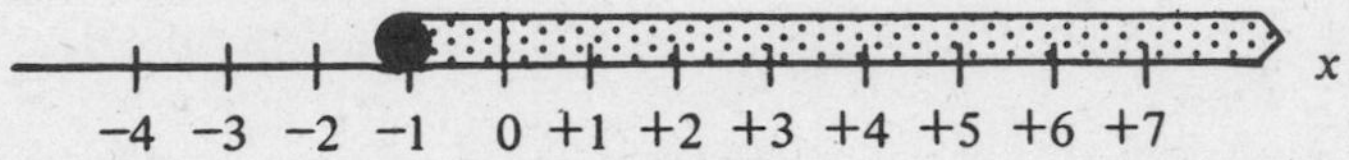
x
−4 −3 −2 −1 0 +1 +2 +3 +4 +5 +6 +7
20. (See frame 10, Chapter 10, for solution.)

Chapter 11

21. 2:5
22. 3:7
23. True
24. $\frac{6}{8}$, or $\frac{9}{12}$, and so on
25. True
26. $2x$; 20
27. 10
28. $a = 4, -4$ (2 solutions because the equation becomes quadratic)
29. $y = 3, -1$
30. 50 feet
31. True
32. 500
33. The area remains constant.
34. True
35. 6
36. True
37. $a = 10$
38. True
39. a and $b:y$
40. True

Chapter 12

41. $n - 3$
42. $24 - 7 = x + 3$
43. $7n - 8 = 15$
44. $2(k-3) - 7 = \frac{1}{2}(k-8)$; $k = 6$
45. 6
46. n = first number; $3n + 2$ = second number; $n + 3n + 2 = 14$; $n = 3$; hence the numbers are 3 and 11.
47. Let x = first number
    y = second number

(1) $y = 3x + 2$
(2) $x + y = 14$

Substituting (1) in (2):

$$x + 3x + 2 = 14$$
$$4x = 12$$
$$x = 3$$
$$y = 11$$

The numbers are 3 and 11.

48. $(18 + 7) + 5$
49. Son's age is 10, father's age is 30.
50. 60%
51. 65%
52. 2 hours
53. 225 miles
54. 4 mph
55. 5 hours
56. Rate of slower is 40 mph; faster is 50 mph.
57. $3\frac{1}{2}$ hours
58. 5 hours (time *out* is 2 hours)
59. 30 mph and 40 mph
60. Let x = pounds of 80¢ tea, $20 - x$ = pounds of \$1.10 tea; $80x + 110(20 - x) = 20 \cdot 92$; $x = 12$ pounds of 80¢ tea, $20 - x = 8$ pounds of \$1.10 tea. (Keep all figures in cents when working the problem.)

A Parting Word

If you have arrived at this point by following the approved path—working the problems in the text as you came to them, making sure you could solve the problems in the Self-Tests, and not taking any unapproved shortcuts—you have done very well indeed. You should by now have acquired a working familiarity with the fundamental topics of basic algebra: the number system, polynomials, factoring, algebraic fractions, exponents and roots, graphing, equations, ratios and proportions, and how to solve word problems. You have also learned something about quadratic equations and inequalities, topics that sometimes are reserved for intermediate or advanced algebra.

If you are interested in pursuing the study of algebra further you will find many good texts available, all of which include additional subject matter and a somewhat more advanced treatment of several of the topics we have covered in this book. Apart from some additional practice with the fundamentals of algebra, you should be ready now for an introduction to the remaining aspects of mathematics leading to the study of calculus. You will find these presented in a companion volume of the Wiley Self-Teaching Guide series titled *Geometry and Trigonometry for Calculus.* Included in this book are the subjects of plane (Euclidean) geometry, plane trigonometry, analytic geometry, the conic sections, polar coordinates, and limits. All are essential to the study of calculus.

Having gotten this far there is no reason you cannot go further. Hopefully, you have found the self-teaching approach a pleasant and fairly painless way to learn. Keep up the good work!

The author

Appendix

GLOSSARY OF TERMS

Absolute value (of a signed number): The numerical value of a number without regard to its sign. Thus $|x| = x,\ x \geqslant 0$ and $|x| = -x,\ x < 0$.
Algebraic representation: Use of mathematical symbols to represent word statements.
Axiom: A statement that is accepted without proof.

Binomial: An algebraic expression of two terms.

Checking (an equation): Substituting solution values for the unknown back in the original equation (or statement of the problem) to verify that it reduces it to an identity.
Clearing (an expression of) fractions: Multiplying all terms in an expression by the LCD in order to establish a common denominator.
Coefficient: The name given to a factor (or group of factors) of a product to describe its relation to the remaining factors.
Common monomial factor: The combined literal and numerical factors common to all terms of a polynomial.
Cross product: The sum of the products of the inner and outer terms of a binomial product.

Denominator: The numbers (or letters) below the fraction bar.
Digit: Arabic symbol for a number.
Dividend: The number being divided.
Divisor: The number by which the dividend is being divided.

Equation: A statement of equality between two algebraic expressions.
Evaluate: Substituting numbers for letters to find the value of an algebraic expression.
Exponent: The small number written to the right of and slightly above another number to indicate how many times the latter is to be used as a factor.
Expression (algebraic): A collection of terms combined by symbols of operation and grouping.

Factor: Any one of two or more numbers or letter which when multiplied together form a product.
Factoring: Expressing a number (or algebraic expression) as a product of prime factors.
Formula: Use of symbols to express a rule in brief form.
Fraction: The quotient of two numbers or expressions; part of any object, quantity, or digit; usually indicated by a numerator and denominator, separated by a fraction bar.

Identity: An unconditional equality.
Integer: A whole number.
Inverse: Having the opposite effect. An inverse operation has the effect of undoing another operation. (For example, subtraction is the inverse of addition, and division is the inverse of multiplication.)
Inversely: Oppositely.
Inversely proportional: Increasing or decreasing oppositely.

Literal factor: A letter used as a factor.
Literal number: A letter used to represent a number.
Lowest common denominator (LCD): The smallest number to which all others of a group may be changed.

Minuend: A number from which another number is being subtracted.
Monomial: An algebraic expression of one term.

Negative number: A number whose value is less than zero.
Number: The concept of quantity, or count.
Number scale (*or line*): A scale along which equally spaced positive and negative numbers represent distances (in opposite directions) from zero.
Numeral: A symbol for a number; a digit from 0 to 9, or a combination of digits.
Numerator: The numbers or letter lying above a fraction bar.
Numerical factor: A numeral used as a factor.

Polynomial: An algebraic expression that has only positive whole numbers for the exponents of the variables.
Positive number: A number whose value is greater than zero.
Power (*of a number*): The product of a number used two or more times as a factor.
Prime number: A whole number greater than one that has no whole number factors except one and itself.
Product: The result of multiplication.
Proportional (*to*): Changing correspondingly with some other quantity or term.

Quotient: The result of division.

Reciprocal: The reciprocal of a number is 1 divided by that number.
Root (*of an equation*): A number which, when substituted for the unknown letter, makes the equation a true statement (that is, both sides of the equation equal).

Satisfy (*an equation*): Finding the numerical values for the literal terms of an equation that will reduce it to an identity or a true statement.
Signed number: A positive or negative number, preceded by a plus sign (or no sign) if it is positive, and by a minus sign if it is negative.
Signs of operation: Plus or minus signs used to indicate addition or subtraction.
Signs of quality: Plus or minus signs used to indicate positive or negative numbers.
Simplify: Combine like terms of an algebraic expression in order to condense it as much as possible.
Solving (*an equation*): Finding values of the unknown which when substituted in the equation will make it a true statement. (See *Root.*)
Square root, principal: The positive square root of a number or algebraic expression.
Square root: One of the two equal factors of a number.
Subtrahend: The number being subtracted.
Symbol (*numerical*): A figure that stands for a number. (See *Numeral.*)

Term: A number, letter, or the product of numbers and letters.
Terms, like: Terms that differ only in their numerical coefficients (that is, that have the same literal part).
Terms, unlike: Terms having different literal coefficients or whose literal factors have different exponents.
Trinomial: An algebraic expression of three terms.

Whole number: The set of counting numbers and zero.

SYMBOLS USED IN THIS BOOK

Symbol	Meaning
$\|a\|$	absolute value
$=$	equals
$>$	greater than
$<$	less than
$\geqslant$	greater than or equal to
$\leqslant$	less than or equal to
$\ldots$	and so on
$\sqrt{\ }$	square root
$+$	addition or positive
$-$	subtraction or negative
$\cdot, \times$	multiplication
$\div$	division
$\neq$	not equal to
$\ngtr$	not greater than
$\nless$	not less than
(x, y)	coordinates of a point
(a, b)	ordered pair
f	function
$f(x)$	function of x
$:$	ratio

TABLE OF POWERS AND ROOTS

NO.	SQ.	SQ. ROOT	CUBE	CUBE ROOT	NO.	SQ.	SQ. ROOT	CUBE	CUBE ROOT
1	1	1.000	1	1.000	**51**	2,601	7.141	132,651	3.708
2	4	1.414	8	1.260	**52**	2,704	7.211	140,608	3.733
3	9	1.732	27	1.442	**53**	2,809	7.280	148,877	3.756
4	16	2.000	64	1.587	**54**	2,916	7.348	157,464	3.780
5	25	2.236	125	1.710	**55**	3,025	7.416	166,375	3.803
6	36	2.449	216	1.817	**56**	3,136	7.483	175,616	3.826
7	49	2.646	343	1.913	**57**	3,249	7.550	185,193	3.848
8	64	2.828	512	2.000	**58**	3,364	7.616	195,112	3.871
9	81	3.000	729	2.080	**59**	3,481	7.681	205,379	3.893
10	100	3.162	1,000	2.154	**60**	3,600	7.746	216,000	3.915
11	121	3.317	1,331	2.224	**61**	3,721	7.810	226,981	3.936
12	144	3.464	1,728	2.289	**62**	3,844	7.874	238,328	3.958
13	169	3.606	2,197	2.351	**63**	3,969	7.937	250,047	3.979
14	196	3.742	2,744	2.410	**64**	4,096	8.000	262,144	4.000
15	225	3.873	3,375	2.466	**65**	4,225	8.062	274,625	4.021
16	256	4.000	4,096	2.520	**66**	4,356	8.124	287,496	4.041
17	289	4.123	4,913	2.571	**67**	4,489	8.185	300,763	4.062
18	324	4.243	5,832	2.621	**68**	4,624	8.246	314,432	4.082
19	361	4.359	6,859	2.668	**69**	4,761	8.307	328,509	4.102
20	400	4.472	8,000	2.714	**70**	4,900	8.367	343,000	4.121
21	441	4.583	9,261	2.759	**71**	5,041	8.426	357,911	4.141
22	484	4.690	10,648	2.802	**72**	5,184	8.485	373,248	4.160
23	529	4.796	12,167	2.844	**73**	5,329	8.544	389,017	4.179
24	576	4.899	13,824	2.884	**74**	5,476	8.602	405,224	4.198
25	625	5.000	15,625	2.924	**75**	5,625	8.660	421,875	4.217
26	676	5.099	17,576	2.962	**76**	5,776	8.718	438,976	4.236
27	729	5.196	19,683	3.000	**77**	5,929	8.775	456,533	4.254
28	784	5.292	21,952	3.037	**78**	6,084	8.832	474,552	4.273
29	841	5.385	24,389	3.072	**79**	6,241	8.888	493,039	4.291
30	900	5.477	27,000	3.107	**80**	6,400	8.944	512,000	4.309
31	961	5.568	29,791	3.141	**81**	6,561	9.000	531,441	4.327
32	1,024	5.657	32,768	3.175	**82**	6,724	9.055	551,368	4.344
33	1,089	5.745	35,937	3.208	**83**	6,889	9.110	571,787	4.362
34	1,156	5.831	39,304	3.240	**84**	7,056	9.165	592,704	4.380
35	1,225	5.916	42,875	3.271	**85**	7,225	9.220	614,125	4.397
36	1,296	6.000	46,656	3.302	**86**	7,396	9.274	636,056	4.414
37	1,369	6.083	50,653	3.332	**87**	7,569	9.327	658,503	4.431
38	1,444	6.164	54,872	3.362	**88**	7,744	9.381	681,472	4.448
39	1,521	6.245	59,319	3.391	**89**	7,921	9.434	707,969	4.465
40	1,600	6.325	64,000	3.420	**90**	8,100	9.487	729,000	4.481
41	1,681	6.403	68,921	3.448	**91**	8,281	9.539	753,571	4.498
42	1,764	6.481	74,088	3.476	**92**	8,464	9.592	778,688	4.514
43	1,849	6.557	79,507	3.503	**93**	8,649	9.644	804,357	4.531
44	1,936	6.633	85,184	3.530	**94**	8,836	9.695	830,584	4.547
45	2,025	6.708	91,125	3.557	**95**	9,025	9.747	857,375	4.563
46	2,116	6.782	97,336	3.583	**96**	9,216	9.798	884,736	4.579
47	2,209	6.856	103,823	3.609	**97**	9,409	9.849	912,673	4.595
48	2,304	6.928	110,592	3.634	**98**	9,604	9.899	941,192	4.610
49	2,401	7.000	117,649	3.659	**99**	9,801	9.950	970,299	4.626
50	2,500	7.071	125,000	3.684	**100**	10,000	10.000	1,000,000	4.642

Index